Springer Remote Sensing/Photogrammetry

The Springer Remote Sensing/Photogrammetry series seeks to publish a broad portfolio of scientific books, aiming at researchers, students, and everyone interested in the broad field of geospatial science and technologies. The series includes peer-reviewed monographs, edited volumes, textbooks, and conference proceedings. It covers the entire area of Remote Sensing, including, but not limited to, land, ocean, atmospheric science and meteorology, geophysics and tectonics, hydrology and water resources management, earth resources, geography and land information, image processing and analysis, satellite imagery, global positioning systems, archaeological investigations, and geomorphological surveying.

More information about this series at http://www.springer.com/series/10182

Vernon Singhroy
Editor

Advances in Remote Sensing for Infrastructure Monitoring

Springer

Editor
Vernon Singhroy
President, EOSPATIAL
Ottawa, ON, Canada

ISSN 2198-0721 ISSN 2198-073X (electronic)
Springer Remote Sensing/Photogrammetry
ISBN 978-3-030-59108-3 ISBN 978-3-030-59109-0 (eBook)
https://doi.org/10.1007/978-3-030-59109-0

This Springer imprint is published by the registered company Springer Nature Switzerland AG
The registered company address is: Gewerbestrasse 11, 6330 Cham, Switzerland

Preface

Infrastructure projects by national, regional, and municipal governments are increasing, and robust monitoring techniques are needed. This is necessary because most civil infrastructure requires frequent repairs and the recent effects of climate change make monitoring even more important. Currently, the national space agencies and some commercial space companies are providing more high-resolution and daily images that are useful for infrastructure characterization and monitoring.

Recent applications on the uses of satellite and airborne images to monitor infrastructure are mainly focused on developing methodologies. This book focuses on modern case studies using high-resolution radar, optical, LIDAR, and UAV images in Canada, Dominica, Guyana, India, Italy, Syria, Taiwan, the USA, and the United Kingdom. The case studies provide operational guidelines and demonstrate how advanced remote sensing techniques are used to monitor civil infrastructure. The book will be of interest to infrastructure consultants and professionals, scientific communities in earth observation and advanced imaging methods, and researchers and professors in earth sciences, climate change, and civil and geoengineering. We also provide a more detailed account on advanced radar interferometry (InSAR), which is a rapidly growing business area for monitoring infrastructure. This book is divided into three parts. Part I focuses on high-resolution, rapid time-series radar interferometry to monitor highways, railways, pipelines, bridges, urban, and water conveyance infrastructures. Part II uses optical images for urban infrastructure characterization. Part III focuses on both radar and optical images to monitor damages to infrastructure.

Chapters 1–9 focus on InSAR applications. The first two chapters provide the background information on the state-of-the-art sensors and InSAR fundamentals as well as case studies. Chapter 1 provides a summary of the most current sensors used for monitoring infrastructure. The case studies in this chapter use time-series radar InSAR to monitor critical highways, railways, and pipelines affected by landslides. This chapter also includes examples of radar change detection and image fusion techniques to explore for minerals and to monitor mining sites, shoreline changes, and oil spills. Chapters 3–6 provide case studies using InSAR produced from different platforms (e.g., UAVSAR, RADARSAT, Sentinel-1, COSMO-SkyMed) at regional and local scales. These studies combine InSAR with LIDAR, UAV, GNSS, and field instrumentation to monitor linear and urban infrastructure. Bridges have different structures and provide challenges for InSAR monitoring. Some are structurally or functionally deficient and in urgent need of rehabilitation and maintenance. Chapters 7 and 8 discuss InSAR case studies of different bridge structures to support the current monitoring techniques. Chapter 9 provides a case of InSAR monitoring of land subsidence impacting a water conveyance infrastructure.

Urban infrastructure is discussed in Chaps. 10 and 11, which focus on ways to characterize urban form using optical images and to extract building footprint from airborne LiDAR. The last three chapters discuss about monitoring the damages to infrastructure. In Chap. 12, we demonstrate the uses of high-resolution radar for rapid flood mapping and for damage assessment to residential infrastructure. Chapters 13 and 14 discuss the use of optical images for spill response to the recycle and repurpose construction materials after destruction.

I would like to thank the 62 authors from 28 agencies for their valuable contribution to this publication. This compilation of selected case studies will provide useful guidelines to the larger community involved in civil infrastructure characterization and monitoring. The examples are clearly global—North America, Europe, Middle East, Asia, Island States, and the Tropics.

President, EOSPATIAL Vernon Singhroy
Ottawa, ON, Canada

Contents

Editor and Contributors

About the Editor

Vernon Singhroy is the President of EOSPATIAL in Ottawa, Canada. He is an internationally recognized expert on remote sensing applications and was the chief scientist of the Canadian Space Agency, RADARSAT Constellation Mission, launched in June 2019. Dr. Singhroy received his Ph.D. in environmental and resource engineering from the State University of New York, Syracuse. He is a professional engineer. As a senior research scientist with the Canada Centre for Remote Sensing at Natural Resources Canada (1995–2020), Dr. Singhroy has published over 300 papers in scientific journals, proceedings, and books. He was the editor-in-chief of the Canadian Journal of Remote Sensing, and he is the co-editor of four books, including the Encyclopedia of Remote Sensing. Dr. Singhroy has been a Professor of Earth Observation at the International Space University in Strasbourg, France (1998–2020), and he is an adjunct professor in Planetary and Space Sciences at the University of New Brunswick in Canada. Dr. Singhroy received the prestigious Gold Medal Award from the Canadian Remote Sensing Society and the Queen Elizabeth Diamond Jubilee Medal for his contributions to Canadian and international remote sensing applications and education.

Contributors

E. Barton National Physical Laboratory, Teddington, UK

Denis Bergeron Natural Resources Canada, Canada Centre for Mapping and Earth Observation, Sherbrooke, Québec, Canada

Silvia Bianchini Earth Sciences Department, University of Firenze, Firenze, Italy

Andrée Blais-Stevens Geological Survey of Canada, Natural Resources Canada, Ottawa, ON, Canada

Peter Bobrowsky Geological Survey of Canada, Sidney, BC, Canada

Mark H. Bulmer Roedown Research R2, Davidsonville, MD, USA Department of Earth Sciences, University College London, London, UK

Nicola Casagli Earth Sciences Department, University of Firenze, Firenze, Italy

Jonathan Chambers British Geological Survey, Keyworth, UK

François Charbonneau Canada Centre for Remote Sensing, Natural Resources Canada, Ottawa, ON, Canada

Subir Chowdhury Alberta Geological Survey, Alberta Energy Regulator, Edmonton, Alberta, Canada

Andrea Ciampalini Earth Sciences Department, University of Pisa, Pisa, Italy

Robert Cocking Geological Survey of Canada, Vancouver, BC, Canada

Daniel Cusson National Research Council Canada, Ottawa, ON, Canada

M. Das Geological Survey of India, Kolkata, India

Vincent Decker Emergency Geomatics Services, Canada Centre for Mapping and Earth Observation, Natural Resources Canada, Ottawa, ON, Canada

Matteo Del Soldato Earth Sciences Department, University of Firenze, Firenze, Italy

Andrea Donnellan Jet Propulsion Laboratory, California Institute of Technology, Pasadena, CA, USA

Shane Donohue School of Civil Engineering, University College Dublin, Dublin, Ireland

Jayson Eppler MDA Geospatial Services Inc., Vancouver, BC, Canada

Catherine Evans Alberta Geological Survey, Alberta Energy Regulator, Edmonton, Alberta, Canada

Pablo Ezquierro Geological Survey of Spain (IGME), Madrid, Spain

Tom G. Farr Jet Propulsion Laboratory, California Institute of Technology, Pasadena, CA, USA

Mary-Anne Fobert Canada Centre for Remote Sensing, Natural Resources Canada, Ottawa, ON, Canada

Charles Fortin Natural Resources Canada, Canada, Centre for Mapping and Earth Observation, Sherbrooke, Québec, Canada

Parwant Ghuman 3v Geomatics Inc., Vancouver, BC, Canada

Fernando Greene Gondi MDA Geospatial Services Inc., Vancouver, BC, Canada

Christian E. Haselwimmer Chevron Energy Technology Company, San Ramon, CA, USA

Michael Hendry Department of Civil and Environmental Engineering, University of Alberta, Edmonton, AB, Canada

Jessica Holmes School of Natural and Built Environment, Queen's University Belfast, Belfast, UK

British Geological Survey, Keyworth, Great Britain, UK

Wei-Kai Huang Disaster Prevention Technology Research Center, Sinotech Engineering Consultants, Inc., Taipei, Taiwan, China

David Huntley Geological Survey of Canada, Vancouver, BC, Canada

Cathleen E. Jones Jet Propulsion Laboratory, California Institute of Technology, Pasadena, CA, USA

Jamel Joseph Geological Survey of Canada, Vancouver, BC, Canada

Ching-Fang Lee Disaster Prevention Technology Research Center, Sinotech Engineering Consultants, Inc., Taipei, Taiwan, China

Brad Lehrbass Emergency Geomatics Services, Canada Centre for Mapping and Earth Observation, Natural Resources Canada, Ottawa, ON, Canada

Jon Leighton 3v Geomatics Inc., Vancouver, BC, Canada

Junhua Li Canada Centre for Remote Sensing, Natural Resources Canada, Ottawa, ON, Canada

Shih-Yuan Lin Department of Land Economics, National Chengchi University, Taipei, Taiwan, China

Zhen Liu Jet Propulsion Laboratory, California Institute of Technology, Pasadena, CA, USA

Renato Macciotta Department of Civil and Environmental Engineering, University of Alberta, Edmonton, AB, Canada

Roger MacLeod Geological Survey of Canada, Sidney, BC, Canada

Heather McGrath Natural Resources Canada, Canada Centre for Mapping and Earth Observation, Sherbrooke, Québec, Canada

Philip Meldrum British Geological Survey, Keyworth, UK

C. R. Middleton Engineering Department, University of Cambridge, Cambridge, UK

Megan M. Miller Jet Propulsion Laboratory, California Institute of Technology, Pasadena, CA, USA

Roberto Montalti Earth Sciences Department, University of Firenze, Firenze, Italy

Victor Neufeld Emergency Geomatics Services, Canada Centre for Mapping and Earth Observation, Natural Resources Canada, Ottawa, ON, Canada

Ian Olthof Emergency Geomatics Services, Canada Centre for Mapping and Earth Observation, Natural Resources Canada, Ottawa, ON, Canada

Istemi Ozkan National Research Council Canada, Ottawa, ON, Canada

Jean-Samuel Proulx-Bourque Natural Resources Canada, Canada Centre for Mapping and Earth Observation, Sherbrooke, Québec, Canada

Federico Raspini Earth Sciences Department, University of Firenze, Firenze, Italy

C. Rossi Satellite Applications Catapult, Didcot, UK

Drew Rotheram-Clarke Geological Survey of Canada, Vancouver, BC, Canada

S. Selvakumaran Engineering Department, University of Cambridge, Cambridge, UK

Monan Shan Earth Sciences Department, University of Firenze, Firenze, Italy

Todd C. Shipman Alberta Geological Survey, Alberta Energy Regulator, Edmonton, Alberta, Canada

Vernon Singhroy EOSPATIAL, Ottawa, ON, Canada

Lorenzo Solari Centre Tecnològic de, Telecomunicacions de Catalunya (CTTC), Castelldefels, Spain

Simon Tolszczuk-Leclerc Emergency Geomatics Services, Canada Centre for Mapping and Earth Observation, Natural Resources Canada, Ottawa, ON, Canada

Paul Wilkinson British Geological Survey, Keyworth, UK

Ying Zhang Canada Centre for Mapping and Earth Observation, Natural Resources Canada, Ottawa, ON, Canada

Part I
Transportation, Energy and Mining Infrastructure

Chapter 1
Advanced Radar Images for Monitoring Transportation, Energy, Mining and Coastal Infrastructure

Vernon Singhroy, Mary-Anne Fobert, Junhua Li, Andrée Blais-Stevens, François Charbonneau, and M. Das

1.1 Introduction

In this book, we focus on the use of advanced radar imaging techniques to monitor some of the main types of critical transportation and energy infrastructure, such as roads, railways, and pipelines. We provide case studies on mining infrastructure to locate and monitor active mining sites. Imaging techniques used to facilitate mineral exploration are discussed. Radar monitoring techniques for shoreline change and ocean oil spills affecting coastal infrastructure are also included. These operational examples will assist in providing guidelines on the use of advanced radar imaging techniques to characterize and monitor civil infrastructure. They will also be useful in the planning of future sustainable climate-resilient infrastructure (InterAmerican Development Bank 2018).

V. Singhroy (✉)
EOSPATIAL, Ottawa, ON, Canada

M.-A. Fobert · J. Li · F. Charbonneau
Canada Centre for Remote Sensing, Ottawa, ON, Canada
e-mail: vern.singhroy@eospatial.com

A. Blais-Stevens
Geological Survey of Canada, Ottawa, ON, Canada

M. Das
Geological Survey of India, Kolkata, India

V. Singhroy (ed.), *Advances in Remote Sensing for Infrastructure Monitoring*,
Springer Remote Sensing/Photogrammetry,
https://doi.org/10.1007/978-3-030-59109-0_1

InSAR Infrastructure

Monitoring surface deformation on civil infrastructures requires high-resolution, spatial, and temporal InSAR imagery for mitigation and safety purposes. An interferometric image represents the phase difference between the backscattered signals of two SAR images obtained from similar positions in space. In the case of spaceborne SAR, the images are acquired from repeat-pass orbits. The phase differences result from topography and changes in the line-of-sight distance (range) to the radar due to surface displacement, or changes in the atmospheric propagation path length. Typical scales for SAR interferometry in land deformation applications are millimeters to centimeters per orbit cycle of the radar satellite.

Radar interferometry (InSAR) techniques that use imagery from several radar satellites are increasingly being used for deformation monitoring (Singhroy et al. 2005, 2008, 2010, 2016; Colesanti and Wasowski 2006; Singhroy, 2008; Calò et al. 2012, 2014; Wasowski and Bovenga 2014a, b, and others). In some cases, ground-based InSAR data (Pratesi et al. 2015) are being used, but these are not discussed in this book. Our examples demonstrate the operational use of space-based InSAR for infrastructure. The application of InSAR techniques to monitor unstable slopes in both rock and soil has been developing rapidly, including its application to other deformation studies, such as subsidence, volcanoes, and earthquakes.

Landslides affecting infrastructure can result in extreme economic and societal costs, despite our increased understanding of the mechanisms of failure and large ground deformation (Singhroy 2012). Current state of the art, real-time monitoring of active slopes developed for early warning of landslides is very expensive. Radar interferometry is increasingly being used to complement real-time monitoring techniques such as GPS and in situ field measurements (Singhroy and Molch 2004; Singhroy 2005; Alasset et al. 2007; Pearse et al. 2014).

Other case studies in this chapter use standard operational image processing techniques. These include multi-date radar change detection to monitor shoreline changes and update mining sites. Enhanced image visualization and filtering were also used for ocean oil spill detection.

Sensors for Infrastructure

This section is written for folks with little knowledge of remote sensing applications. We therefore summarize some relevant information related to infrastructure. Tables 1.1 and 1.2 provide the specifications and applications of the current and upcoming radar satellites. Table 1.3 provides a list of the current radar and optical sensors that can be used to characterize and monitor transportation, energy, and mining infrastructure. This introduction is a background to support the images and techniques related to the case studies described in this chapter.

RADAR is an acronym for **RA**dio **D**etection **A**nd **R**anging, which essentially characterizes the function and operation of a radar sensor. The sensor installed on

Table 1.1 Current and planned radar satellites 2007–2025

Satellite	Radarsat-2	Radarsat Constellation Mission (RCM)	TerraSAR-X (TSX)	TanDEM-X (TDX)	ALOS-2/ ALOS-4	Cosmo/ Skymed	Sentinel-1A/–1 B	NISAR	Capella X-SAR Constellation	ICEYE Constellation	KOMPSAT 5/6	NovaSAR-1	PAZ (aka SEOSAR)
Space agency	CSA	CSA	DLR	DLR	JAXA	ASI	ESA	NASA/ ISRO	Palo alto	ICEYE	KARI	UKSA	HISDESAT
Launch	2007	2019	2007	2010	2014/ 2020	2007–2010	2014/2016	2022	2018–2021	2018–2020	2013/2025	2018	2018
Band	C	C	X	X	L	X	C	L, S	X	X	X	S	X
Wavelength (cm)	5.7	5.7	3.1	3.1	22.9	3.1	5.6	24, 12	3.1	3.1	3.2	10	3.1
Polarization	Single, Dual, Quad	Single, Dual, Quad, CP	Single, Dual, Quad	Single, Dual, Quad	Single, Dual, Quad, (experimental CP)	Single, Dual	HH, Dual	All	HH	VV	Single	Single, Dual, Tri (HH+VV +HV)	Single, Dual
Incidence angle (deg)	10–60	10–60	15–60	20–55	8–70	Variable	20–46	33–47	10–50	10–35	20–55	16–73	15–60
Resolution range (m)	3–100	1–100	1–16	1–16	3–100	1–100	2–93	3–48, 3–24	0.3–10	1–20	1–20	6–35	1–6
Resolution azimuth (m)	3–100	3–100	1–16	1–16	1–100	1–100	5–89	7	0.3–1	1–20	1–20	6–35	1–16
Scene width (km)	50–500	5–500	5–100	5–100	25–490	10–200 (up to 1300)	20–410	>240	10–50	5–30	5–100	20–400	5–100
Repeat cycle (days) InSAR	24	4–12	2–11	2–11	14	5–16	12	12	0.13–0.25 (3–6 h)	1	28	16	11
Orbital elevation (km)	798	592	514	515	628	620	693	747	485–525	570	550	583	514

Table 1.2 Application potential of RADARSAT-1, RADARSAT-2[1], and RCM

	Application	Satellite		
		RADARSAT-1	RADARSAT-2	RCM
Agriculture	Crop type	–/+	++P	++CP
	Crop condition	–/+	++P	++CP
	Crop yield	–	–/+	–/+
Cartography	DEM interferometry	+	+	++
	DEM stereoscopy	++	++	++
	DEM polarimetry	N.A.	–/+	–
	Cartographic feature extraction	+	++	++
Disaster management	Floods	++	++	++
	Geological hazards	–/+	++	++
	Hurricanes	+	+	+
	Oil spills	++	++	++
	Search and rescue	–/+	+	+
Forestry	Forest type	–/+	–/+	+
	Clearcuts	–/+	++	++
	Fire scars	–/+	++	++
	Biomass	–	–/+	–/+
Geology	Terrain mapping	–/+	+	+
	Structure	++	++	++CP
	Lithology	–/+	–/+	–/+
Hydrology	Soil moisture	–/+	+	+CP
	Snow	–/+	+	+
	Wetlands	–/+	++	++CP
Oceans	Winds	+	++	++
	Ships	+	++	++CP
	Waves	–/+	–/+	+
	Currents	–/+	+	+
	Coastal zones	–/+	++	++
Sea and land ice	River ice	+	++	++
	Sea ice edge and ice concentration	+	++	++CP
	Sea ice type	+	++	++CP
	Sea ice topography and structure	–/+	++	++CP
	Icebergs	–/+	++	++
	Polar glaciology	–/+	++	++
InSAR motion applications	Subsidence		++	++
	Slope stability/landslides		++	++
	Co-seismic deformation		++	++
	Oil Sands mine heaving		++	++
	Land motion		++	++

Key: "1"" use of single data images assumed, "–" minimal, "–/+" limited, "+" moderate, "++" strong, "CP" compact pol, "P" polarimetric. van der Sanden (2004) modified by Singhroy

Table 1.3 Potential application of advanced remote sensing for infrastructure

Application	Stereo Air photos	Drone	Lidar	High-Resolution Optical	Radar	InSAR Land Motion
Buildings Urban/Rural	++	++	++	++	+	++
Transportation corridor roads/ Highways, railway	++	++	++	++	+	++
Dykes/Dams	+	++	++	++	+	++
Mine wall motion	–	+	+	–	–	++
Flood extent	++	++	+	+	++	–
Damaged buildings	++	++	+	++	++	
Bridge motion	–	–	–	–	++	++
Fire damage	++	++	–	++	+	–
Land motion	++	++	+	++	+	++
Construction materials, Metals (mining), Geological structures	++	–	+	+	++	–

Key: "–" minimal, "–/+" limited, "+" moderate, and "++" strong

airplanes and satellites transmits a microwave (radio) signal toward the target on Earth and receives the backscattered portion of the signal. The strength of the backscattered signal is measured to discriminate between different targets.

Operational applications of radar imagery have been developed over the past 30 years. Currently, radar satellites provide high-resolution (1–10 m) and medium to low resolution (10–100 m), day and night, all-weather imagery for many research applications. These include geoscience and climate change research, environmental and disaster monitoring, topographic mapping, change detection, security-related applications, and planetary exploration (Moreira et al. 2013; Singhroy 2013). Table 1.1 provides a summary of current and future radar satellite missions, and their technical specifications. The major space agencies focused on launching C-, X-, S-, and L-band frequency satellites with different resolutions, viewing geometries, and polarizations.

There are a number of high-resolution radar constellation satellites being planned (Table 1.1). For example, the recently launched (June 2019) RADARSAT Constellation Mission (RCM) is designed as a scalable constellation of three small satellites. RCM is optimized for coherent change detection (CCD), a compact polarimetry mode (Charbonneau et al. 2010), special imaging modes optimized for ship detection, and a combined SAR and automatic identification system (AIS) (Thompson 2015). The frequent revisits of RCM will allow 1-day global monitoring for applications such as maritime surveillance, disaster management, and ecosystem monitoring.

Table 1.2 provides a summary and the ratings for the overall applications of RADARSAT-1 & 2 and the potential applications of the RADARSAT Constellation Mission. Other radar satellites can provide similar applications.

RADARSAT-2 has improved multipolarization and higher resolution capabilities in comparison to RADARSAT-1 and ENVISAT. This improvement has allowed for further developments in agricultural applications, particularly crop-type mapping and condition assessment, soil tillage and crop residue mapping, and soil moisture estimation. The additional polarizations have increased the information and understanding of the scattering mechanisms and target interactions. Integrating C-band SAR with other radar frequencies, or with data acquired by optical sensors, has provided many operational applications (McNairn and Brisco 2004; Singhroy and Pilkington 2012 among others).

SAR polarimetry is a widely used technique to derive qualitative and quantitative physical information for land and water. This is based on the measurement of the polarimetric signatures of natural, agricultural, and man-made scatterers. The scattering mechanism allows for their identification and separation, and are based on the shape, orientation, and dielectric properties of the target (Cloude 2010; Moreira et al. 2013; Lopez-Sanchez et al. 2014). For example, the RADARSAT Constellation Mission (RCM) consists of three small synthetic aperture radar (SAR) satellites flying in a constellation configuration. It has fully polarimetric (FP) capabilities, in addition to single-polarization (HH, HV, VV), conventional (HH-HV, VV-VH, and HH-VV), and hybrid (i.e., compact) dual polarization. Recent results have shown that the multi-frequency and compact polarimetric images from RCM, when combined with fully polarimetric data, were useful for estimating soil moisture conditions, and improved ship detection and classification, and the mapping of geological structures (Merzouki and McNairn 2015; Touzi and Vachon 2015; Fobert et al. 2018).

Radar interferometry (InSAR) represents the phase differences between the backscatter signals in two or more time-series SAR images obtained from similar positions in space. Radar satellites acquire images from repeat-pass orbits. These orbit cycles, from current radar satellites, can range from 1 to 6 days for Cosmo-SkyMed, 4 days for RCM, 11 days for TerraSAR-X, and 12 days for Sentinel-1. Table 1.1 provides a summary of the technical specifications of these satellites.

The phase differences between two repeat-pass images at a fixed line of site (LOS) can provide information on topography, small surface displacements, or changes in the atmospheric propagation path length. Land deformation measured from SAR interferometry ranges from millimeters to centimeters per orbit cycle. InSAR techniques are used to monitor deformation under specific conditions, provided that coherence is maintained over the respective orbit cycle. Using data pairs with short perpendicular baselines, short time intervals between acquisitions, and correcting for the effect of topography and atmospheric effects, reliable measurements of surface deformation can be achieved.

InSAR techniques using images from several radar satellites are increasingly being used for landslide deformation monitoring (Colesanti and Wasowski 2006; Singhroy 2012; Calò et al. 2014; Singhroy et al. 2016; Bianchini et al. 2018). InSAR can assist in characterizing and monitoring geohazards and can be used as a screening tool for mitigation and remediation for the protection of assets, property, and life (Singhroy et al. 2016; Gee et al. 2019). InSAR techniques are also being used to monitor co-seismic deformation (Boncori 2019), volcanic eruption (Dumont et al. 2018 and others), and glacier retreat (Brancato et al. 2020 and others). Some examples are described below as they relate to infrastructure monitoring.

Optical

Table 1.3 provides a summary of the capabilities of other imaging sensors for mapping and monitoring infrastructure.

Optical images use the visible, near-infrared, and short wave infrared parts of the electromagnetic spectrum. The images are acquired from aircraft and satellites. Aerial photos are still being used for mapping and interpreting landforms and surficial materials related to civil engineering work. All civil construction is built on solid bedrock and surficial soil materials, therefore, knowing their distribution and properties are fundamental. Mollard and Janes (1984) provided useful keys and examples of air photo interpretation for various terrain and surficial materials that can be used for transportation, energy corridor, and mining infrastructure. Table 1.3 summarizes some of the ways air photos are applied for mapping attributes for infrastructure.

High-resolution optical systems (0.5–5 m spatial resolution) are generally the most useful in identifying and interpreting the fine details required for monitoring civil engineering work. Table 1.3 provides a summary of the capabilities of high-resolution optical sensors for infrastructure. Several case studies on the use of high-resolution optical images, and other spatial data related to monitoring infrastructure and site characterization, are described in this book and by Singhroy (1996).

Hyperspectral images provide high spatial and spectral information that are particularly useful in infrastructure site characterization, geological mapping, and mineral exploration. They can provide sub-pixel compositional information to identify surface minerals of exposed rocks. Spectral libraries and image enhancement techniques are now used in mineral identification, mine tailings, and oil contamination sites as part of the techniques associated with mining infrastructure (Singhroy 1996; van der Meer et al. 2012; Scafutto et al. 2016). Spectral signatures from the field and airborne sensors were also used to identify surface mineralization and stressed vegetation associated with high concentrations of base metals in the surficial material (Singhroy and Kruse 1991; Feng et al. 2018).

The occurrence of oil spills indicates damage to the pipelines. Finding these on-land spills can guide cleanup and repair operations. Hyperspectral images were used to characterize the type of oil contamination in mine tailings (Scafutto et al. 2016).

UAV

Unmanned aerial vehicle (UAV) systems are generally understood as drones with RGB cameras. Current UAV systems are providing multispectral, hyperspectral, longwave infrared range cameras (thermal) and light-weight LiDAR (light detection and ranging) data (Yao et al. 2019). UAV sensors and platforms are used for Structural Health Monitoring (SHM) related to the integrity of engineering structures. Research has shown that UAVs are increasingly being used to monitor infrastructure for early warning, situational assessment, and decision support applications (Flammini et al. 2016). The short response time and low mobility cost of UAVs will be useful for monitoring civil engineering works. These high-resolution drone images are also used in several other applications, such as agriculture, forestry, mining, and environmental, among others (Fernandes et al. 2018; Yao et al. 2019; Poley and McDermid 2020). Drone images will continue to provide a powerful tool for infrastructure monitoring because of their ultrahigh spatial resolution, flexibility, and ease of integration with other high-resolution radar, LiDAR, and optical images. Table 1.3 provides some of the applications of drone images for infrastructure monitoring and characterization.

LIDAR

Light Detection And Ranging (LiDAR) provides accurate three-dimensional (3D) point cloud data that are used to characterize, monitor, and plan civil infrastructure developments. LiDAR is used for topographic mapping because of its capability to provide 3D measurements. As the light wave passes through the vegetation, the backscatter wave provides information about the vegetation parameters. Airborne Laser Scanning (ALS) are mainly used to generate detailed bare earth digital terrain models (DTM) and to estimate forest height, density, etc. ALS systems emit laser pulses with footprints ranging from 0.1 m to 2 m near the surface for submeter measurement accuracy of the terrain surface height (White et al. 2016; Shang et al. 2019). Terrestrial Laser Scanners (TLS) and Mobile Laser Scanning (MLS) provide millimeter-level accuracy and have point densities of a few thousand points/m^2. MLS has many applications, including the monitoring of urban transportation infrastructure (Wang et al. 2019; Sánchez-Rodríguez et al. 2019). TLSs are used for measuring and monitoring specific infrastructure and buildings with high levels of detail (e.g., Nowak et al. 2020). Multispectral laser scanning is a rapidly growing technology. Images are acquired at different wavelengths allowing for the recording of a diversity of spectral reflectance. The multispectral LiDAR data, using intensity and height images created from LiDAR points, provide detailed land cover classification in urban environments (Morsy et al. 2017). Table 1.2 shows some of the main applications of lidar data for measuring attributes related to civil infrastructure.

1.2 Transportation Corridors

InSAR Monitoring of a Major Highway Affected by Coastal Landslide

Canada has the longest coastline in the world. Coastal hazards are frequent because of various combinations of topographic, geomorphic, and geological conditions, which are conducive to mass failure (Mosher 2008). In addition, the stability of the coastal environment is impacted by climatic, geographic, oceanographic, and anthropogenic factors (Irvine 2012). Coastal landslides represent a specific threat on the coastlines of populated areas in southern British Columbia and the Atlantic provinces because of the high tsunami potential. These landslides are triggered by seawater intrusion, coastal scouring, high intensity/frequency rainfall and storm events, human activity, and rapid snowmelt. They are aided by favorable geological and geotechnical conditions, such as fractured bedrock on steep slopes or earthquake tremors. They usually occur without warning and with little time lag between failures. Therefore, high-resolution, spatial, and rapid temporal images generated from the RADARSAT Constellation Mission are required as part of the integrated monitoring of these coastlines.

We acquired RADARSAT-2 InSAR images in ultrafine beam mode (U17, ~3 m resolution) with 43° incidence angle and HH polarization for every 24-day orbital cycle from March to October 2010. During the summer of 2010, we acquired TerraSAR-X InSAR images every 11 days for 3 months to fill the gaps between the RADARSAT-2 acquisitions. The TerraSAR-X data were in spotlight (1 m) mode with 43.1° incidence angle and HH polarization.

Our InSAR method used the GAMMA software to perform standard interferometric processing (Gamma Remote Sensing 2015). This includes image co-registration and re-sampling, interferogram generation, topographic phase removal using a 1 m Lidar DEM to get a set of co-registered unwrapped differential interferograms for both RADARSAT-2 and TerraSAR-X data. An improved SBAS (Small Baseline Subset) algorithm (Samsonov et al. 2011) was applied to the unwrapped differential interferograms. The SBAS method minimizes the effects of the atmosphere and the lack of resolution in the DEM on the accuracy of InSAR measurements. It produces a deformation time-series using interferograms with short time acquisitions, which are usually more coherent. A high-pass filter with a Gaussian window was applied to remove the residual orbital and long-wavelength atmospheric signals. A Singular Value Decomposition (SVD) inversion was applied to simultaneously solve for the individual deformation rates and the residual topographic errors.

During the spring of 2008, the coastal landslide at Daniel's Harbour destroyed several homes, as well as the only highway corridor connecting all the settlements along the northeastern coast. After the event, the wet landslide debris continued to move about 25 mm during the summer and early fall months, as shown in Fig. 1.1.

Our field investigation and earlier geotechnical investigations by Batterson et al. (1999) suggested that the Daniel's Harbour landslide was triggered by coastal erosion from high winter ocean waves combined with spring rainfall and groundwater seeps from the nearby reservoir. In other coastal areas of Newfoundland, sea-level rise, wave action, slope angle, sediment type, storm surge, and human activities are the main factors triggering landslides (Irvine 2012). The high-resolution InSAR images are effective in characterizing differential motion on this coastal landslide, especially during active wet spring periods. Because these are spontaneous events, a 4 day revisit from RCM will provide an improvement to current RADARSAT-2, Sentinel-1, TerraSAR-X, and ALOS-2 acquisitions over the active spring and summer months in these coastal areas. In tropical areas, similar methods can be applied during the hurricane and typhoon seasons.

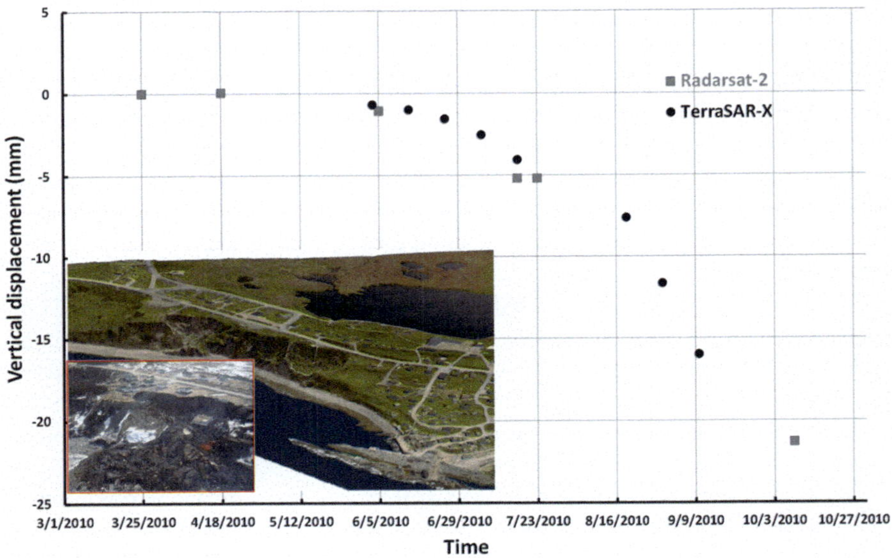

Fig. 1.1 InSAR motion at Daniel's Harbour

InSAR Monitoring of a Coastal Railway

This case study reports on a rock failure affecting a nearby railway line. Field measurements of fissures show crack widths of tens of centimeters to meters, which indicate significant rock slope movement (Locat et al. 2010). We use InSAR techniques to monitor the rock slope movement as part of a continuous monitoring program. Our results show 30 mm of motion using RADARSAT-2 images over a 2 year period.

The site is highly vegetated with mature trees which strongly reduces the InSAR coherence at C-band. We installed 23 metallic reflectors (13 trihedrals and 10 dihedrals) within and outside the landslide area to monitor the landslide motion. The reflectors serve as permanent coherent targets with high backscattering intensity and accurate phase (Fig. 1.2). The locations of the reflectors within and outside the landslide area were based on prior knowledge of the unstable rock slope.

Seven targets are oriented and inclined to maximize the line of sight (LOS) backscattering return of RADARSAT-2 Spotlight (SLA3, at $\theta = 32°$) for ascending passes. Sixteen targets are oriented for descending passes and inclined to optimize the return for RADARSAT-2 Spotlight SLA 76 and SLA19 ($\theta = 27°$ and $44°$ respectively).

Our InSAR displacement estimates show that the landslide is characterized by a more complex series of active blocks than previously estimated from traditional in situ instruments. All these blocks are moving at different rates, from 15 to 55 mm a year, as shown in Fig. 1.3. Results from ground-based monitoring instrumentation confirmed our displacement measurements and have assisted in the understanding of the failure mechanisms of this complex rockslide (Cloutier et al. 2010).

Fig. 1.2 Two metallic trihedrals installed on the retaining wall along the railway (1&2) and one on the tilted block (3). Reflector 1 shows a differential GPS instrument which provided measurements to validate the InSAR results

At the Gascons site, we observed that a combination of favorable geological conditions, combined with fractured bedrock, produced multiple failure mechanisms that require high-resolution rapid monitoring throughout the year. High-resolution data, and the 4 day repeat pass provided by the RADARSAT Constellation Mission, would maintain high coherence and detailed resolution to monitor the moving blocks within the rock slide.

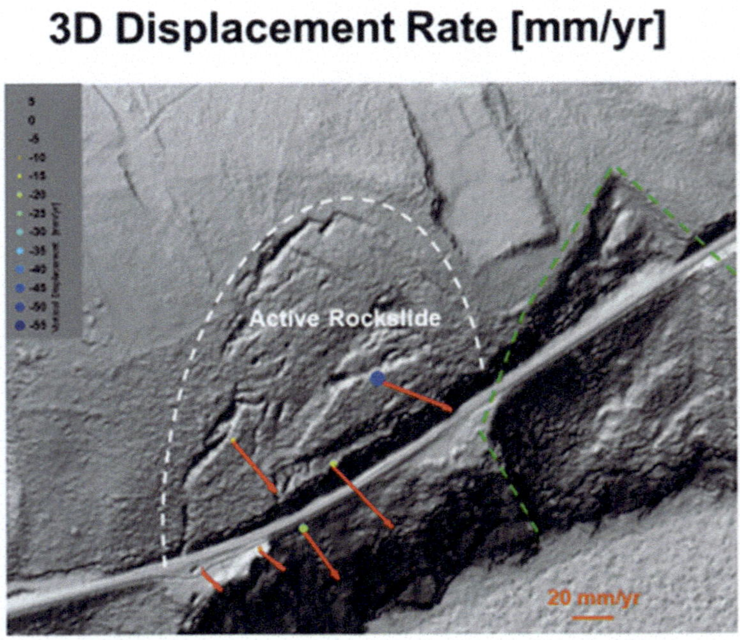

Fig. 1.3 RADARSAT InSAR results showing rockslide motion at the Gascons Landslide, Quebec. The LIDAR image shows the outline (white stippled line) of the rockslide and the railway line. Red arrows shows displacement (direction and annual horizontal rate) of the different rockslide blocks. The colored circles indicate reflector location and their vertical displacement rate. Locations with ascending and descending configurations are shown

InSAR Monitoring of a Himalayan Highway and Railway Affected by Monsoons

In this study, we used RADARSAT-2 InSAR to assist in developing an early warning system for Himalayan landslides. In Darjeeling district, West Bengal, there are about 3000 landslides, many of which are active and retrogressive. They are affecting settlements and transportation corridors, and in some cases, have caused fatalities (Ghosh et al. 2011; Das et al. 2018). The movement of these landslides is slow as evident from ground cracks and cracks in the walls on some of the buildings. It is difficult to monitor all of these landslides because of the steep Himalayan terrain and incoherent vegetation cover, including tea plantations. We focus our investigation on the Gayabari landslide (Fig. 1.4) which is one of the many active landslides on these Himalayan slopes. The Gayabari landslide is a rockslide located on a moderately steep (25–35°) to steep (35–45°) slope at the right bank of Shiva Khola River. The landslide covers an area of 0.19 km^2, and measures 647 m in length and 258 m in width. The only strategic railway (Darjeeling Himalayan Railway (DHR)), and National Highway (NH-55), connecting the scenic towns and tea plantations of Darjeeling Himalayas, run along the middle of the

Fig. 1.4 View of the Gayabari Landslide. The Giddapahar village is situated just above the crown of the landslide. The Darjeeling Himalaya Railway (DHR)/NH-55 connecting Darjeeling district passes through the middle of the Gayabari landslide

slide (Fig. 1.4). The geomorphology of this terrain represents a complicated interaction of erosional and gravitational processes (mass wasting) which are sometimes triggered by active tectonic effects of the Himalayas and monsoonal heavy rainfall.

We used 20 RADARSAT-2 ultrafine (3 m) scenes with downslope viewing acquired between September 2016 and January 2018. Five trihedral corner reflectors (CR) were installed to provide coherent signals for accurate time-series measurements. One CR was installed on a stable site away from the landslide. The measurement on the stable CR, shown in Fig. 1.5, was compared with InSAR measurements taken from the active vegetated slopes. We used the GAMMA software to generate unwrapped differential interferograms and used the SBAS method (Casu et al. 2006) to produce time-series displacement. We measured 2–5 cm motion over the year, with the maximum motion (shown in red in Fig. 1.5) occurring during the wet monsoon season. The time-series displacement profiles for the CRs are also shown in Fig. 1.5. The movement data from the pillars installed in and around the landslide area, as well as the prevalent ground cracks, coincide with the seasonal deformation measured from the InSAR analysis. Our study shows that InSAR can be used to develop an early warning system for monitoring active Himalayan landslides affecting critical infrastructure within the region.

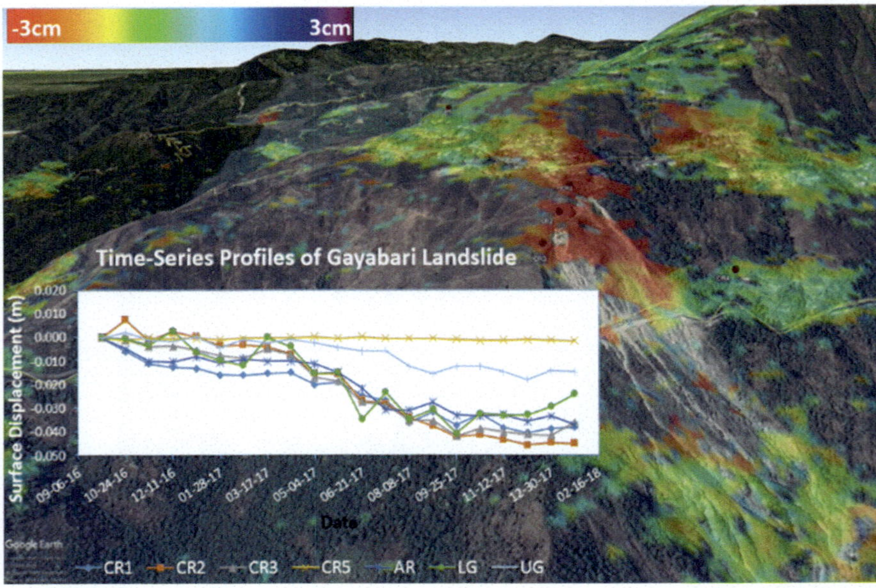

Fig. 1.5 Gayabari landslide displacement motion from RADARSAT-2 InSAR. The black dots on the image show the location of the installed corner reflectors (CRs)

InSAR Deformation Monitoring Before and After Hurricane Maria, Dominica

Dominica is a small forested volcanic island (47 × 29 km) in the eastern Caribbean. Its highest peak is 1447 m asl, and about 60% of its slopes are steeper than 30% gradient (World Bank, Dominica 2004). Over 80% of the island is soaked by an average rainfall of 2500 mm throughout the year (Benson et al. 2001), especially in the hurricane season from June to November. The island's volcanic bedrock is overlain by weak wet clayey soils, which are prone to landslides, especially on steep slopes (De Graff et al. 2012). A national-scale landslide susceptibility map and research on triggering mechanisms for Dominica were conducted by van Westen (2016).

In September 2017, category 5 hurricane Maria triggered about 10,000 landslides throughout the island (van Westen and Zhang 2018). It resulted in total damages of USD931 million, which amount to 226% of Dominica's 2016 gross domestic product (GCD Government of the Commonwealth of Dominica 2017). The need to reduce risk and build resilience for future hazards in Dominica was therefore identified (GCD Government of the Commonwealth of Dominica 2017). A community-based approach to landslide risk reduction to vulnerable communities has also been recommended (Anderson et al. 2011).

We used RADARSAT-2 InSAR to identify post landslide activity after hurricane Maria. We analyzed 40 RADARSAT-2 multi-look fine (3 m) InSAR images from January 2014 to June 2018. Our results focused on bare slopes with good InSAR

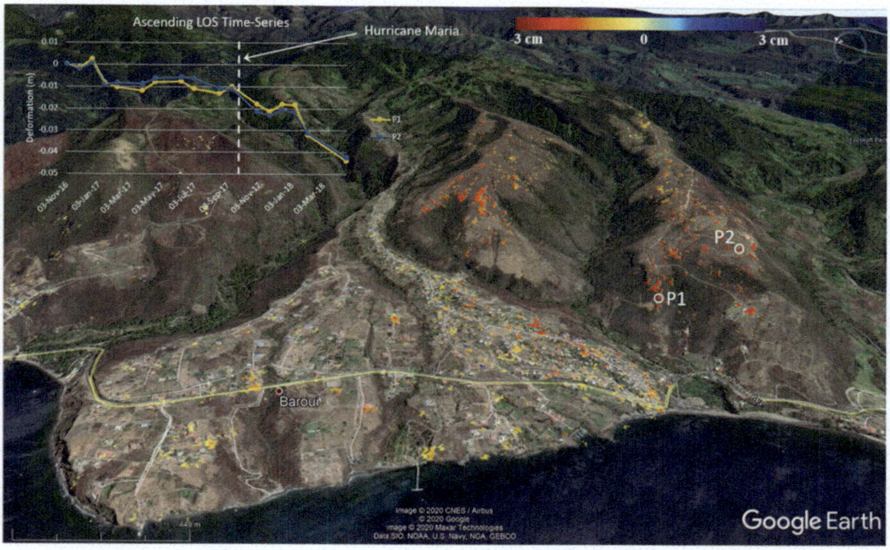

Fig. 1.6 Landslide motion from RADARSAT-2 InSAR over Dominica before and after Hurricane Maria (September 2017)

coherence and where landslides were expected to occur. Our results, for a small area in the eastern part of Dominica (Fig. 1.6), is representative of the island. We show that there is considerable landslide motion on the crest of some of the bare slopes. Further work needs to be done because some bare slopes show landslide motion, but not all. We recommend more continuous rapid revisit InSAR monitoring throughout the year, especially during the hurricane season. This will assist in understanding the differential motion of the slopes within Dominica. This case study shows InSAR can be integrated with existing techniques to develop early warning and mitigation techniques for other hurricane-prone Caribbean island states.

1.3 Energy Corridors

InSAR Monitoring of Pipeline Routes

In Canada and elsewhere, discussions are ongoing regarding pipeline routing, monitoring, construction, and planning related to the distribution of oil and gas for processing and export. Detailed characterization of the terrain and ground movement along a pipeline corridor is critical for pipeline integrity (Porter et al. 2014). We provide three case studies using InSAR deformation time-series analysis as part of an early warning technique to monitor the safety of oil pipelines on different terrain. We analyzed 157 RADARSAT-2 InSAR images for the three sites discussed below. We focused on the use of InSAR to identify potential risks in the planning of pipeline routes within (1) a steep mountain valley and on (2) permafrost terrain. Our third case study used InSAR to monitor a pipeline built on a soft foundation within a wetland area where an oil spill has occurred. Our results show that InSAR deformation monitoring can be used to investigate areas of potential risk along pipeline routes.

Pipeline on Permafrost Terrain

The study was conducted along the proposed Mackenzie Valley pipeline in northern Canada on permafrost terrain. The pipeline route was planned, many geotechnical studies were conducted, but the pipeline was never built. Many important lessons were learnt however, related to large engineering projects planned on permafrost terrain. In addition, the socioeconomic impacts of permafrost degradation can be costly particularly for engineering construction in these areas (Larsen et al. 2008). This case study is only one example. Our objective here is to emphasize the importance of InSAR for monitoring current and planned infrastructure on permafrost terrain affected by its potential degradation.

Permafrost is defined as frozen ground that remains at or below 0 °C for two or more years. It is estimated to occupy about 24% of the northern hemisphere land surface (Zhang et al. 1999). The active layer of surficial material, approximately 1 m above the permafrost, freezes and thaws seasonally and therefore, affects the foundations of engineering and transportation infrastructures. Monitoring and mapping of recent permafrost degradation have been documented by Lawrence and Slater (2005), Lantuit and Pollard (2008), Short et al. (2011), Fraser et al. (2018), and others.

The Mackenzie Valley is experiencing one of the highest rises in mean annual temperature for any region in Canada. This increase in temperature is triggering climate-driven thaw in the permafrost and subsequent landslides (Dyke and Brooks 2000; Fraser et al. 2018). In areas where the thin forest cover is burned by wildfires, permafrost thaw and landslides are accelerated (Couture and Riopel 2008). There are approximately 2000 landslides, mainly triggered by climate-driven permafrost thaw, along the proposed Mackenzie Valley pipeline route (Aylsworth et al. 2000).

This case study focuses on the Thunder River landslide within the pipeline corridor. The study area is characterized by permafrost with a thickness reaching up to 100 m. Relief is relatively low, with an elevation difference of 402 m. Vegetation cover is dominated by sparse-to-open coniferous forests, shrubs, and exposed land produced from natural fire scars.

The Thunder River landslide is a retrogressive thaw flow where the active layer is detached at the contact with the ground ice (Fig. 1.7 insert). The slide debris consists of thawed permafrost with a slurry of sediments and melted water, which can flow at an angle as low as 1°–2°. Such retrogressive thaw flows may start from a small-scale slope failure that can be as small as a few meters wide and retrogress over a distance as large as a few hundred meters (Wang et al. 2009). The deep thaw is caused by current warm summers combined with the destruction of the insulating surface from forest fires. The retrogressive failures may remain active for many

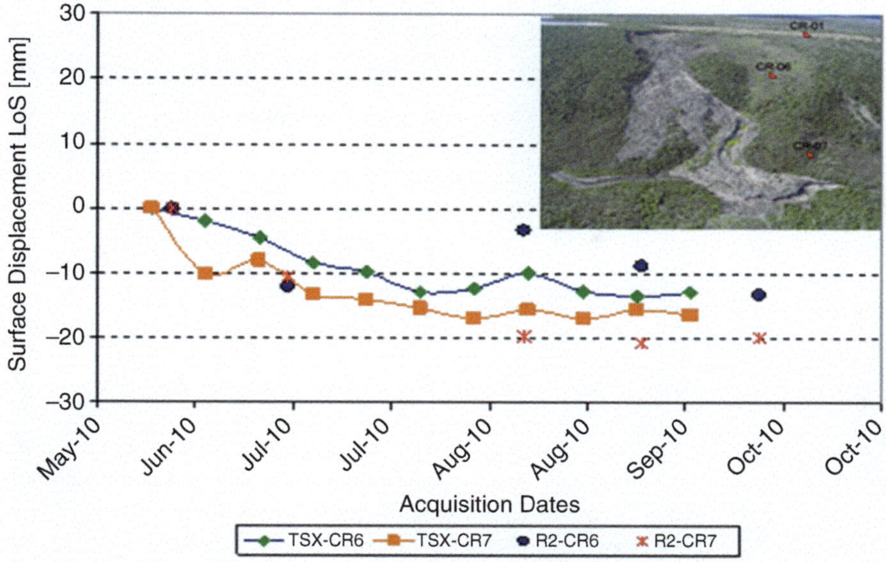

Fig. 1.7 RADARSAT 2 and TerraSAR InSAR time-series showing surface deformation resulting from permafrost thaw within the active layer due to warmer summer temperatures at the Thunder River landslide

years as long as the ice-rich materials are exposed (Wang et al. 2009). Understanding the failure processes and their rates of movement are important in developing appropriate remedial measures related to the proposed Mackenzie Valley pipeline route.

Figure 1.7 shows typical ground motion from RADARSAT-2 and TerraSAR InSAR data during the summer months over permafrost terrain within the Thunder River area. Both InSAR observations presented similar temporal variations of deformation, with about 20 mm of downslope motion recorded from the corner reflectors installed near the retrogressive thaw flow landslide. However, TerraSAR provides more dynamic motion within the time-series because it has a shorter revisit time compared to RADARSAT-2. In Fig. 1.8, the colors represent motion resulting from the active layer freeze-thaw cycles, and warmer temperatures in the summer. The area in red shows a small ground heave of about +20 mm and the areas in blue show thaw of about −20 mm over the summer months within the proposed pipeline corridor. Earlier long term InSAR monitoring studies using corner reflectors near the landslide (Fig. 1.7) showed the downslope motion of 20–30 cm from 2006 to 2010 (Singhroy 2008; Singhroy et al. 2010). The maximum motion was observed during the summer months. The thin veneer of thermally conductive mineral soils are linked to the active layer responses to summer warming (Fraser et al. 2018). Farquharson et al. (2019) also reported ground subsidence totaling tens of centimeters within the active layer in some areas. Research on active layer response to summer warming is still ongoing as the responses are variable, and depend on a number of factors, such as local snow accumulation and active layer thickness (Zhang 2005; Lawrence and Slater

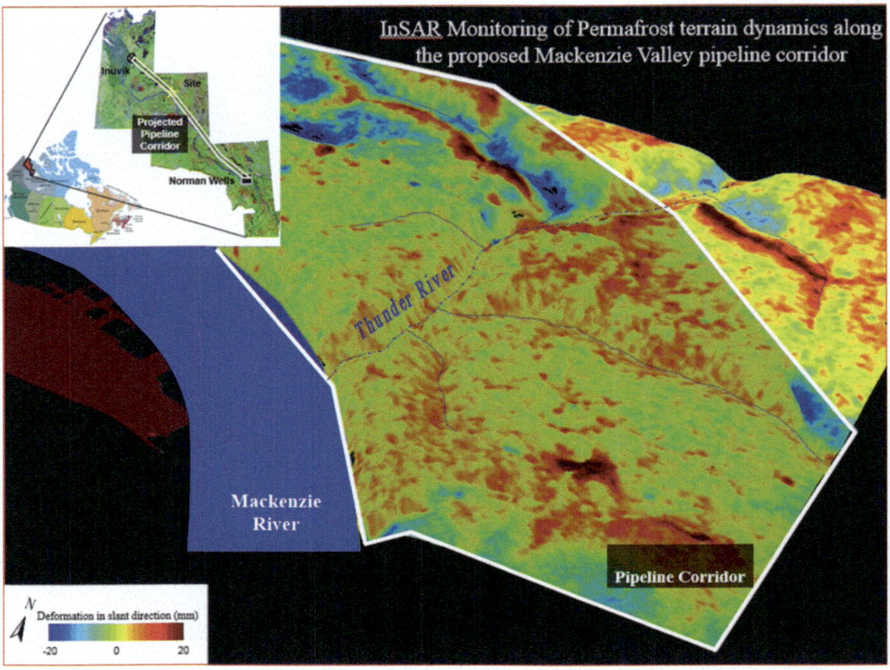

Fig. 1.8 InSAR deformation monitoring on a proposed pipeline route over permafrost terrain. Red areas show where the ground is heaving (swells) and the blue areas show where the ground is sinking because of thawing

2010). To understand the impact of this dynamic motion on existing and future infrastructure, integrated monitoring would be required, including high-resolution rapid revisit InSAR provided by the RADARSAT Constellation Mission and other radar missions.

Pipeline Routes in Steep Mountainous Areas

This study focused on a proposed pipeline route within a narrow mountain valley in the Canadian Cordillera. The frequency of landslides on these mountain slopes are increasing and may be linked to climate change (Cloutier et al. 2017). The steep valley slopes produce debris flows which can affect pipeline routes. Identifying potential debris flows, and areas of downslope sediment accumulation are critical for pipeline safety within the valley.

InSAR monitoring was carried out following the methods outlined by Singhroy and Li (2017). We processed 17 RADARSAT-2 ultra-fine (3 m) scenes acquired from December 2015 to January 2017 in an area close to Nimbus Mountain, British Columbia, Canada. We selected a downslope view that corresponded to a slope angle of 35° in ascending orbit. We also selected short perpendicular baselines

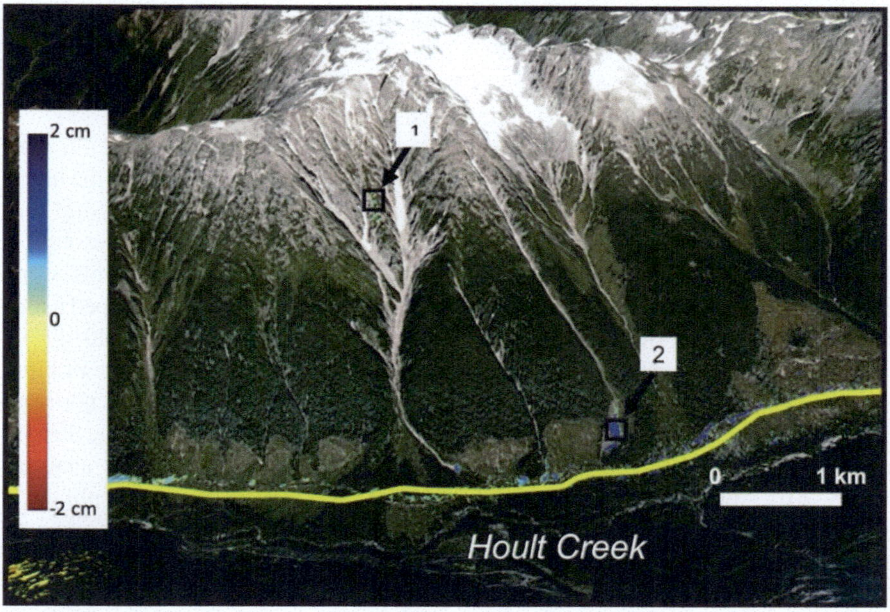

Fig. 1.9 Ground displacement results from InSAR monitoring of slopes within the Holt Creek valley. Blue tones indicate accumulation of sediment and yellow and red tones indicate depletion and potential debris flow zones. Points 1 and 2 reflect active debris flow channels. The yellow line shows a previously proposed pipeline route (Blais-Stevens et al. 2018)

(<100 m) with short time intervals between acquisitions and corrected for topography and atmospheric conditions (Singhroy et al. 2012).

Our InSAR results show an area where debris flows occur. The area at Point 1 shows slope deformation at higher elevations in the debris flow channel. This corresponds to the area classified as having a high likelihood of landslide initiation in the geological map published by Maynard et al. (2018). Point 2 at the base of the slope is on a debris flow fan where sediment accumulation increases over time (Blais-Stevens et al. 2018; Fig. 1.9). Furthermore, most of the steep stream channels also show sediment accumulation (blue) at their base. In the lower left area of Fig. 1.9, the yellow indicates sediment depletion on the opposite side of the valley at high elevations. Identifying these areas of sediment motion in the upper and lower parts of the valley from InSAR data before a pipeline is constructed will assist in building safe pipelines and developing the appropriate mitigation measures.

Pipelines on Wetland Areas

Wetland areas are particularly challenging for pipeline construction. Seasonal displacement and swelling often cause structural damage to the pipes. The high compressibility of the peat results in consolidation, as well as lateral and horizontal displacement (Corrales et al. 2017). In the boreal forest regions of Canada, wetlands are also subjected to the freeze/thaw cycles. This seasonal process describes the transition of the landscape between frozen and thawed conditions which can be measured from ground reflected GNSS signals (Chew et al. 2017).

Our InSAR analysis over the wetland site uses 9 years of RADARSAT-2 ultrafine (134 images) data from August 2008 to November 2017. The resolution of the ultrafine beam mode is ~1.6 m in range and ~ 2.8 m in azimuth. InSAR processing was done using the same methods outlined by Singhroy and Li (2017). A 5 m resolution DEM was used for topographic correction. The flattened interferograms were filtered using an adaptive noise filter (Goldstein and Werner 1998), unwrapped using a minimum-cost algorithm (Costantini 1998) and geocoded (Pearse et al. 2014).

In July 2015, a pipeline spilled five million liters of emulsion (a mixture of bitumen, sand, and water) at the Long Lake site in the oil sands region of northern Alberta (Singhroy et al. 2012). The spill occurred in the wetland area near the Steam-Assisted Gravity Drainage (SAGD) site. Our 9 year RADARSAT InSAR analysis (2008–2017) over this area shows a −10 cm deformation over the spill area (Fig. 1.10). Investigation by the company suggests that the cause of the rupture was a thermally driven upheaval, along with an incompatible pipeline design for the muskeg ground conditions (Price 2017). Our InSAR time-series results show seasonal

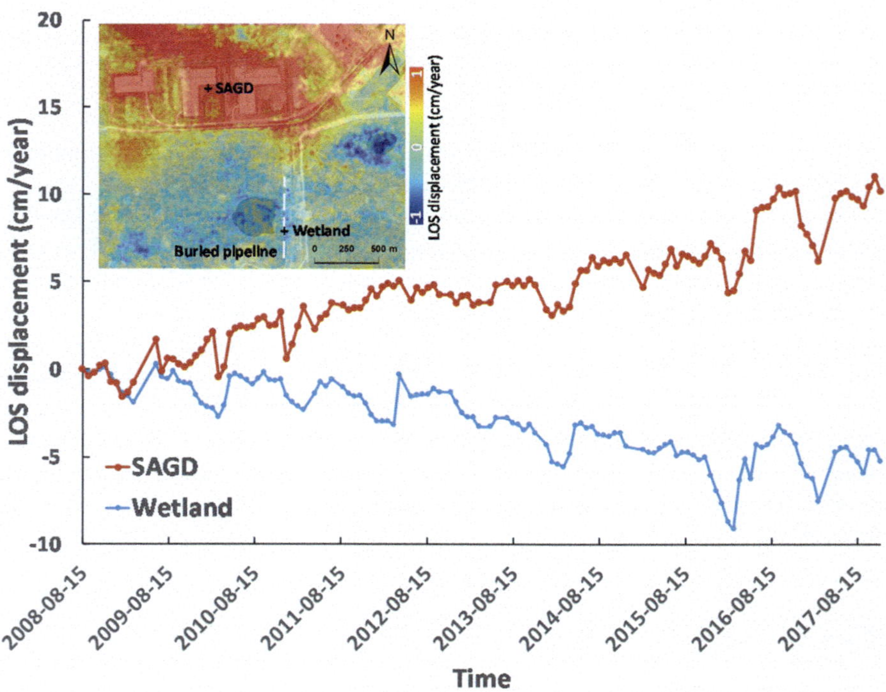

Fig. 1.10 InSAR deformation due to a pipeline leak on a wetland (blue) near SAGD site (Red) using RADARSAT InSAR time-series

fluctuations of 5 cm along the InSAR line of sight (LOS) before and after the oil spill within the wetland (muskeg) area. This amount of fluctuation can result in significant mechanical stress on the pipeline and indicates that InSAR land motion monitoring on muskeg ground (and other wetland areas) is essential when developing an early warning system to prevent potential pipeline rupture.

1.4 Mining Infrastructure

InSAR Monitoring of Oil Extraction in Alberta

Thermal viscous oil exploration requires a sound understanding of geomechanics to enhance the recovery process and reduce environmental impacts. Surface deformation measurements, as part of the geomechanical monitoring process, are critically important. They provide a reliable way to calibrate and verify Thermal-Hydro-Mechanical (THM) models dealing with heat transfer, fluid flow, and mechanical stress and strain within a reservoir. The Alberta oil sands in northern Alberta contain one of the world's largest deposits of crude oil, in the form of viscous bitumen embedded within uncemented sand (Mossop 1980; Mossop and Flach 1983). The four major oil sand deposits extend over a large area of northern Alberta, with total initial in-place reserves estimated at 1.8 trillion barrels (Oil Sands 2010).

The oil sands differ from other crude oil deposits in that they are heavier and have a viscous tar-like consistency. They are unable to flow under reservoir conditions and must be recovered using unconventional methods. Mining reserves deeper than about 65 m is not economical. The bitumen must be produced in situ using enhanced recovery techniques. This relies on a thermal method of heating the bitumen to reduce its viscosity. This thermal method is known as the steam-assisted gravity drainage (SAGD) method. In the SAGD process, high-pressure steam is injected into the bitumen reservoirs at close to, or higher than, initial reservoir pressures. Groups of horizontal wells are drilled into the bitumen reservoir, side by side, while the heated oil drains from around the growing steam chambers, driven by gravity, to lower horizontal wells (Butler 1994; Collins 2007; Peacock 2011). The combined withdrawal and injection of material into the subsurface is known to cause surface deformation, which has sometimes led to the failure of well casings. The causes of failure depend on a variety of factors including the rate of the injection and withdrawal, temperature and pressure of injected material, properties of the reservoir and overburden, and local geology (ERCB 2010; Peacock 2011). We provide an example of InSAR monitoring of surface deformation at the Mackay SAGD site to identify and monitor potential over-pressure zones.

Glacial and recent deposits with different overburden and lithological thickness vary from 50 to 360 m above the bitumen productive layer. Singhroy and Charbonneau (2014) have shown that the overburden thickness is not influencing the surface heave. About 82% are considered to be recoverable mainly by thermal in situ methods and 18% by surface mining (Alberta Energy Utilities Board 2005; Hein and Cotterill 2006).

Several techniques are being used to measure surface deformation in the oil sands. These include leveling, gravimeters, tilt meters, inclinometers, GPS, and InSAR (Stancliffe and van der Kooij 2001; Dusseault 2011; Nur Khakin et al. 2013; Shen et al. 2014). We processed over 60 InSAR RADARSAT-2 ultrafine mode (3 m) scenes starting from summer 2011 to fall 2013 at the Mackay site. Surface deformation is a response to changes in pressure and temperature in the subsurface, so heave measurements can be used as a tool for monitoring reservoir pressures indirectly. The geomechanical process of SAGD, where constant pressure is used to achieve production through fluid gravity segregation, results in an increase in temperature, pressure, porosity, permeability, dilation, and surface deformation (Dusseault 2011; Shen et al. 2014). The InSAR observations show persistent surface heave occurring at rates of 2 cm/year over a 2 year monitoring period from 2011 to 2013 (Fig. 1.11). We found that heaving above the horizontal injector wells is strongly correlated with rates of steam injection, even though there is a net fluid loss from the reservoir pore space as oil and water are withdrawn through the production wells. There are also minor swelling and ponding on the infill rehabilitation area near the tailings pond shown on (Fig. 1.11).

Monitoring Mining Activities in Guyana

Mining in tropical countries significantly contributes to the global supply of minerals. Monitoring environmentally destructive illegal mining is a global problem, especially in the remote parts of South America and Africa (particularly Ghana and Congo) (Ansner et al. 2014; Boadi et al. 2016). The practice of unregulated and illegal mining in these countries is sometimes associated with deforestation, loss of habitat and biodiversity, mercury pollution, and excessive river sediments. Small-scale gold mining in the remote, densely forested areas in the interior of Guyana are numerous, mobile, and difficult to reach. It is estimated that the gold mining sector produces in excess of USD300 million in gold annually. The government is aware of the environmental problems produced by mining, but controlling them is beyond its capabilities (Roopernarine 2014; Hilso and Laing 2017). The majority of the country's gold mining is conducted by small and medium-scale operators. The mining properties of large international companies are known and regulated.

In this example, we used ALOS PALSAR (AP) imagery to monitor small-scale mining sites in one of the remote areas in Guyana. ALOS L-band images have been very useful for geological mapping in similar densely forested areas (Pour and Hasim 2017). We focused our study on the use of change detection techniques applied to radar imagery to monitor small-scale gold mining activities along with

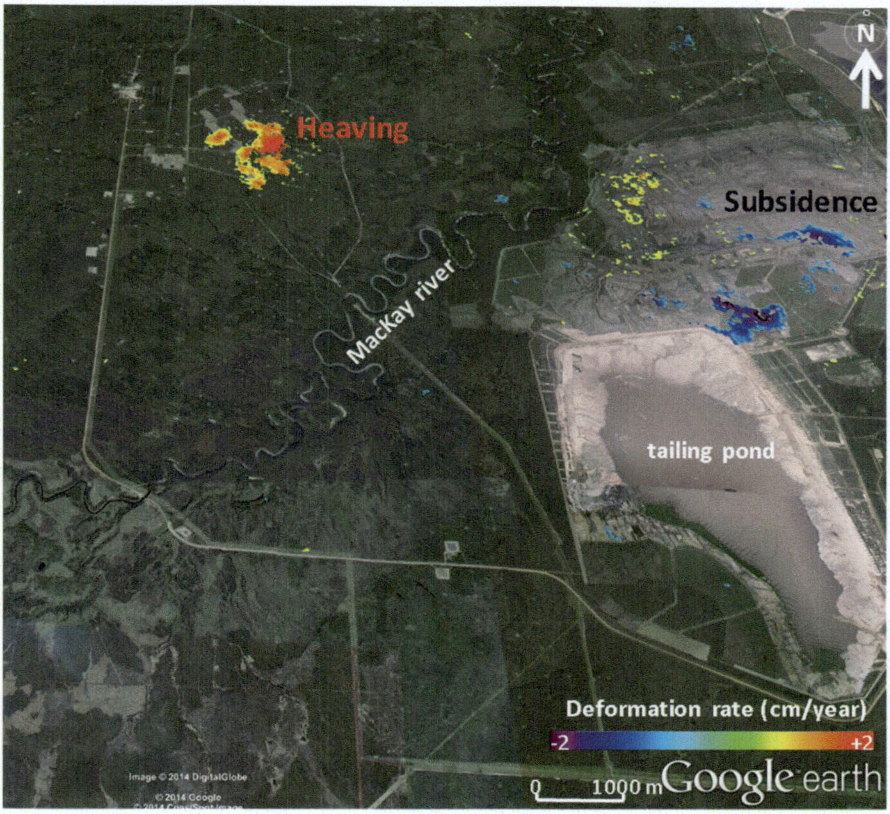

Fig. 1.11 RADARSAT-2 InSAR time-series showing surface deformation at the Mackay River SAGD extraction site

parts of the Mazaruni River in Guyana (Figs. 1.12 and 1.13). The geometric and radiometrically corrected AP images were produced globally by JAXA for 2008 and 2016 (Global PALSAR-2/PALSAR/JERS-1 Mosaic and Forest/Non-Forest map https://www.eorc.jaxa.jp/ALOS/en/palsar_fnf/fnf_index.htm; Singhroy et al. 2019). Based on the intensity level of the backscatter in the radar images, JAXA separated the data into three classes; water, forest, and non-forest. By identifying the areas that changed from forest in 2008 to non-forest in 2016, we were able to observe the changes in placer mining operations along rivers within this time period shown in yellow in Fig. 1.12. Within the study area, our results show an increase of 370% in gold mining activities over this 8 year period. The monitoring of mining activities of other areas can be done using this radar change detection technique. The change maps will assist these countries in monitoring legal and illegal mining activities.

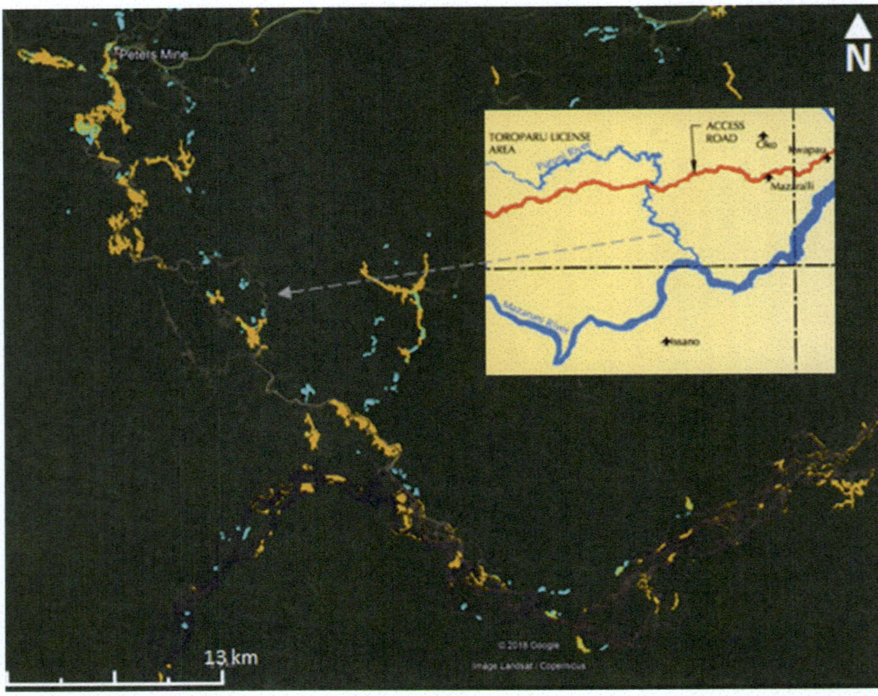

Fig. 1.12 Placer gold mining along river banks in a cloudy, dense tropical forest identified from ALOS PALSAR images. The yellow color represents the changes in mining operations from 2008–16. Comparing these areas with registered mining leases will help identify illegal mining operations. Inset map helps to identify the location of the small-scale mining

Mineral Exploration Using Radar Image Fusion

The search for minerals and engineering construction materials is an important part of mining infrastructure. In areas where the surficial materials and bedrock are covered by vegetation, there is a need to improve the interpretation of remote sensing images that indirectly provide useful information on minable construction materials and minerals. Image enhancement and data fusion techniques have been used to facilitate the search for minable surficial materials and minerals in vegetated areas (Singhroy 1996; Barnett et al. 2004; LaRocque et al. 2011; Harris et al. 2011, and others).

In this section, we present an example of using simple image fusion techniques to assist in delineating geological faults, as well as lithological contacts which are used in mineral exploration. Our example is focused within an active mining area in Sudbury Canada.

The Sudbury impact structure is one of the world's largest producer of nickel. The area has about 90 Ni-Cu-PGE (nickel, copper, platinum-group metals) deposits, and includes several operating mines. The mineralization is mainly associated with

Fig. 1.13 Field photo of local small-scale unregulated placer gold mining in the tropical forest areas in Guyana (Guyana Geology and Mines, 2019)

the basal contact of the Sudbury Igneous Complex (SIG) (Ames et al. 2008). Norite and gabbro are two of the four major lithologies of the SIG shown in Fig. 1.14. Tuba et al. (2014), and others, have suggested that large-scale brittle fault structures are conduits for PGE enriched fluids in Sudbury and several other locations around the world. Mapping the surface expression of faults using radar images has been extensively reported by Lowman (1991), Singhroy et al. (1993a, b), Singhroy and Lowman (2012), and others.

The Sudbury area consists of glaciated undulating terrain (400 m) covered by trees within the boreal forest. Unlike images/air photos from which topographic maps are produced or a lidar image which looks down perpendicular at the terrain, we selected a 35° viewing angle to enhance topographically controlled structures. In addition, we added three different polarizations of RADARSAT-2 imagery (VH, HH, and VV) to enhance the moisture captured in the faults (Fig. 1.14). Radar is a moisture seeker, and recent research has shown (Fobert et al. 2018) that polarimetric radar imagery are particularly useful in mapping moisture trapped faults in vegetated terrains. Optical satellite images such as Landsat are less useful because their optical and infrared bands are not as sensitive to moisture-filled linear features as the SAR backscatter.

Using the biased viewing angle and the polarimetric radar images we provide an example where a polarimetric composite image (Fig. 1.14c) provides additional information on lithologic contacts and faults to assist in mineral exploration

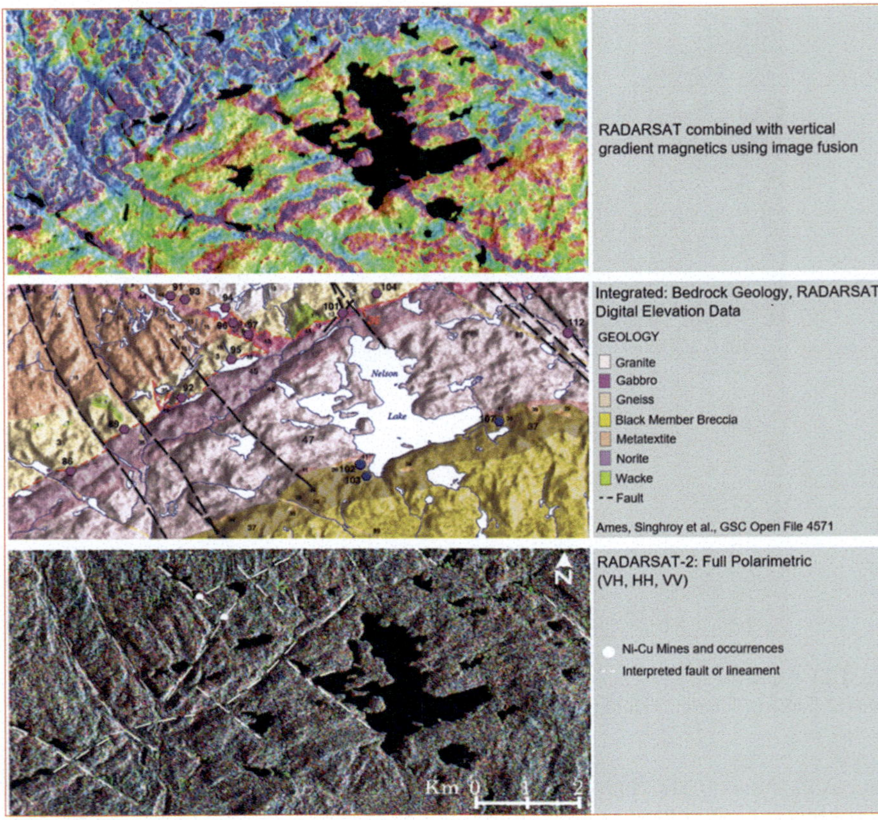

Fig. 1.14 Shows image fusion techniques used to identify faults, rock types, and contact zones which control the nickel copper (Ni Cu) mineralization (Ames et al. 2006) in the Nelson Lake area, Sudbury, Ontario. (**a**) shows the fusion of vertical gradient magnetics and a RADARSAT image. Faults are shown as NW-SE red linear. (**b**) shows the published rock types, faults and mineral map overlain on a RADARSAT image. (**c**) shows a polarimetric RADARSAT-2 image (VH, HH, VV) used to identify additional NW-SE faults and SW-NE contacts not clearly shown on the geological map and magnetic image

programs. The fusion of radar and airborne magnetic images (Fig. 1.14a) has enhanced the interpretation of the magnetic images by confirming the magnetic signatures related to faults, dykes, or contacts, and add new topographically expressed or moisture-filled faults. Figure 1.14b shows an example where there is a need to update an existing geological map. The radar image overlay can also confirm or add lithological and structural information. These integrated radar image maps provide additional and up-to-date structural and lithologic contact information to assist in targeting drilling areas for exploration, education, and research.

1.5 Coastal Infrastructure

Monitoring Coastal Change Using Radar Time Series Images

Coastal areas are the most vulnerable to climate change due to sea-level rise (SLR). This includes coastal flooding and changes to the littoral processes (i.e., sediment transport patterns) (Nicholls et al. 2011; Sánchez-Arcilla et al. 2011).

All of the productive agricultural lands, and most of Guyana's population, reside in the narrow and fertile coastal plain (425 × 50 km), which is below sea level. The coastal dikes, and concrete and earth dams constructed for sea defense, need constant monitoring and improvement to control flooding and to protect both coastal infrastructure settlements and irrigation for agricultural areas.

This study shows that parts of the Guyana coastline are undergoing significant changes. Recent research has shown that changes of a few meters to half of a kilometer have occurred from satellite image observations over the past 50 years (Singhroy 1996a, 1999). Similar coastline changes observed from radar imagery are occurring along the coast of French Guyana and NE Brazil (Barbosa et al. 1999; Trebossen et al. 2005). This rate of change is variable and occurred in different parts of the coastline. The changes are due to SLR, sand mining, mangrove depletion, and natural shoreline erosion and accretion processes (Singhroy 1995, 1997; Temitope et al. 2019). Ahmad and Lakhan (2012) utilized a Geographical Information System (GIS)-based approach to model coastline changes, i.e., its linkages to recurring episodes of coastal accretion, erosion, and mudbank development.

An example of shoreline change is shown in Fig. 1.15. We observe a change of 0–400 m from 2007 to 2018 using ALOS L-band radar imagery. The results show areas of stable, eroded, and accreted shoreline. There appear to be more areas of erosion than accretion which may be the result of increasing storm events due to SLR. Our example is a snapshot of similar processes that are occurring along the entire Guyana coastline. This has serious implications for sea defense, mangrove protection and restoration, and flood protection for infrastructure settlements and agriculture. Measuring the shoreline changes from satellite radar imagery will also assist in locating areas for near-shore gas pipeline facilities being planned because of the recent offshore oil and gas discoveries. In addition, the potential of oil spills along the Guyana coast requires the need to understand the dynamics of long and short term littoral processes, and requires the ability to monitor and measure the impacts of oil spills on coastal ecosystems.

Oil Spill Monitoring

Ocean oil spills can be caused by tanker ruptures, illegal oil discharges by ships, or natural oil seepage. The International Tanker Owners Pollution Federation (ITOPF) estimated that between 1970 and 2016 approximately 5.73 million tons of oil were

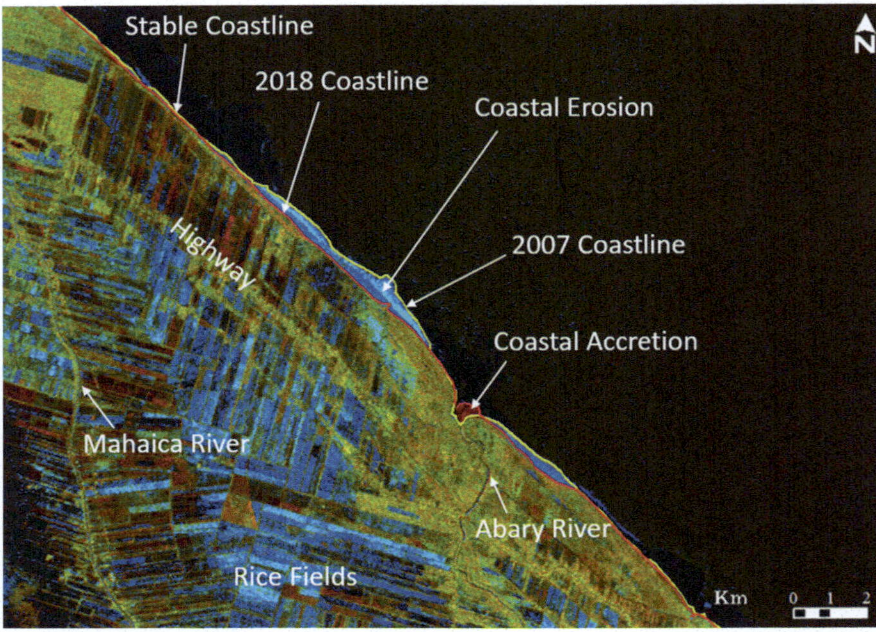

Fig. 1.15 Coastal changes of the Guyana coastline over an 11 year period (2007–2018) using cloud free radar images

lost as a result of tanker incidents (Roser 2020). However, these incidents have decreased substantially since 1970. While the majority of these spills are minor (less than 7 tons) (ITOPF), both large and small spills can cause serious damage to coastal ecosystems and can harm coastlines and near-shore waters.

Radar imagery is highly useful in the detection of oil spills because of its day and night imagining capability and the clear distinction of oil on the sea surface. Oil spills floating on the sea surface appear dark on the radar image (Fig. 1.16). The oil slicks dampen the surface waves resulting in a low radar backscatter (dark areas). However, not all dark areas on the ocean surface are oil spills. The dark areas on the open ocean surface can sometimes have lookalikes. These include cold water upwelling, flow regimes associated with internal waves and oceanic eddies, changes in the stability of the air–sea interface, and from natural surface films produced by large areas of plankton or fish (Alpers et al. 2017). Manual and automatic pattern recognition techniques are used to discriminate between oil slicks and lookalikes (Brekkea and Solbergb 2005; Alpers et al. 2017). In addition, polarimetric SAR observations have been used to improve the oil slick characterization and separation from lookalikes (Migliaccio et al. 2015).

In this case study, we report on a recent oil spill off the coast of Newfoundland, Canada, and the use of RADARSAT imagery to discriminate the spill. On July 17th, 2019, approximately 12,000 L of oil and water leaked from a storage tank aboard the Hibernia offshore oil platform (C-NLOPB 2019). The data used were

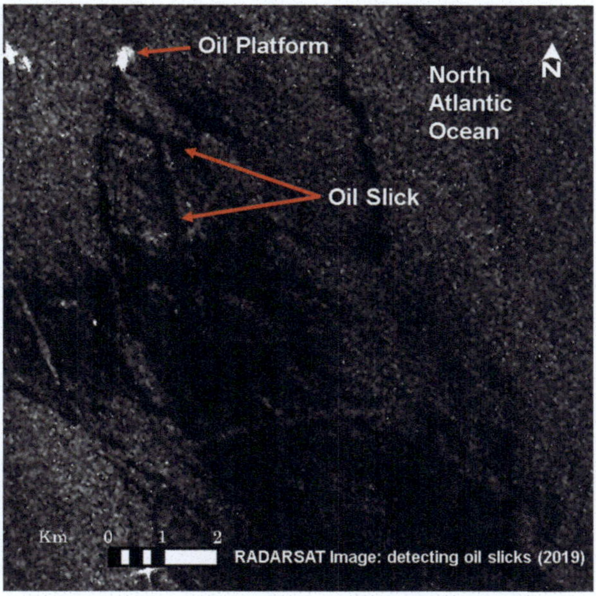

Fig. 1.16 RADARSAT-2 image showing the extent of the oil spill from July 2019 (dark areas) near the Hibernia oil platform in the North Atlantic (2019)

RADARSAT-2 Standard mode 5 acquired on July 21, 2019. The polarization selected was VV as it is well suited for marine applications such as oil spill detection (Li et al. 2018). The data were obtained in an ascending right-looking orbit, with incidence angles of 36°–42° and 12.5 × 12.5 m spatial resolution. Subsequently, the data were geometrically corrected and filtered with a 5 × 5 Enhanced Lee filter (Hu et al. 2012). Normalization was then applied to enhance visualization.

Our results show that the oil spills appear dark on the image. This was a relatively calm sea and as such the separation of the oil slicks from the background ocean was very clear (Fig. 1.16). The oil platforms act as corner reflectors and appear as bright targets on the image. There were no cleanup efforts for this particular spill because the oil was very thin and therefore not suitable for cleanup.

Although oil spills can be easily detected during relatively calm ocean conditions, continuing research and verification methods are being developed for the estimation of oil spill thickness and on the discrimination of oil spills from lookalikes (Fingas 2018).

1.6 Conclusion

We provided a review of the application potential of the common sensors that are being used for monitoring infrastructure. We focused on case studies using high-resolution radar and InSAR images for critical transportation and energy corridors, mining, and coastal infrastructure. These case studies will assist in developing integrated early warning systems and mitigation strategies for climate-resilient infrastructure.

Our case studies show that:

- High-resolution InSAR images are effective in monitoring landslides along highways and railways in coastal and mountainous Himalayan communities. Land motion is triggered during active wet spring periods (Newfoundland and Quebec), monsoons (Indian Himalayan), and the hurricane season (Dominica). Landslides are spontaneous and require timely InSAR images from the RADARSAT Constellation Mission (4 days), and other radar constellations. In the tropical island states, and on steep Himalayan slopes, more rapid InSAR monitoring is needed during hurricane, typhoon, and monsoon events.
- As part of safe mining practices, InSAR is used to monitor uplift rates produced by steam injection in the mining of the oil sands in Canada. The surface heaving measurements will assist in understanding the dynamic changes of the reservoir properties.
- InSAR is used to monitor pipeline corridors in high risk mountainous, wetland, and permafrost terrain. This can assist in targeting geotechnical investigations and develop an early warning system to prevent potential oil spills.
- Change detection techniques of multi-temporal radar images are effective in monitoring mining infrastructures in remote vegetated tropical areas. This can help in identifying illegal mining in cloudy remote tropical areas.
- Coastal erosion and accretion change low lying shorelines. Radar time-series and change detection techniques are useful to monitor these shifting coastlines. This is significant for monitoring the effects of the rising sea level on low lying coastlines and nearby coastal infrastructure.
- The fusion of high-resolution vertical gradient magnetics and multipolarization radar images is providing additional information for the improvement of the fault and rock unit delineation in vegetated areas. This will assist in mineral exploration.
- Radar images are very useful in detecting oil spills in the ocean. Research is currently ongoing to differentiate oil spills from lookalikes.

Acknowledgments The case studies described above are an extension and a reinterpretation of the published works conducted by the authors at the Canada Centre for Remote Sensing. The methodologies are simplified for a general audience.

References

Ahmad, S.R., and C. Lakhan. 2012. GIS-based analysis and modeling of coastline advance and retreat along the coast of Guyana. *Marine Geodesy* 35 (1): 115. https://doi.org/10.1080/01490419.2011.637851.

Alasset, P.J., V. Poncos, V. Singhroy, and R. Couture. 2007. InSAR monitoring of a landslide in a permafrost environment: Constraints and results. In *100th Canadian Institute of Geomatics/3rd*

International Symposium on Geo-Information for Disaster Management, Toronto (ON), CD-ROM.

Alberta Energy and Utilities Board. 2005. Alberta's Reserves 2004 and Supply Demand Outlook 2005–2014: Statistical Series. ST98-2005, 1–1–A-29.

Alpers, W., B. Holt, and K. Zeng. 2017. Oil spill detection by imaging radars: challenges and pitfalls. *Remote Sensing of Environment* 201 (2017): 133–147. https://doi.org/10.1016/j.rse.2017.09.002.

Ames, D.E., V. Singhroy, J. Buckle, and K. Molch. 2006. *Geology, Integrated Bedrock Geology RADARSAT - Digital Elevation Data of Sudbury*. Ontario: Geological Survey of Canada, Open File 4571, 2006, 1 sheet; 1 CD-ROM, https://doi.org/10.4095/222241 (Open Access).

Ames, D.E., A. Davidson, and N. Wodicka. 2008. Geology of the giant Sudbury polymetallic mining camp, Ontario. *Canada Economic Geology and the Bulletin of the Society of Economic Geologists* 103 (5): 1057–1077.

Anderson, M.G., E. Holcombe, J.R. Blake, F.D. Ghesquire, N. Holm-Nielsen, and T. Fisseh. 2011. Reducing landslide risk in communities: evidence from the Eastern Caribbean. *Applied Geography* 31: 590–599.

Ansner, G.P., W. Llactayo, R. Tupayachi, and E.R. Luna. 2014. Elevated rates of gold mining in the Amazon revealed though high-resolution monitoring. *Proceedings of the National Academy of Sciences of the United States of America* 110: 18454–18459.

Aylsworth, J.M., A. Duk-Rodkin, T. Robertson, and J.A. Traynor. 2000. Landslides of the Mackenzie valley and adjacent mountainous and coastal regions. *Geology Survey Canada Bulletin* 547: 167–176.

Barbosa, M.P., V. Singhroy, and R. Saint-Jean. 1999. Mapping coastal erosion in Southern Paraíba, Brazil from RADARSAT-1. *Canadian Journal of Remote Sensing* 29 (3): 323–328.

Barnett, P.J., V.H. Singhroy, J. Shirota, and S. Leney. 2004. Methods for remote engineering geology Terrain analysis in boreal forest regions of Ontario. *Canada Environmental and Engineering Geology* X (3): 229–241.

Batterson, M., D.G. Livermam, J. Ryan, and D. Taylor. 1999. The assessment of Geological Hazards and Disasters in Newfoundland: An Update". In *Current Research Newfoundland Dept of Mines and Energy Geological Survey Report 99–1*, pp 95–123.

Benson, C., E. Clay, F.V. Michael, and A.W. Robertson. 2001. *Dominica: Natural Disasters and Economic Development in a Small Island State. Disaster Risk Management Working Paper Series No. 5, Disaster Management Facility*. Washington, DC: World Bank.

Bianchini, S., F. Raspini, L. Solari, M. Del Soldato, A. Ciampalini, A. Rosi, and N. Casagli. 2018. From picture to movie: twenty years of ground deformation recording over tuscany region, Italy, with satellite InSAR front. *Earth Science* 6 (177): 8 p. https://doi.org/10.3389/feart.2018.00177.

Blais-Stevens, A., D.E. Maynard, I. Weiland, V. Singhroy, J. Li, P. Behnia, and M.A. Fobert. 2018. Landslides in the Kitimat-Morice River Corridor, northwest British Columbia, Canada, Paper 604. In *GeoEdmonton Conference Proceedings, Canadian Geotechnical Society Annual Meeting, Edmonton*, 8 p.

Boadi, S., C. Ayine Nsor, O. Owusu Antobre, and E. Acquah. 2016. An analysis of illegal mining on the Offin shelterbelt forest reserve, Ghana: implications on community livelihood. *Journal of Sustainable Mining* 15: 115–119.

Boncori, J.P.M. 2019. Measuring coseismic deformation with spaceborne synthetic aperture radar: a review volume 7 | Article 16. *Frontiers in Earth Science* 7: 20p. https://doi.org/10.3389/feart.2019.00016.

Brancato, V., E. Rignot, P. Milillo, M. Morlighem, J. Mouginot, L. An, B. Scheuchl, S. Jeong, P. Rizzoli, J.L. Bueso Bello, and P. Prats-Iraolaa. 2020. Grounding line retreat of Denman glacier, East Antarctica, measured with COSMO-SkyMed radar interferometry data. *Geophysical Research Letters* 47: e2019GL086291. https://doi.org/10.1029/2019GL086291.

Brekkea, C., and A.H.S. Solbergb. 2005. Oil spill detection by satellite remote sensing. *Remote Sensing of Environment* 95: 1–13. https://doi.org/10.1016/j.rse.2004.11.015.

Butler, R.P. 1994. Steam-assisted gravity drainage: concept, development, performance and future. *Journal of Canadian Petroleum Technology* 33 (2): 8 p. https://doi.org/10.2118/94-02-05.

Calò, F., D. Calcaterra, A. Iodice, M. Parise, and M. Ramondini. 2012. Assessing the activity of a large landslide in southern Italy by ground-monitoring and SAR interferometric techniques. *International Journal of Remote Sensing* 33 (11): 3512–3530.

Calò, F., F. Ardizzone, R. Castaldo, P. Lollino, P. Tizzani, F. Guzzetti, R. Lanari, M.-G. Angeli, F. Pontoni, and M. Manunta. 2014. Enhanced landslide investigations through advanced DInSAR techniques: the Ivancich case study, Assisi, Italy. *Remote Sensing of Environment* 142: 69–82. https://doi.org/10.1016/j.rse.2013.11.003.

Casu, F., M. Manzo, and R. Lanari. 2006. A quantitative assessment of the SBAS algorithm performance for surface deformation retrieval from DInSAR data. *Remote Sensing of Environment* 102 (3–4): 195–210.

Charbonneau, F., B. Brisco, R.K. Raney, H. McNairn, C. Liu, P.W. Vachon, J. Shang, R. De Abreu, C. Champagne, A. Merzouki, and T. Geldsetzer. 2010. Compact polarimetry overview and applications assessment. *Canadian Journal of Remote Sensing* 36 (2): S298–S315. https://doi.org/10.5589/m10-062.

Chew, C., S. Lowe, N. Parazoo, S. Esterhuizen, S. Oveisgharan, E. Podest, C. Zuffada, and A. Freedman. 2017. SMAP radar receiver measures land surface freeze/thaw state through capture of forward-scattered L-band signals. *Remote Sensing of Environment* 198: 333–344.

Cloude, S. 2010. *Polarisation: Applications in Remote Sensing*. 1st ed. Oxford: Oxford University Press.

Cloutier, C., J. Locat, R. Couture, and P.E. Lord. 2010. Caractérisation des instabilités côtières dans le secteur de Port-Daniel-Gascons, Gaspésie, Québec. In *Proceedings of the 63rd Canadian Geotechnical Conference & 6th Can. Permafrost Conf., Calgary*, 71–79. ESS contribution No. 20100061.

Cloutier, C., J. Locat, M. Geertsema, M. Jakob, and M. Schnorbus. 2017. Potential impacts of climate change on landslide occurrence in Canada. In *Slope Safety Preparedness for Impact of Climate Change*, ed. K. Ho, S. Lacasse, and L. Picarelli, 71–104. London: Taylor and Francis Group. https://doi.org/10.1201/9781315387789-5.

C-NLOPB. (2019). Canada-Newfoundland and Labrador Offshore Petroleum Board 2019 Incident Disclosure 2019 Hibernia Oil Spill.

Colesanti, C., and J. Wasowski. 2006. Investigating landslides with space-borne synthetic aperture radar (SAR) interferometry. *Engineering Geology* 88 (3–4): 173–199.

Collins, P.M. 2007. The false lucre of low-pressure SAGD. *Journal of Canadian Petroleum Technology* 46 (1): 20–27.

Corrales, J., A. Garcia, and A.M. Ordonez. 2017. OCENSA Oil Pipeline Damage due to the Construction of an Embankment for a Highway on Soft Soils: Case Study. In *Proc. ASME 2017 International Pipeline Geotechnical Conference. Lima. Peru.*

Costantini, M. 1998. A novel phase unwrapping method based on network programming. *IEEE Transactions on Geoscience and Remote Sensing* 36 (3): 813–821.

Couture, R., and S. Riopel. 2008. Regional landslide susceptibility mapping, Mackenzie Valley, Northwest Territories. In *Proceedings of the 4th Canadian Conference on Geohazards: From Causes to Management. Québec, May 20–24, 2008*, 375–382. Quebec City: Presses de l'Université Laval.

Das, M.K., S. Bhowmik, R. Karmakar, S.B. Dasgupta, and A.B. Ekka. 2018. Monitoring of Chibo and Gayabari (14th Mile) landslides, Darjeeling and Kalimpong districts, West Bengal using InSAR. Open File report Geological Survey of India, 1–80.

De Graff, J.V., H.C. Romesburg, R. Ahmad, and J.P. McCalpin. 2012. Producing landslide-susceptibility maps for regional planning in data-scarce regions. *Natural Hazards* 64: 729–749. https://doi.org/10.1007/s11069-012-0267-5.

Dumont, S., F. Sigmundsson, M.M. Parks, V.J.P. Drouin, G.B.M. Pedersen, I. Jónsdóttir, A. Höskuldsson, A. Hooper, K. Spaans, M. Bagnardi, M.T. Gudmundsson, S. Barsotti, K. Jónsdóttir, T. Högnadóttir, E. Magnússon, A.R. Hjartardóttir, T. Dürig, C. Rossi, and

B. Oddsson. 2018. Integration of SAR data into monitoring of the 2014–2015 Holuhraun eruption, Iceland: contribution of the Icelandic volcanoes supersite and the FutureVolc projects. *Frontiers in Earth Science* 6: 231. https://doi.org/10.3389/feart.2018.00231.

Dusseault, M.B. 2011. Geomechanical challenges in petroleum reservoir exploitation. *Journal of Civil Engineering* 15 (4): 669–678.

Dyke, L.D., and G.R. Brooks. 2000. The physical environment of the Mackenzie Valley, Northwest Territories: a baseline for the assessment of environmental change. In *Geological Survey of Canada, Bulletin*, ed. L.D. Dyke and G.R. Brooks, vol. 547, 7–10; 1 CD-ROM, https://doi.org/10.4095/211901. (Open Access).

ERCB Energy Resources Conservation Board. 2010. Total E&P Canada Ltd. surface steam release of May 18, 2006 Joslyn Creek SAGD thermal operation: ERCB staff review and analysis.

Farquharson, L.M., V.E. Romanovsky, W.L. Cable, D.A. Walker, S.V. Kokelj, and D. Nicolsky. 2019. Climate change drives widespread and rapid thermokarst development in very cold permafrost in the Canadian high Arctic. *Geophysical Research Letters* 46: 6681–6689. https://doi.org/10.1029/2019GL082187.

Feng, J., D. Rogge, and B. Rivard. 2018. Comparison of lithological mapping results from airborne hyperspectral VNIR-SWIR, LWIR and combined data. *International Journal of Applied Earth Observation and Geoinformation* 64: 340–335. https://doi.org/10.1016/j.jag.2017.03.003.

Fernandes, R., C. Prevost, F. Canisius, S.G. Leblanc, M. Maloley, S. Oakes, K. Holman, and A. Knuby. 2018. Monitoring Snow Depth Change Across a Range of Landscapes with Ephemeral Snow Packs Using Structure from Motion Applied to Lightweight Unmanned Aerial Vehicle Videos. May 2018 The Cryosphere Discussions.

Fingas, M. 2018. The challenges of remotely measuring oil slick thickness. *Remote Sensing* 10: 319.

Flammini, F., C. Pragliola, and G. Smarra. 2016. Railway infrastructure monitoring by drones. 2016 international conference on electrical systems for Aircraft, railway, ship propulsion and road vehicles & international transportation electrification conference (ESARS-ITEC). *Toulouse* 2016: 1–6.

Fobert, M.A., J.G. Spray, and V. Singhroy. 2018. Assessing the benefits of simulated RADARSAT constellation mission polarimetry images for structural mapping of an impact crater in the Canadian shield. *Canadian Journal of Remote Sensing* 44 (4): 321–336. https://doi.org/10.1080/07038992.2018.1517022.

Fraser, R.H., S.V. Kokelj, T.C. Lantz, M. McFarlane-Winchester, I. Olthof, and D. Lacelle. 2018. Climate sensitivity of high Arctic permafrost terrain demonstrated by widespread ice-wedge thermokarst on Banks Island. *Remote Sensing* 10: 954.

Gamma Remote Sensing. 2015. http://www.gamma-rs.ch/software/ipta-interferometric-point-target-analysis.html.

GCD Government of the Commonwealth of Dominica. 2017. Post Disaster Assessment, Hurricane Maria, A Report by the Government of the Commonwealth of Dominica. 145pp.

Gee, D., A. Sowter, S. Grebby, G. de Lange, A. Athab, and S. Marsh. 2019. National geohazards mapping in Europe: interferometric analysis of the Netherlands. *Engineering Geology* 256: 1–22. https://doi.org/10.1016/j.enggeo.2019.02.020.

Ghosh, S., E.J.M. Carranza, C.J. van Westen, V.G. Jetten, and D.N. Bhattacharya. 2011. Selecting and weighting spatial predictors for empirical modeling of landslide susceptibility in the Darjeeling Himalayas, India. *Geomorphology* 131: 35–56.

Global PALSAR-2/PALSAR/JERS-1 Mosaic and Forest/Non-Forestmap. https://www.eorc.jaxa.jp/ALOS/en/palsar_fnf/fnf_index.htm.

Goldstein, R.M., and C. Werner. 1998. Radar interferogram filtering for geophysical applications. *Geophysical Research Letters* 25 (21): 4035–4038.

Harris, J., L. Wickhert, T. Lynds, P. Behnia, R. Rainbird, and E. Grunsky. 2011. Remote predictive mapping 3. Optical remote sensing – a review for remote predictive geological mapping in Northern Canada. *Geoscience Canada* 38 (2): 49–83.

Hein, F.J., and D.K. Cotterill. 2006. The Athabasca oil sands – a regional geological perspective, Fort McMurray area, Alberta, Canada. *Natural Resources Research* 15 (2): 85–102. https://doi.org/10.1007/s11053-006-9015-4.

Hilso, G., and T. Laing. 2017. Guyana gold: a unique resource curse? *Journal of Development Studies* 53 (2): 229–248. https://doi.org/10.1080/00220388.2016.1160066.

Hu, Z., L. Wei, and M. Guo. 2012. Method and implementation of oil spill detection in SAR image. In *Frontiers in Computer Education. Advances in Intelligent and Soft Computing*, ed. S. Sambath and E. Zhu, vol. 133. Berlin: Springer. https://doi.org/10.1007/978-3-642-27552-499.

InterAmerican Development Bank. 2018. What is Sustainable Infrastructure? A Framework to Guide Sustainability Across the Project Cycle Technicnal Note 1388. 38pp. https://doi.org/10.18235/0001043.

Irvine, M.L. 2012. Coastal Monitoring in Newfoundland and Labrador. In *Current Research, Newfoundland and Labrador Department of Natural Resources Geological Survey, Report 12–1*, 191–197.

Lantuit, H., and W. Pollard. 2008. Fifty years of coastal erosion and retrogressive thaw slump activity on Herschel Island, southern Beaufort Sea, Yukon territory, Canada. *Geomorphology* 95: 84–102.

LaRocque, A., B. Leblon, J. Harris, C. Jefferson, V. Tschirhart, and Y. Shelat. 2011. Surficial materials mapping in Nunavut, Canada with multibeam RADARSAT-2 dual-polarization C-HH and C-HV, LANDSAT-7 ETM+, and DEM data. *Canadian Journal of Remote Sensing* 38 (3): 281–305. https://doi.org/10.5589/m12-020.

Larsen, P.H., S. Goldsmith, O. Smith, M.L. Wilson, K. Strzepek, P. Chinowsky, and B. Saylor. 2008. Estimating future costs for Alaska public infrastructure at risk from climate change. *Global Environmental Change* 18: 442–457. https://doi.org/10.1016/j.gloenvcha.2008.03.005.

Lawrence, D.M., and A.G. Slater. 2005. A projection of severe near-surface permafrost degradation during the 21st century. *Geophysical Research Letters* 32: L24401., pp. 1–5. https://doi.org/10.1029/2005GL025080.

———. 2010. The contribution of snow condition trends to future ground climate. *Climate Dynamics* 34: 969–981. https://doi.org/10.1007/s00382-009-0537-4.

Li, X., F. Nunziata, and O. Garcia. 2018. Oil spill detection from single- and multipolarization SAR imagery. In *Reference Module in Earth Systems and Environmental Sciences, Comprehensive Remote Sensing*, ed. S. Liang, vol. 8, 231–248. Rockville: The Oceanography Society. https://doi.org/10.1016/B978-0-12-409548-9.10407-5.

Locat, J., C. Cloutier, R. Couture, F. Charbonneau, L. Danisch, S. Gravel, D. Hébert, M. Jaboyedoff, C. Jacob, P.E. Lord, K. Murnaghan, A. Nadeau, A. Pedrazzini, P. Therrien, and V. Singhroy. 2010. An integrated mass movement monitoring system for rockslide hazard assessment at Gascons, Gaspé peninsula, Québec: an overview. In *Proceedings of the 63rd Canadian Geotechnical Conference & 6th Can. Permafrost Conf., Calgary*, 35–43. ESS Contribution No. 20100060.

Lopez-Sanchez, J.M., F. Vicente-Guijalba, J.D. Ballester-Berman, and S.R. Cloude. 2014. Polarimetric response of rice fields at C-band: analysis and phenology retrieval. *IEEE Transactions on Geoscience and Remote Sensing* 52 (5): 2977–2993.

Lowman, P.D.J.R. 1991. Original shape of the sudbury structure, Canada: a study with airborne imaging radar. *Canadian Journal of Remote Sensing* 17 (2): 152–161. https://doi.org/10.1080/07038992.1991.10855288.

Maynard, D.E., I.C. Weiland, and A. Blais-Stevens. 2018. Surficial Geology, middle Kitimat River – Hoult Creek, Kitimat – Morice River corridor, British Columbia, parts of NTS 103-I/1 and 8, Geological Survey of Canada, Canadian Geoscience Map 319. scale 1:25 000.

McNairn, H., and B. Brisco. 2004. The application of C-band polarimetric SAR for agriculture: a review. *Canadian Journal of Remote Sensing* 30 (3): 525–542. https://doi.org/10.5589/m03-069.

Merzouki, A., and H. McNairn. 2015. A hybrid multi-angle and multipolarization approach to soil moisture retrieval using the integral equation model: preparing for the RADARSAT constellation Mission. *Canadian Journal of Remote Sensing* 41 (5): 349–362. https://doi.org/10.1080/07038992.2015.1104629.

Migliaccio, M., F. Nunziata, and A. Buono. 2015. SAR polarimetry for sea oil slick observation. *International Journal of Remote Sensing* 36: 3243–3273. https://doi.org/10.1080/01431161.2015.105730.

Mollard, J.D., and J.R. Janes. 1984. Airphoto interpretation and the Canadian landscape; Mollard, J, 422 sheets. https://doi.org/10.4095/299395. (Open Access).

Moreira, A., P. Prats-Iraola, M. Younis, G. Krieger, I. Hajnsek, and K.P. Papathanassiou. 2013. A tutorial on synthetic aperture radar. *IEEE Geoscience and Remote Sensing Magazine* 1: 6–43. https://doi.org/10.1109/MGRS.2013.2248301.

Morsy, S., A. Ahmed Shaker, and A. El-Rabbany. 2017. Multispectral LiDAR data for land cover classification of urban areas. *Sensors* 17: 958. https://doi.org/10.3390/s17050958.

Mosher, D.C. 2008. Submarine mass movement in Canada: Geohazard with far reaching implications. In *Proceedings of the 4th Canadian conference on Geohazards*, ed. J. Locat, D. Perret, D. Turmel, D. Demers, and S. Leroueil, 55–62. Richmond: Canadian Geotechnical Society.

Mossop, G.D. 1980. Geology of the Athabasca oil sands. *Science* 207 (4427): 145–152.

Mossop, G.D., and P.D. Flach. 1983. Deep channel sedimentation in the lower cretaceous McMurray formation, Athabasca Oil Sands, Alberta. *Sedimentology* 30: 493–509. https://doi.org/10.1111/j.1365-3091.1983.tb00688.

Nicholls, R.J., N. Marinova, J.A. Lowe, S. Brown, D. Gusmão, J. Hinkel, and R.S.J. Tol. 2011. Sea-level rise and its possible impacts given a 'beyond 4°C world' in the twentyfirst century. *Philosophical Transactions of the Royal Society A* 369: 161–181.

Nowak, R., R. Orłowicz, and R. Rutkowski. 2020. Use of TLS (LiDAR) for Building Diagnostics with the Example of a Historic Building in Karlino. Buildings, 10 (2), art. no. 24.

Nur Khakin, M.Y., T. Tsuji, and T. Matsuoka. 2013. Detection of localized surface uplift by differential SAR interferometry at the Hangingstone oil sand field, Alberta, Canada. *IEEE, Journal of Selected Topics in Applied Earth Observation and Remote Sensing* 6 (6): 2344–2354.

Oil Sands. 2010. *A Strategic Resource for Canada, North America and the World*, 4p. Ottawa: Government of Canada.

Peacock, M.J. 2011. Athabasca oil sands: reservoir characterization and its impact on thermal and mining opportunities. In *Petroleum Geology: From Mature Basins to New Frontiers – Proceedings of the 7th Petroleum Geology Conference*, ed. B.A. Vining and S.C. Pickering, 1141–1150. London: Petroleum Geology Conferences Ltd. Published by the Geological Society. https://doi.org/10.1144/0071141.

Pearse, J., V. Singhroy, S. Samsonov, and J. Li. 2014. Anomalous surface heave induced by enhanced oil recovery in northern Alberta: InSAR observations and numerical modeling. *Journal of Geophysical Research - Solid Earth* 119 (8): 6630–6649. https://doi.org/10.1002/2013JB010885.

Poley, L.G., and G.J.A. McDermid. 2020. Systematic review of the factors influencing the estimation of vegetation aboveground biomass using unmanned aerial systems. *Remote Sensing* 12: 1052. https://doi.org/10.3390/rs12071052.

Porter, M., M. Leir, and G. Ferris. 2014. Integrating terrain and geohazard knowledge into the pipeline lifecycle. In *6th Canadian GeoHazards Conference., Kingston*, 8p. Kingston: Canadian Geotechnical Society.

Pour, A.B., and M. Hasim. 2017. Application of Landsat-8 and ALOS-2 data for structural and landslide hazard mapping in Kelantan, Malaysia. *Natural Hazards and Earth System Sciences* 17: 1285–1303. https://doi.org/10.5194/nhess-17-1285-2017.

Pratesi, F., T. Nolesini, S. Bianchini, D. Leva, L. Lombardi, R. Fanti, and N. Casagli. 2015. Early warning GBInSAR-based method for monitoring Volterra (Tuscany, Italy) City walls. *IEEE Journal of Selected Topics in Applied Earth Observations and Remote Sensing* 8 (4): 1753–1762. https://doi.org/10.1109/JSTARS.2015.2402290.

Price, B. 2017. Nexen Energy ULC- Media Advisory.

Roopernarine, L. 2014. Small scale gold mining and environmental policy challenges in Guyana: protection or pollution. *Canadian Journal of Latin America and Caribbean Studies* 31 (61): 115–143. https://doi.org/10.1080/08263663.2006.10816892.

Roser, M. 2020. *Oil Spills*. OurWorldInData.org. "https://ourworldindata.org/oil-spills".

Samsonov, S., M. Kooij, and K. Tiampo. 2011. A simultaneous inversion for deformation rates and topographic errors of DInSAR data utilizing linear least square inversion techniques. *Computers and Geosciences* 37: 1083–1091.

Sánchez-Arcilla, A., C. Mösso, J.P. Sierra, M. Mestres, A. Harzallah, M. Senouci, and M.E. Raey. 2011. Climatic drivers of potential hazards in Mediterranean coasts. *Regional Environmental Change* 11: 617–636.

Sánchez-Rodríguez, A., M. Soilán, M. Cabaleiro, and P. Arias. 2019. Automated inspection of railway tunnels' power line using LiDAR point clouds. *Remote Sensing* 11: 2567. https://doi.org/10.3390/rs11212567.

Scafutto, R.D.P.M., C.R. deSouza Filho, and B. Rivard. 2016. Characterization of mineral substrates impregnated with crude oils using proximal infrared hyperspectral imaging. *Remote Sensing of Environment* 179: 116–130.

Shang, C., P. Treitz, J. Caspersen, and T. Jones. 2019. Estimation of forest structural and compositional variables using ALS data and multi-seasonal satellite imagery. *International Journal of Applied Earth Observation and Geoinformation* 78: 360–337.

Shen, L., V. Singhroy, and S. Samsonov. 2014. In *Forward Modeling of SAGD-Induced Heave and Caprock Failure Analysis Proceedings, SPE, Heavy oil conference. Calgary*, 10 p.

Short, N., B. Brisco, N. Couture, W. Pollard, K. Murnaghan, and P. Budkewitsch. 2011. A comparison of TerraSAR-X, RADARSAT-2 and ALOS-PALSAR interferometry for monitoring permafrost environments, case study from Herschel Island. *Canada Remote Sensing of Environment* 115: 3491–3506.

Singhroy, V. 1995. SAR integrated techniques for geohazard assessment. *Advances in Space Research* 15 (11): 67–78.

———. 1996. Environmental and geological site characterization in vegetated areas: image enhancement guidelines. In *Remote Sensing and GIS: Applications and Standards. ASTM Special Technical Publication #1279*, ed. V. Singhroy, D. Nebert, and A. Johnson, 5–17. West Conshohocken: American Society for Testing and Materials.

———. 1996a. Interpretation of SAR images for coastal zone mapping in Guyana. *Canadian Journal of Remote Sensing* 22 (3): 317–328.

———. 1997. Interpretation of SAR images for coastal zone mapping in Guyana. In *Land use Land Degradation and Land Management in Guyana*, ed. Parry Eden Williams, 32–41. Montreal: Mc.Gill. Uni Press.

———. 1999. Monitoring and mapping areas affected by water control projects in coastal Guyana. In *Hydrology in the Humid Tropic Environment*, 81–93. Oxfordshire: IAHS publication # 253.

———. 2005. Remote sensing for landslide assessment (Chapter 16). In *Landslides Hazard and Risk*, ed. T. Glade, M. Anderson, and M.J. Crozier, 469–490. Hoboken: Wiley Press.

———. 2008. Satellite remote sensing applications for landslide detection and monitoring (chap. 7). In *Landslide Disaster Risk Reduction*, ed. Kioji Sassa and Paolo Canuti, 143–158. Berlin: Springer.

———. 2012. In *Landslides: Encyclopedia of Remote Sensing*, ed. E.G. Njoku, M.J. Abrams, G. Asrar, F.S. Marzano, P. Minnett, V.V. Salomonson, V. Singhroy, and J. Turk. Berlin: Springer, ISBN 978-0-387-36698-2.

———. 2013. In *Operational Applications of Radar Images in Handbook of Satellite Applications*, ed. Pelton, Madry, and Camacho-Lara, 739–756. Berlin: Springer Press. ISBN: 978-1-4419-7670-3.

Singhroy, V., and F. Charbonneau. 2014. RADARSAT: Science and Applications. *Physics in Canada* 70 (4): 212–217.

Singhroy, V., and F.A. Kruse. 1991. Detection of Metal Stress in Boreal Forest Species Using the 670 UM Chlorophyll Absorption Band. In *ERIM 8th Thematic Conference on Geologic Remote Sensing*, 361–372.

Singhroy, V., and J. Li. 2017. Rapid Revisit InSAR Time Series Monitoring of Surface Deformation at Long Lake, Alberta. GeoOttawa, Ottawa, 6pp.

Singhroy, V., and P. Lowman. 2012. In *Remote Sensing and Geologic Structure. Encyclopedia of Remote Sensing*, ed. E.G. Njoku, M.J. Abrams, G. Asrar, F.S. Marzano, P. Minnett, V.V. Salomonson, V. Singhroy, and J. Turk, 681–683. Berlin: Springer, ISBN 978-0-387-36698-2.

Singhroy, V., and K. Molch. 2004. Characterizing and monitoring rockslides from SAR techniques. *Advances in Space Research* 33 (3): 290–295.

Singhroy, V., and M. Pilkington. 2012. In *Geological Mapping using Earth Magnetic Field: Encyclopedia of Remote Sensing*, ed. E.G. Njoku, M.J. Abrams, G. Asrar, F.S. Marzano, P. Minnett, V.V. Salomonson, V. Singhroy, and J. Turk, 232–237. Berlin: Springer, ISBN 978-0-387-36698-2.

Singhroy, V., R. Slaney, P. Lowman, J. Harris, and W. Moon. 1993a. RADARSAT and radar geology in Canada. *Canadian Journal of Remote Sensing* 14 (4): 329–332.

———. 1993b. RADARSAT and radar geology in Canada. *Canadian Journal of Remote Sensing* 19 (4): 338–351. https://doi.org/10.1080/07038992.1993.10874569.

Singhroy, V., R. Couture, and K. Molch. 2005. InSAR monitoring of frank slide. In *Landslide Risk Management*, ed. Hungr Fell and Eberhardt Couture, 611–614. Rotterdam: Balkema Publishers.

Singhroy, V., P.J. Alasset, R. Couture, and C. Froese. 2008. InSAR monitoring of landslides in Canada. In *Proceedings IEEE-IGARSS 08, Boston (MA)*, 202–205.

Singhroy, V., K. Murnaghan, and R. Couture. 2010. InSAR monitoring of a retrogressive thaw flow at Thunder River, lower Mackenzie. In *Geo2010, Proceedings of the 63rd Annual Canadian Geotechnical Conference and the 6th Canadian Permafrost Conference*, 1317–1322. (ESS Cont. # 20100181).

Singhroy, V., F. Charbonneau, C. Froese, and R. Couture. 2012. Guidelines for InSAR monitoring of landslides in Canada. In *Landslides and Engineered Slopes*, ed. E. Eberhardt, C. Froese, K. Turner, and S. Leroueil, vol. 2, 1281–1287. Boca Raton: CRC Press.

Singhroy, V., J. Li, and F. Charbonneau. 2016. High Resolution Rapid Revisit InSAR Monitoring of Surface Deformation. *Canadian Journal of Remote Sensing* 41 (5): 458–472. https://doi.org/10.1080/07038992.2015.1104638.

Singhroy, V., J. Li, M. Fobert, C.F. Lee, and M.K. Das. 2019. Monitoring Post Landslide Activity from RADARSAT Constellation Mission IGARSS, Yokohama, 4p.

Stancliffe, R.P.W., and M.W.A. van der Kooij. 2001. The use of satellite-based radar interferometry to monitor production activity at the cold Lake heavy oil field, Alberta, Canada. *AAPG Bulletin* 85 (5): 781–793.

Temitope, D., T. Oyedotun, and L. Johnson-Bhola. 2019. Shoreline change identification and analysis along the coast of Guyana, South America. *Interdisciplinary Environmental Review* 20 (1): 7–31.

Thompson, A.A. 2015. Overview of the RADARSAT constellation mission. *Canadian Journal of Remote Sensing* 41 (5): 401–407. https://doi.org/10.1080/07038992.2015.1104633.

Touzi, R., and P.W. Vachon. 2015. RCM polarimetric SAR for enhanced ship detection and classification. *Canadian Journal of Remote Sensing* 41 (5): 473–484. https://doi.org/10.1080/07038992.2015.1110010.

Trebossen, H., B. Deffontaines, N. Classeau, J. Kouame, and L.P. Rudant. 2005. Monitoring coastal evolution and associated littoral hazards of French Guiana shoreline with radar images. *Comptes Rendus Geoscience* 337: 1140–1153.

Tuba, G., F. Molnár, and D.E. Ames. 2014. Multi-stage hydrothermal processes involved in "low-sulfide" Cu (–Ni)–PGE mineralization in the footwall of the Sudbury igneous complex (Canada): Amy Lake PGE zone, east range. *Mineralium Deposita* 49: 7–47. https://doi.org/10.1007/s00126-013-0468-1.

van der Meer, F.D., H.M.A. van der Werff, F.J.A. van Ruitenbeek, C.A. Hecker, W.H. Bakker, M.F. Noomen, M. van der Meijde, E.J.M. Carranza, J.B. de Smeth, and T. Woldai. 2012. Multi- and hyperspectral geologic remote sensing: a review. *International Journal of Applied Earth Observation and Geoinformation* 14 (1): 112–128.

van der Sanden, J. 2004. Anticipated applications potential of RADARSAT-2 data. *Canadian Journal of Remote Sensing* 30 (3): 369–379.
van Westen, C.J. 2016. National Scale Landslide Susceptibility Assessment for Dominica. The World Bank Caribbean Handbook on Risk Information Management Technical Report. 91, https://doi.org/10.13140/RG.2.1.4313.2400.
van Westen, C.J., and J. Zhang. 2018. Tropical Cyclone Maria. Inventory of landslides and flooded areas UNITAR Map Product ID: 2762 GLIDE: TC20170918DMA.
Wang, B., B. Paudel, and H. Li. 2009. Retrogression characteristics of landslides in fine-grained permafrost soils, Mackenzie Valley, Canada. *Landslides* 6: 121–127.
Wang, Y., Q. Chen, Q. Zhu, L. Liu, C. Li, and D. Zheng. 2019. A survey of mobile laser scanning applications and key techniques over urban areas. *Remote Sensing* 11: 1540. https://doi.org/10.3390/rs11131540.
Wasowski, J., and F. Bovenga. 2014a. Investigating landslides and unstable slopes with satellite multi temporal interferometry: current issues and future perspectives. *Engineering Geology* 174: 103–138. https://doi.org/10.1016/j.enggeo.2014.03.003.
———. 2014b. Remote sensing of landslide motion with emphasis on satellite multitemporal interferometry applications: an overview. In *Landslide Hazards, Risks and Disasters*, ed. T. Davies, 345–403. Berlin: Springer. https://doi.org/10.1016/B978-0-12-396452-6.00011-2.
White, J.C., N.C. Coops, M.A. Wulder, M. Vastaranta, T. Thomas Hilker, and P. Tompalski. 2016. Remote sensing technologies for enhancing forest inventories: a review. *Canadian Journal of Remote Sensing* 42 (5): 619–641. https://doi.org/10.1080/07038992.2016.1207484.
World Bank, Dominica. 2004. Natural disasters and economic development of a small island state. Disaster risk management No. 2. Disaster Management Facility, Washington, DC.
Yao, H., R. Qin, and X. Chen. 2019. Unmanned aerial vehicle for remote sensing applications—a review. *Remote Sensing* 11: 1443. https://doi.org/10.3390/rs11121443.
Zhang, T. 2005. Influence of the seasonal snow cover on the ground thermal regime: an overview. *Reviews of Geophysics* 43: RG4002. https://doi.org/10.1029/2004RG000157.
Zhang, T., R.G. Barry, K. Knoles, J.A. Heginbottom, and J. Brown. 1999. Statistics and characteristics of permafrost and ground-ice distribution in the Northern Hemisphere. *Polar Geography* 23: 132–154.

Chapter 2
InSAR Satellite Surveys: Key Considerations for Monitoring Infrastructure

Jon Leighton, Parwant Ghuman, and Christian E. Haselwimmer

2.1 Fundamental Concepts

This section defines InSAR and provides key concepts, necessary for evaluating a site's suitability for InSAR.

What Is InSAR?

Interferometric Synthetic Aperture Radar (InSAR) is a technique for monitoring millimetre to centimetre-level displacement of the Earth's surface through repeated imaging of an area over time (Hanssen 2001). During each pass, a radar satellite transmits microwaves towards the Earth's surface, and records the returned echoes as a radar image. Each pixel in the image contains amplitude (target brightness, typically expressed as intensity, which is the square of the amplitude) and more importantly phase. The phase value for a pixel is expressed as an angle, and proportional to the distance to the ground. Figure 2.1 shows how phase measurements over time relate to ground displacement. Note that the satellite acquires from a side-looking perspective, this is an important aspect of SAR, discussed in SAR Geometry on page 9.

When phase data from the first image is subtracted from another, an interferogram is formed, which contains the phase difference for each pixel.

J. Leighton (✉) · P. Ghuman
3v Geomatics Inc., Vancouver, BC, Canada
e-mail: jleighton@3vGeomatics.com; pghuman@3vGeomatics.com

C. E. Haselwimmer
Chevron Energy Technology Company, San Ramon, CA, USA
e-mail: cehaselwimmer@chevron.com

V. Singhroy (ed.), *Advances in Remote Sensing for Infrastructure Monitoring*, Springer Remote Sensing/Photogrammetry,
https://doi.org/10.1007/978-3-030-59109-0_2

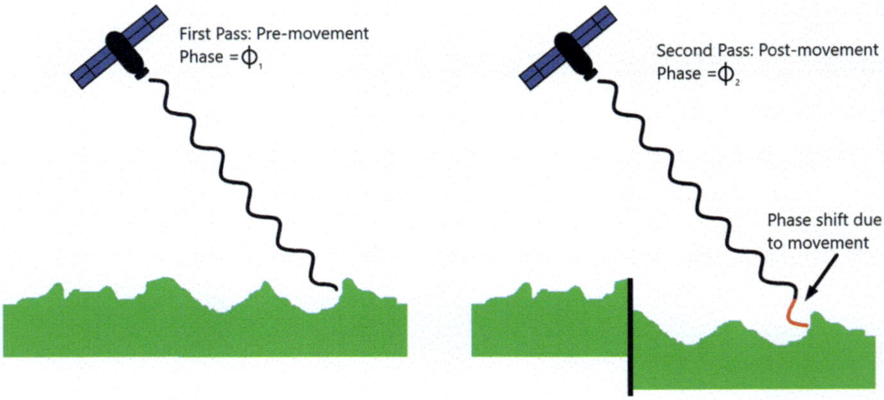

Fig. 2.1 Left: The SAR acquires an image of the ground, which contains phase information. Right: The ground has displaced prior to a subsequent image being acquired, and the phase measurement is consequently longer

Because phase is related to the ground distance, the difference in phase is therefore related to the change in ground distance that has occurred in the time between the two image acquisitions. Unfortunately phase differences also contain other unwanted contributions, relating to topography, atmospheric conditions, and noise. The goal of InSAR processing is usually to correctly separate displacement from other unwanted effects that also influence the phase (Woodhouse 2006).

This is most effectively performed when a series of 15 or more repeated radar images is considered. Common approaches utilise the spatio-temporal statistics of the data to derive displacement from an interconnected network of interferograms that overlap in time (Ferretti et al. 2001; Berardino et al. 2002; Mora et al. 2003; Hooper 2008). Despite best efforts, some areas will not yield useful results, discussed further in Coherence on page 7.

End products typically map displacement over time for areas up to thousands of square kilometres on the ground, which often equates to millions of measurable targets. InSAR, therefore, is essentially a simultaneous satellite survey of millions of targets on the ground, repeated every few days or weeks. Figure 2.2 shows InSAR displacement results over Seattle.

InSAR can be a difficult technology to apply for a few reasons:

- During each acquisition, microwaves are emitted by the radar and travel to the Earth and back. Along the way, these signals encounter the ionosphere, the troposphere, topography, pressure and temperature changes, displacement, and even soil moisture and snow (Hanssen 2001). These effects vary from pixel to pixel and over time, and, therefore, manifest in the data as such.
- Phase values for each pixel are bounded between 0° and 360°. They must be integrated spatially in order to recover higher levels of displacement.
- Each pixel over time has varying amounts of information. Some are fully information bearing, some contain partial information, and others are complete noise.

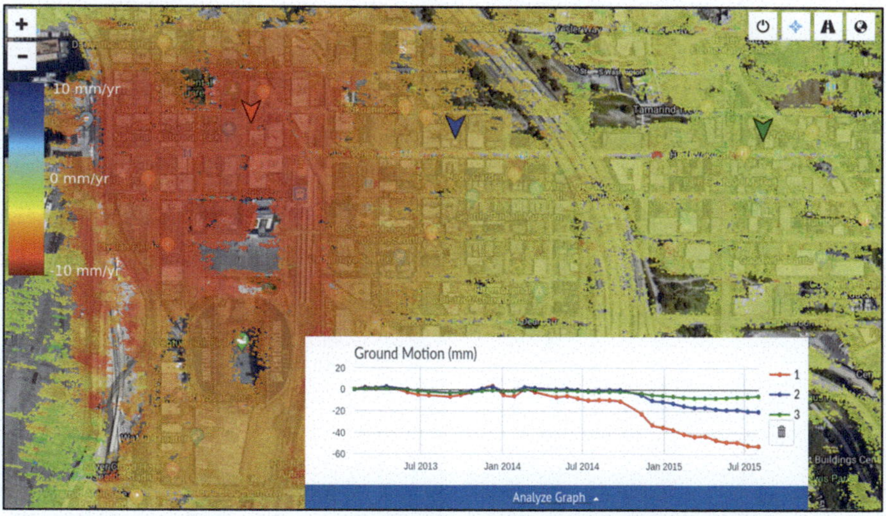

Fig. 2.2 A sample web map of InSAR displacement data generated over Seattle. Tens of millions of targets were tracked over a 3-year time period to map tunnelling-related subsidence. Optical image credit: Google Earth, Landsat/Copernicus

Figure 2.3 shows a single interferogram on the left; the per-pixel phase difference between two radar images, represented by a blue-pink-yellow colour map. The phase values vary across the image for several reasons. The right image shows displacement results, following a complex processing chain to remove unwanted signals, noise, and in this case, non-moving areas.

Targets

The measurable pixels in an InSAR analysis are often referred to as *targets*. InSAR tracks repeatable targets over time. Man-made infrastructure, exposed rocks, and bare ground are examples of targets that are repeatable from one image acquisition to the next. Water, snow, leaves, and crops are not trackable because they change too much between images. Targets that do not change for the duration of a *stack* of images are called *Permanent Targets* (Ferretti et al. 2001), such as built infrastructure or dry deserts. Displacement may or may not be occurring, but the target itself has the same structure. Targets that stay unchanged only for some portion of images, or are only available for monitoring during specific seasons (e.g. due to snow) are called *Temporary Targets* (Hooper et al. 2012).

The quality of targets also varies spatially, with some exhibiting point-like behaviour with low noise, termed *Point Targets*. Here, the phase response from solid objects such as buildings or boulders dominates over all other objects in the pixel. Point Targets tend to provide excellent InSAR measurements. Pixels containing no

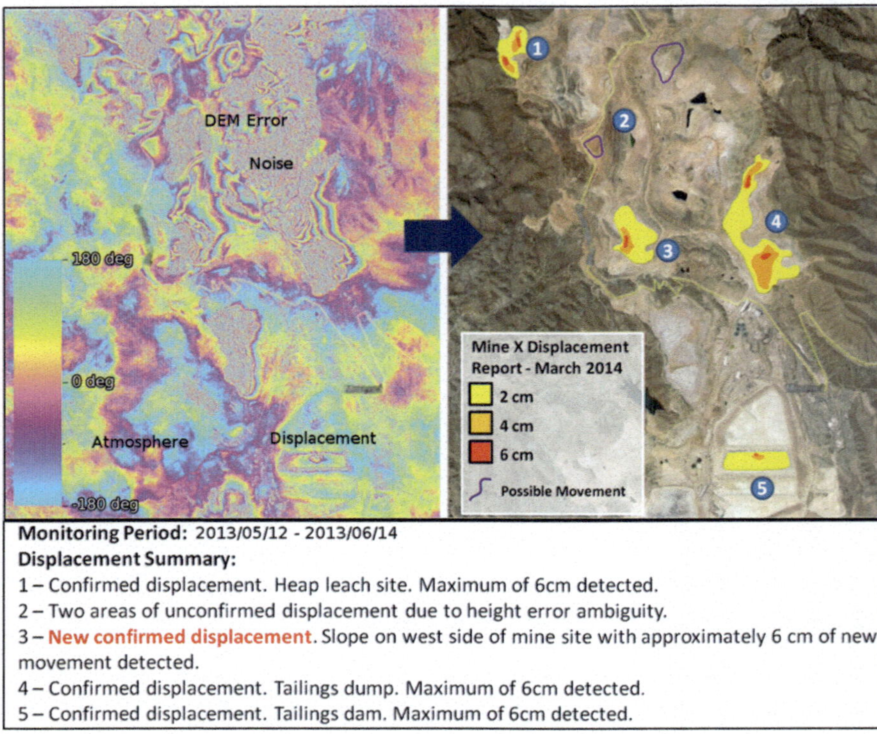

Fig. 2.3 Differencing two radar images produces an "interferogram" (left) containing mixed signals and noise. A derived product (right) indicates when, where, and how much displacement has occurred. Additional context is provided to aid interpretation (bottom). Optical image credit: Google Earth, Landsat/Copernicus

solid dominant object are referred to as *Distributed Targets;* these might contain bare ground or a complex structure. Figure 2.4 shows the various types of targets, with permanent point targets offering the highest quality measurements, and distributed temporary the lowest.

Coherence

For any given pixel, a successful InSAR measurement between a pair of images relies on the target maintaining some continuity over time. If the target is disturbed between acquisitions, then that interferometric measurement is lost. This characteristic of target quality in an interferogram is quantified by the metric of *coherence* (Cattabeni et al. 1994). The coherence of a target may come and go over time (see Temporary Targets discussed in the section above), but continuously changing areas such as water produce very low coherence and cannot be monitored using InSAR.

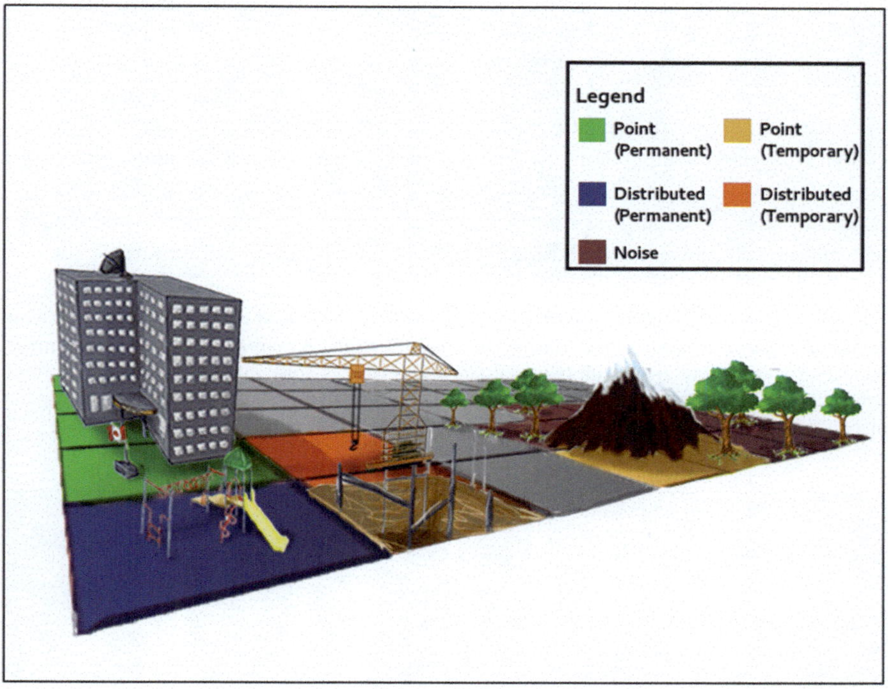

Fig. 2.4 Measurable pixels in InSAR are often referred to as targets. Their quality can vary based on ground cover (point vs. distributed) and whether they persist over time (permanent vs. temporary)

InSAR coherence captures measurement quality, and without it there is no means of discerning information from noise. A target's coherence is expressed as a continuum between zero (no coherence) and one (perfect coherence). Many InSAR processes employ coherence estimates, such as target identification, as weights for algorithms, and for masking out low quality data.

Figure 2.5 left shows an optical image of an oil pumping site bordering a town; vegetation dominates elsewhere. One coherence image can be generated for each interferogram, and Fig. 2.5 right shows an example of this. Here, coherence is being measured spatially; targets are attributed high coherence if they are similar to their neighbours. Processing plants and nearby urban infrastructure provide many permanent point targets with high coherence (closer to 1), whereas vegetated areas are less coherent (closer to 0).

Coherence over time is also important. Figure 2.6 shows coherence images from successive interferograms over a tailings area. Sporadic activity, the lack of hard targets, and changing ground conditions provide targets that are neither point like, nor permanent. Temporary and Distributed Targets must be considered if all the available displacement information is to be extracted.

Fig. 2.5 Left: A residential area (north east), an area of natural ground with some vegetation (west and industrial infrastructure all lie close to one another). Right: A corresponding coherence image indicating many bright areas with potential targets, as well as noise in vegetated areas. A useful coherence image such as this successfully attributes measurement potential, and is a key metric for many stages of InSAR analyses. Optical image credit: Google Earth, Landsat/Copernicus

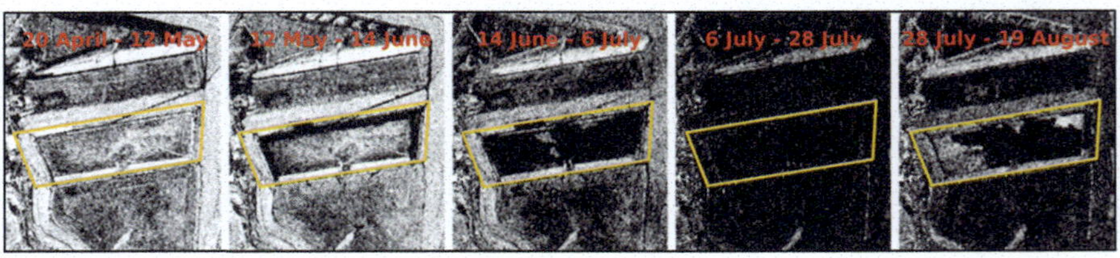

Fig. 2.6 Dynamic areas such as mine tailings have variable coherence over time; temporary target analysis is necessary to monitor such areas. The colour legend shown in Fig. 2.5 applies here also

SAR Geometry

SAR satellites are necessarily side-looking sensors. Microwave pulses emitted and received by the radar are ordered in the image by range. Their range is based on their travel time; pulses that take longer to return are assumed to be further away. A radar pointing straight down would result in many pulses returning simultaneously, which would be superimposed in the image. A side-looking arrangement, as shown in Fig. 2.1, therefore, provides greater opportunity for targets to be separated in range, resulting in greater detail in the image. This is always true for flat areas, but the presence of topography affects measurement coverage and measurement sensitivity (Woodhouse 2006). Effects due to topography can be summarised as follows:

- *Foreshortening*. Moderate slopes facing towards the satellite return pulses that arrive at similar times, and this section of the image is, therefore, compressed providing little opportunity for target measurements (*foreshortening*). Sensitivity to downslope displacement is also poor for these slopes, as displacement is occurring nearly perpendicular to the line of sight (LOS). No target measurements occur on slopes perpendicular to the satellite look direction, as all pulses return simultaneously and are superimposed (maximum foreshortening).
- *Layover*. If the top of the slope is closer to the radar than the base (*layover*), the feature appears reversed in the image as pulses are ordered in reverse. This is

expected for buildings, and with the correct approach, useful displacement information can still be extracted from the data.

- *Ideal Slopes*. Moderate slopes facing away from the satellite provide the best opportunity for targets and good sensitivity to downslope displacement. Coherence can be adversely affected on these slopes because more pulses are reflected away (*specular reflection*), but as long as reasonable surface roughness is present, this tends not to be an issue.
- *Shadow*. Slopes facing away that are steeper than the look angle (angle between nadir and the look direction) are completely obscured to the satellite (*shadow*).

Figure 2.7 illustrates the effects of foreshortening, layover, and shadow in a radar intensity image.

Such geometric effects are easy to simulate using an accurate Digital Elevation Model (DEM). These are available as open source or proprietary data products. Prior to an InSAR monitoring project, a DEM can be used to model all available satellite passes to evaluate which provide the best coverage. A coverage and downslope displacement sensitivity map can be provided to the client.

InSAR Geometry

In addition to the side-looking geometry of single SAR passes, the relative geometry between passes is also important for InSAR. Recall that interferograms are the per-pixel phase difference between two radar images and constitute the fundamental observable for InSAR analyses (Bamler and Hartl 1998). The satellite position can

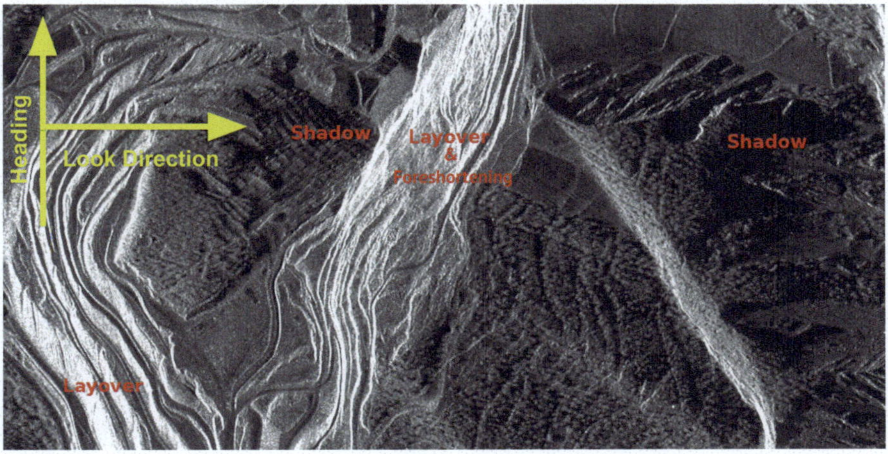

Fig. 2.7 Regions of foreshortening, layover, and shadow manifest in areas with topography as bright areas. They are bright because of simultaneous signal returns and more pulses returning to the sensor. Dark areas facing away produce fewer returns or no returns if the slope is steep enough to be occluded from view

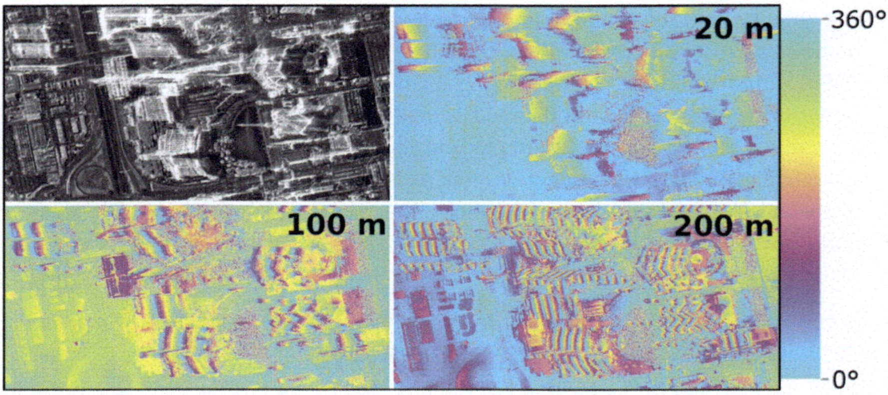

Fig. 2.8 Skyscrapers along the Las Vegas Boulevard appear as bright structures in a radar intensity image (top left). As these buildings are not represented in the DEM, they generate a response in interferograms seen as phase cycles termed *fringes*. The number of fringes present is proportional to the interferometric baseline. Three examples are shown with baselines of 20 m, 100 m, and 200 m

change by hundreds of metres between satellite passes. Surface topography (defined here to include buildings or structures) manifests in interferograms as spatial phase changes, and by an amount proportional to the distance between the two sensor positions at acquisition time (the interferometric baseline). Figure 2.8 shows how larger baselines are more sensitive to topographic effects. The main purpose of the DEM in InSAR is to model and remove this unwanted effect in each interferogram. DEMs are never perfect, and a residual effect typically remains, especially if the DEM is dated or coarser resolution than the SAR data. In the case shown, an excellent correction to the DEM can be estimated because the images are coherent, the topography is unchanging, and the effect is highly systematic; it scales linearly with the size of the baseline.

For any given pixel, if coherence is high and the surface topography is unchanging, a good correction to the DEM is trivial to determine using linear regression, if ten or more images are available. However, surface topography in places like open pit mines not only changes, but the interferograms are less coherent in those areas, because they are changing. Solutions to this problem are more challenging, but data redundancy, rejecting large baseline interferograms, and restricting the timespan of the stack all help.

Atmospheric Effects

Phase differences in interferograms arising from ground displacement are of primary interest, but may be contaminated with phase differences due to atmosphere, ionosphere, snow, and soil moisture. If the displacement signal dominates over sources of error, it can usually be isolated and quantified with high precision. If the displacement is subtle or the contamination is high, isolating displacement becomes more challenging.

Phase changes due to non-homogeneous water vapour in the troposphere are common (Hanssen 2001), but stormy conditions produce more extreme phase turbulence. In these cases, atmosphere can mask the onset of new displacement, and more time is required for the displacement signal to raise above the higher noise levels. Figure 2.9 shows three examples of phase turbulence due to tropospheric water vapour, shown for both C- and X-band interferograms. The X-band data, with its shorter wavelength, is more easily perturbed.

In addition to water vapour in the troposphere (wet or hydrodynamic atmosphere), phase delays also occur due to pressure changes with altitude, known as *hydrostatic atmosphere*, hereafter referred to as dry atmosphere. This is easier to remove as it is highly correlated with height and can be modelled using the DEM (Bekaert et al. 2015).

Other sources of error are now discussed.

- Phase effects due to the ionosphere are typically only seen with longer wavelength sensors in data covering high latitudes (Woodhouse 2006).
- Consistent snow cover between winter images may still provide good coherence if the snow is dry, but the resulting data is biased due to variations in snow density, depth, and ice content. The most common solution is to simply exclude data

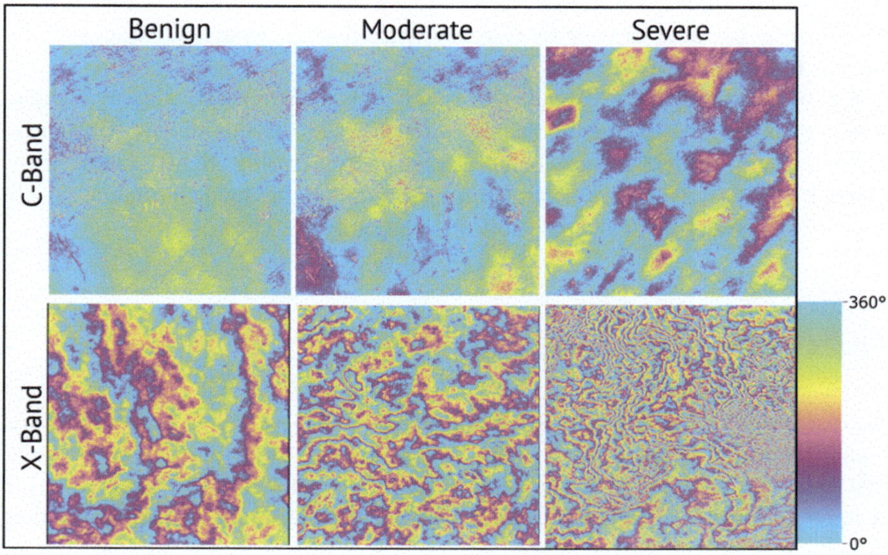

Fig. 2.9 Non-homogeneous atmospheric water vapour manifests as phase variance in interferograms. Phase variance increases as sensor wavelength decreases. As variance increases, the onset of subtle displacement becomes harder to detect

during snow-covered periods. Coherence is very low during snow fall and snow melt due to the absence of repeatable targets.

- Soil moisture signals due to irrigation or water drainage are a nuisance and are most notable in agricultural areas.

Wavelength

SAR satellites operate in the microwave portion of the electromagnetic spectrum; the peak-to-peak length of the wave is known as its wavelength and represents the measuring stick for estimating ground displacement. Wavelengths are grouped into lettered bands for ease of referencing. The most common wavelengths used are 3.1 cm (X-band), 5.6 cm (C-band), and 23.6 cm (L-band). InSAR can capture tiny movements from a thousand kilometres away through the analysis of per-pixel phase values. These phase values can be conceptualised as fractions of a wavelength (see Fig. 2.1), which are subtracted between images to produce an interferogram. Shorter wavelengths are more sensitive to displacement, but also more sensitive to other effects including scattering from vegetation and atmospheric turbulence. Strong displacement over a small area can cause *aliasing*; this is analogous to contours of displacement densifying to the point where they become superimposed. This leads to underestimation in the results, and longer wavelengths may be desirable in these cases.

Areas containing trees cannot be monitored with X-band because returning pulses are randomised by foliage; therefore, the targets are not repeatable. Longer wavelengths are able to penetrate vegetation, reflecting from underlying targets such as tree trunks or bare soil.

Choosing an appropriate wavelength is very important when developing a monitoring plan, but the availability of satellites also affects this choice. Currently, X-band satellites are the most numerous and commercially viable, followed by C-band; there is only one L-band satellite. The number of redundant satellites available affects the risk level and data collection capacity for a project.

Phase Unwrapping

As discussed in What is InSAR, the phase difference values for each pixel are bounded between 0° and 360° degrees. This means that the displacement estimates that result are also inherently bounded; between 0 and 1.5 cm for satellites with X-band wavelengths. Figure 2.10 left shows an X-band interferogram with a displacement zone with an associated colourmap scaled between 0° and 360°. Three clear phase cycles or *fringes* can be seen, with a red transect line passing through the displacement zone. Phase values for pixels along the transect are plotted on the right

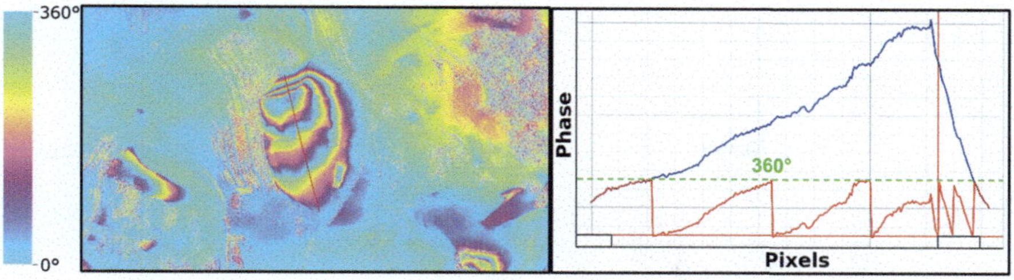

Fig. 2.10 Left: An interferogram with a displacement zone in the centre, and a red transect line passing through it. Right: Phase values along the red transect line produce a sawtooth plot (also red) because the values are natively bounded between 0° and 360°. Successful unwrapping integrates phase values spatially to recover higher magnitude displacement (blue line)

in red; values steadily grow in magnitude, but abruptly return to 0° once they reach 360°. This is termed as *wrapped phase*.

It is trivial here for a human observer to estimate an unbounded displacement maximum, simply by counting the fringes and multiplying by 1.5 cm. To do this algorithmically, for all pixels, given an assumed starting point, is known as *phase unwrapping* (Ghiglia and Pritt 1998). This two-dimensional counting process works to recover the unbounded phase values shown by the blue line in Fig. 2.10 right.

Despite the simplistic example shown, accurate phase unwrapping represents the biggest challenge for InSAR processing. Displacement zones can be complex, with discontinuities or entirely segregated regions due to incoherent areas, and the process must be conducted in two dimensions.

Resolution

For optical imagery, spatial resolution is defined by the smallest resolvable object. For InSAR, however, it is better defined as the minimum distance between adjacent targets. Note that neither of these involves pixel spacing, which could be arbitrarily oversampled, without affecting the level at which objects can be resolved or the actual distance between targets. Pixel spacing is, however, often used synonymously with spatial resolution, and usually their values are close. In reality, spatial resolution for InSAR changes depending on the angle of the terrain (see SAR Geometry) and the distribution of measurable objects within the bounds of each pixel.

Figure 2.11 illustrates spatial resolution by comparing InSAR results derived from 3 m and 20 m SAR data; optical imagery is shown for reference.

Spatial resolution has progressively improved over time; typical resolution was over 20 m in the 1990s, 8 m in the 2000s, and 3 m currently. Metre and sub-metre resolutions are also available on current satellites, but coverage decreases with increasing resolution (similar to a zoom lens on an optical camera).

Resolution is also used in a temporal context; the temporal resolution of a SAR satellite, also known as its revisit period, indicates how long before the satellite can

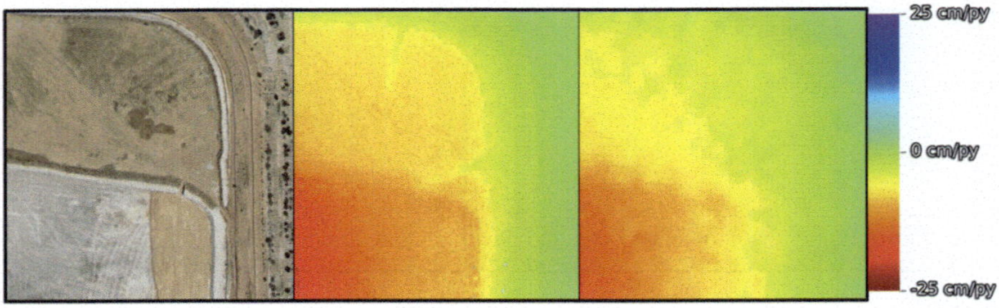

Fig. 2.11 Left: optical imagery of a tailings storage facility. Centre: a displacement heat map from 3 m TerraSAR-X data. Right: the same result for the same period using 20 m Sentinel-1 data. Whilst magnitudes are similar, some smaller features are lost in the Sentinel-1 results. Optical image credit: Google Earth, Landsat/Copernicus

collect data over the same position on the ground. Revisit has also improved over the decades, starting from 35-day revisit in the 1990s, to 24-day in the 2000s, down to 4-day currently. Many modern satellite missions (TSX, Cosmo, Sentinel, RCM) operate in constellations of identical sensors that are interchangeable and can, therefore, revisit more frequently. For example, the Cosmo constellation from the Italian Space Agency comprises 4 satellites with 1-, 3-, 4-, and 8-day revisit, for an average revisit of 4 days. Higher temporal resolution provides opportunities to maintain coherence over dynamic land cover, and also better supports operations at active sites by providing higher quality and more timely monitoring data.

Displacement Sensitivity

InSAR is inherently one dimensional; displacement is measured in the line of sight of the satellite. Displacement for a target moving down a slope that happens to exactly coincide with the look direction, can be fully captured. In general, however, this will rarely be the case, and InSAR will underestimate the real displacement (Pepe and Calò 2017). This is not particular to InSAR, but true for any dimensionally restricted technology, such as levelling or Electronic Distance Measurement (EDM). Figure 2.12 illustrates the issue with a simplified sliding block scenario. Despite usual underestimation, slopes that face away from the look direction provide good opportunities for InSAR monitoring, slopes that face the radar, on the other hand, do not (see SAR Geometry on page 9).

The best workaround for capturing better displacement estimates, as well as providing better coverage, is to image an area of interest from two directions, one looking east (usually ascending passes) and one looking west (usually descending passes). From these two independent LOS measurements, a direction of displacement can be solved in the east-west-up-down plane. The ascending and descending displacement vectors can then be co-projected along the solved vector, not only providing directional information, but also increasing the displacement magnitude

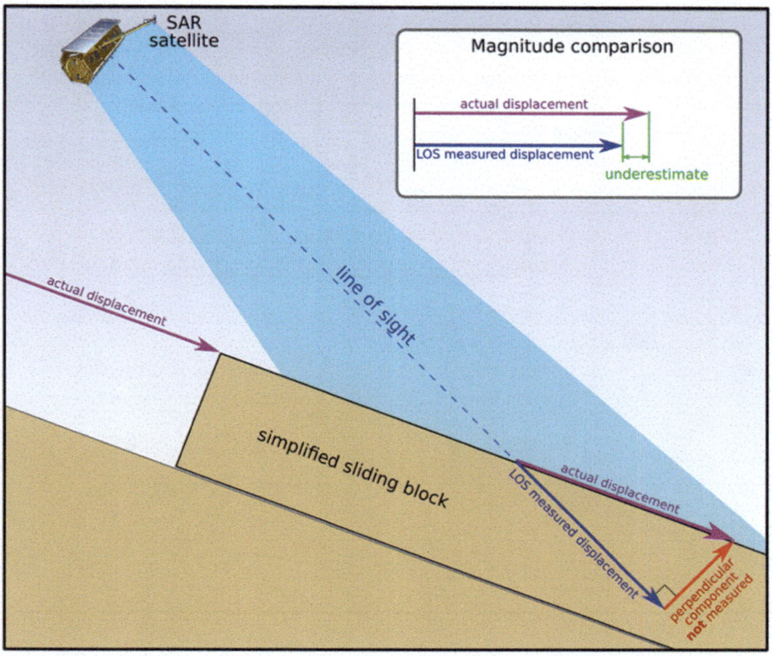

Fig. 2.12 A simplified sliding block displacement scenario. InSAR underestimates the actual displacement, because it only captures the component that is along the LOS

appropriately. Note that the north-south direction is not considered because this is nearly the flight path of the satellite and there is no displacement sensitivity along this axis.

Satellite Options

InSAR monitoring programs are typically conducted in two stages, a historical archive analysis to check site suitability and establish a baseline, followed by ongoing monitoring to highlight emerging risks. Spaceborne SAR satellites have existed since the 1990s; many satellites had good default data collection worldwide, enabling historical analysis for very little cost as older data are generally free. Figure 2.13 shows all of the SAR satellites that are currently viable for operational monitoring. The planned lifetime of a SAR satellite is generally 7 years, and actual lifetimes vary around this number. The data from a given satellite is self-contained due to variations in wavelength and look direction between satellites. Exceptions to this exist when satellites are specifically cloned from one another and launched along the same orbit track (e.g. TSX and PAZ). Results from different sensors *can* be combined, but issues can occur if the resolutions differ significantly.

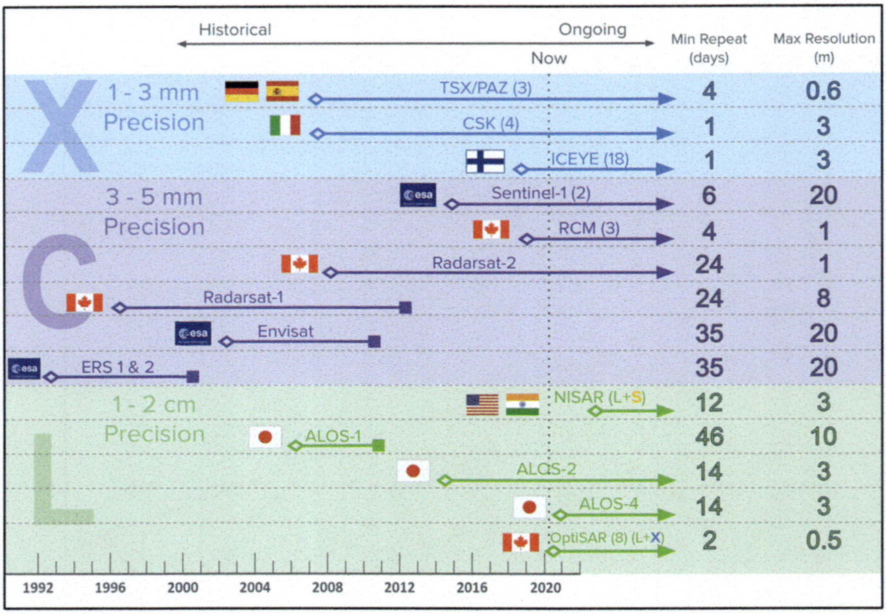

Fig. 2.13 Key specifications of historic and currently operational satellites. Lower precision satellites in L-band wavelengths are the only option for densely vegetated areas. Higher precision satellites in C- and X-band wavelengths can achieve good results over open ground and urban areas

2.2 Case Studies

This section draws on the InSAR characteristics already presented to show how they can influence site evaluation. A number of case studies are discussed, starting with an overview of the project, followed by a site evaluation. Key results are illustrated at the end of each case study.

Seattle Tunnelling

An engineering construction project in Seattle, WA, USA, involved extensive tunnelling through the downtown area using the largest tunnel boring machine (TBM) in the world, named Bertha (nicknamed Big Bertha by the media). The TBM stalled in December 2013 after hitting a steel pipe and InSAR results captured up to 60 mm of displacement. This radiated outward for more than a kilometre, characteristic of displacement caused by groundwater pumping, which was necessary to repair Bertha. The widespread displacement affected local terrestrial benchmarks used as reference stations to monitor subsidence, which resulted in an underestimate from terrestrial survey results. Table 2.1 provides an overview of the case study.

Table 2.1 Overview of the Seattle tunnelling case study

Characteristic	Comment
Targets	The Seattle metropolitan area has dense urban infrastructure such as houses and roads, and superstructures such as skyscrapers and bridges. Urban infrastructure provides excellent permanent targets for InSAR monitoring over years and decades. Snow is rare, making uninterrupted monitoring the norm
Topography	The area has very little topographic relief, but dense skyscrapers in the downtown core cause blind spots due to layover and shadow. The topography is static, thus errors in the DEM (e.g. over infrastructure) can be readily corrected once 15–20 images have been collected
Sources of error	• Dry atmosphere is a non-issue due to flat terrain. Wet atmosphere is not expected to be overly turbulent, consistent with coastal mid-latitude areas • Soil moisture effects are not expected (no agriculture/drainage) • Seattle experiences little to no snow, thus year-round monitoring is expected
Wavelength	Urban infrastructure provides excellent targets for any radar wavelength; C-band and X-band are the best choices for higher precision. Smaller wavelength data will generate noisy measurements over parks and golf courses, and no measurements over areas with trees
Resolution	High spatial resolution (3 m or better) is necessary in order to generate distinct returns from densely packed infrastructure. Weekly imaging is desirable to detect the onset of accelerations due to groundwater extraction, and also to amass the data required to mitigate atmospheric and DEM errors
Known displacement	Information regarding previous displacement is not available
Satellite options	• ALOS-2 (L-band) is unsuitable for detecting subtle displacement • Sentinel resolution is too coarse for monitoring infrastructure • TSX, CSK, and Radarsat-2 are good choices for monitoring infrastructure, but provide no coverage over vegetation. Radarsat-2 has a long revisit (24 days) which may not suffice during the most dynamic time periods

Historical high-resolution Radarsat-2 data was available prior to and during the tunnelling activities, which was ideal for establishing a baseline and characterising the impact of tunnelling and associated groundwater extraction. 35 images were available between February 2013 and August 2015 and results are shown in Fig. 2.14.

Aqueducts and Aquifers

The California Aqueduct transports water over 640 km and is the backbone of the California State Water Project. The aqueduct crosses through the Tulare Lake Basin, in which water is pumped from underground aquifers. Government agencies regulate aquifer pumping as well as monitoring the condition of the aqueduct. Table 2.2 provides an overview of the case study.

TSX was chosen due to the availability of pre-existing archive data. Forty images covering two 1500 km^2 areas during 2015 and 2016 were analysed to produce a

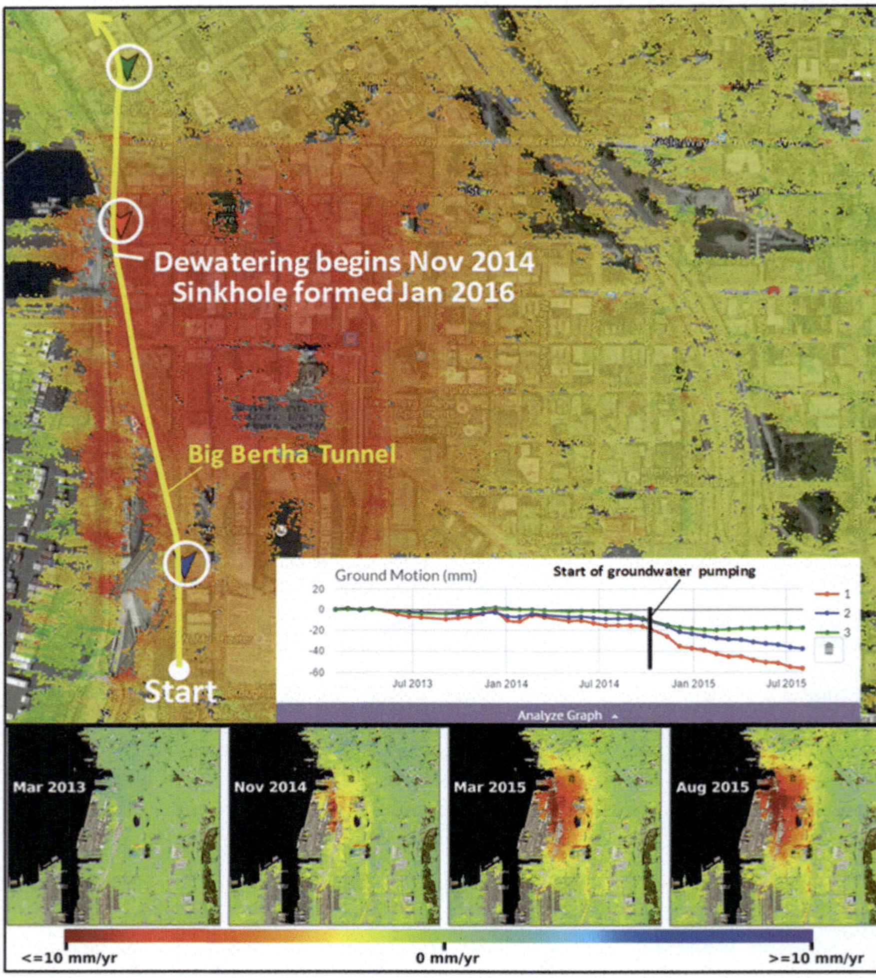

Fig. 2.14 Top: InSAR results over Seattle indicate the onset of displacement coinciding with groundwater pumping following the stall of the Bertha TBM. Bottom: From December 2013 to October 2014, the ground displaced 10–20 mm along the route of the TBM. In November 2014, the ground suddenly began sinking significantly, with up to 60 mm of line-of-sight displacement occurring over the next 9 months. Optical image credit: Google Earth, Landsat/Copernicus

high-resolution (3 m) displacement map over 85 km of the California Aqueduct. A low-resolution map was also produced at 30-m resolution to reveal large-scale displacement across the entire 3000 km^2 area. Five large displacement zones were identified, each with displacement rates exceeding 5 cm/year and a combined area greater than 500 km^2. Results for an example displacement zone are shown in Fig. 2.15. Almost 20 km of the California Aqueduct is within these large

Table 2.2 Overview of the Seattle tunnelling case study

Targets	The aqueduct traverses hundreds of kilometres over mostly agricultural fields but some arid land as well. Permanent Targets are expected from arid land as well as isolated lampposts, fences, roads, canal infrastructure (e.g. pumping stations, bridges, canal banks, etc.), and scattered small buildings. Targets over agricultural fields, which are subject to crop cover, resetting by soil turnover and soil moisture. The area does not receive snow
Topography	The agricultural areas are relatively flat and significant geometric effects are not expected. Single look monitoring is, therefore, adequate. The topography is unchanging, hence a recent publicly available DEM suffices
Sources of error	• Significant soil moisture effects are expected over agricultural fields • Dry atmosphere effects should be insignificant, except for the hilly areas. Wet atmosphere has turbulence typical for mid-latitudes • Ionospheric effects can manifest at mid-latitudes; some ionospheric changes may appear if long-wavelength L-band images are used
Wavelength	L-band would achieve the best coverage by penetrating vegetated areas. However, X-band or C-band will be more sensitive to subtler displacement at the expense of some coverage
Resolution	Low resolution is sufficient for aquifer related subsidence, but 3 m or better spatial resolution is needed for monitoring small displacement features on the aqueduct itself. Biweekly images are sufficient as gradual displacement over time is expected (see next)
Known displacement	No site-specific information was available, but displacement magnitudes and gradients are typically low for aquifer applications
Satellite options	• C-band options are not ideal; Sentinel-1 with 20-m spatial resolution is too coarse; Radarsat-2 has high-resolution modes, but its 24-day repeat is limiting • ALOS-2 L-band data would make a good option archive data where already available to build upon • Both CSK and TSX are good options for generating a perforated displacement map with coverage over infrastructure and bare land with high spatial resolution and displacement sensitivity

displacement zones. The high-resolution map revealed two additional small (100 m) localised areas of the aqueduct that were subsiding.

Northern Alberta: Pipelines

Pipelines in Northern Alberta traverse hundreds of kilometres of challenging terrain subject to landslides and seasonal changes. Operators are interested in characterising geohazards throughout the area that threaten pipeline integrity, and planning routes for future pipelines. Sprawling pipeline infrastructure in remote areas is difficult to monitor effectively and efficiently using ground-based techniques. Table 2.3 provides an overview of the case study.

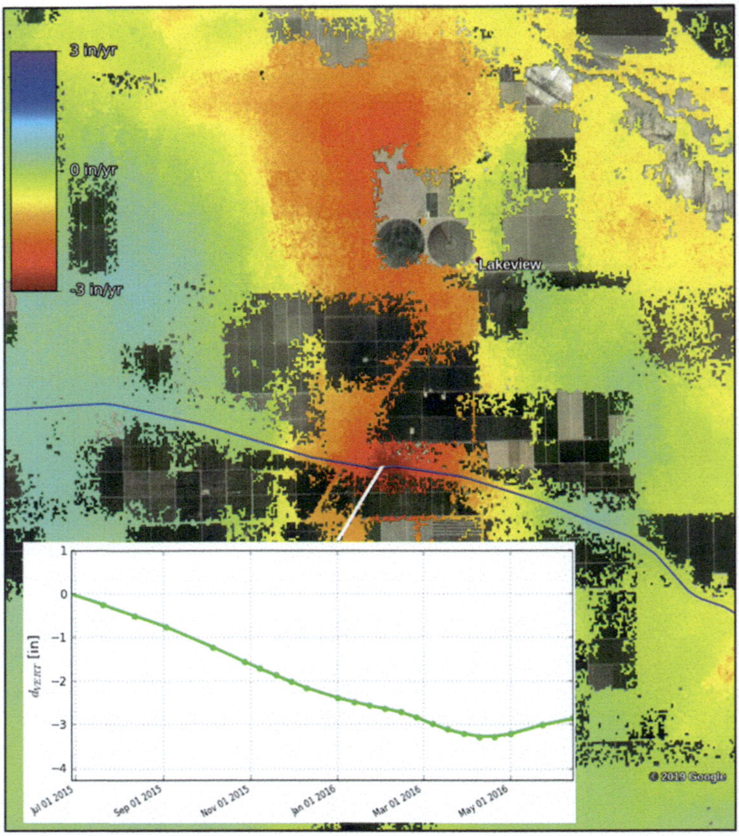

Fig. 2.15 Ground displacement along the California Aqueduct (blue line). Good coverage was achieved over bare ground and man-made infrastructure including the aqueduct itself using TSX data. Agricultural fields exhibited lower coverage due to crop cover and soil turnover, and also higher noise due to soil moisture effects. The results cover an area measuring 3000 sq. km and indicate multiple large displacement regions. Optical image credit: Google Earth, Landsat/ Copernicus

ALOS-2 was not commercially operational at the time of this project. 3vG analysed 17 ALOS-1 images, acquired between February 2007 and March 2011. Despite its lower resolution (15 metres), and its long temporal repeat (46 days minimum), displacement of 2–10 cm/year was evident over pipeline assets and surrounding terrain. Displacement histories for 18 million points were generated.

Seven thousand square kilometres of terrain were almost fully mapped for displacement, identifying numerous geohazards along the pipeline right of way. Excluding water bodies, spatial coverage was near complete. A landslide that had impacted the pipeline was clearly identified in images prior to the event; the known extent of this landslide matched well with InSAR results (see Fig. 2.16). Another

Table 2.3 Overview of the Alberta pipelines case study

Targets	The terrain in the area of interest is highly vegetated and dotted with lakes and swamps. Permanent Targets from processing facilities and other infrastructure are few and far between. The area receives heavy snow annually, limiting the InSAR monitoring period. Ground conditions following snow do not necessarily re-establish, limiting long-term measurements
Topography	The area is mostly flat and geometric effects are, therefore, not expected. The topography is unchanging and hence a recent publicly available DEM is sufficient to remove topographic effects from interferograms
Sources of error	• Dry atmosphere variations should be insignificant as the area is flat. Wet atmosphere can be turbulent if short wavelength sensors (X- or C-band) are used • At these latitudes, significant ionospheric changes may appear if long-wavelength L-band images are used • Significant soil moisture effects due to drainage, lakes, and agriculture are common in such terrain • Snow signals are expected in images with partial or full snow cover
Wavelength	L-band data is the only viable option for monitoring. The longer wavelength penetrates vegetation, possibly even interlinking measurements across the snow season each year. Sub-cm level precision will not be possible as precision is a function of the wavelength (see wavelength section). In agricultural areas, farming activity can reset targets and limit coverage even with L-band imaging
Resolution	High-resolution imaging (3 m or better) is needed as local fast-moving displacement is likely present (see next)
Known displacement	Pipelines in the area have been previously affected by small landslides. Seasonal soil dynamics over water-rich ground were also known to stress the pipeline
Satellite options	• TSX, CSK, Radarsat-2, and Sentinel are not useful due to short wavelengths • ALOS and ALOS-2 are the only options for historical and ongoing monitoring, respectively

landslide that was being monitored with ground-based methods was also accurately mapped. Seasonal freeze/thaw cycles in water-rich areas manifested as uplift/subsidence hotspots in the InSAR data; strong displacement gradients exist at the boundaries of such areas and can stress the pipeline.

Southwest USA Copper Mine

This open pit copper mine in southwest USA is very typical for the region, with large copper reserves of medium to low grade. Mining activity such as blasting, excavation, dumping, leeching produce some of the highest levels of displacement for any InSAR application, commonly reaching 100 cm/year or more. Much of the displacement is expected, but surprise events can occur also and geotechnical engineers often use InSAR to warn of concerning displacement before can affect safety or production. Table 2.4 provides an overview of the case study.

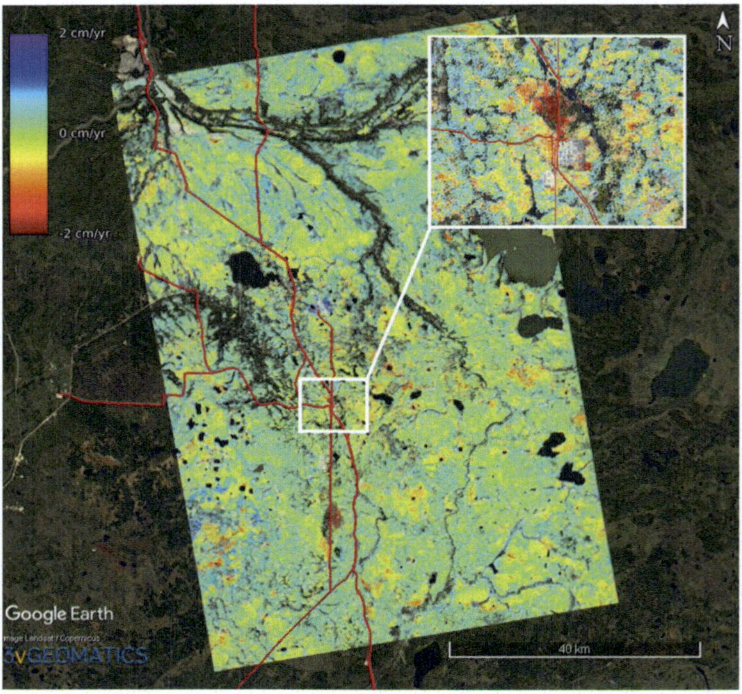

Fig. 2.16 Displacement results along pipeline rights of way in Northern Alberta using archived L-band ALOS-1 SAR data. The results provided a proof of concept for the pipeline operators, and the inset area shown is one example of a known geohazard captured by the InSAR results. Optical image credit: Google Earth, Landsat/Copernicus

TSX was chosen due to the availability of pre-existing archive data. Ninety-two ascending and ninety descending images were available, from July 2016 to April 2019, both data sets were utilised in order to derive displacement direction on slopes. A high-resolution (3 m) displacement map over the mine was produced (Fig. 2.17 left), as well as solved direction estimates for all pixels covered by both directions and with significant displacement (Fig. 2.17 right).

Following this initial baseline report, regular ongoing reports were provided. These were in the form of Rapid Reports, which provide a snapshot of displacement since the previous image, as well as quarterly Comprehensive Reports, providing displacement time series information and directional information as shown.

2.3 Summary

InSAR can be an effective tool for operational monitoring; its biggest advantage lies in mapping vast areas that would be prohibitive to monitor by more traditional means. A careful consideration of factors on the ground, such as the type of cover, the characteristics of the displacement activity, and the topography of the area are

Table 2.4 Overview of the southwest USA copper mine case study

Targets	The mine property covers 100 km^2 over arid terrain with practically no vegetation. Permanent targets are expected from boulders and infrastructure while many other areas will provide high quality distributed targets. Many targets will be temporary, due to mining activity. The area does not receive snow, but a turbulent atmosphere in late summer is likely to leave high residual noise in the data, despite attempts to remove it, and this might bias target selection
Topography	Many slopes exist, some steep enough to provide foreshortening, layover, and shadow effects, especially the pit. Elevation for some areas is expected to change over time, especially on stockpiles, dumps, and the two adjoined pits. The DEM should have resolution comparable with the chosen SAR data and be contemporary with it also. This is often available from the site personnel, but coverage may need to extend with the best open-source DEM data available
Sources of error	• Errors in the DEM are expected to grow significantly over time —Solving errors in the DEM using the InSAR data is straightforward for areas that are coherent over time —For less coherent, DEM updates every 6 months will aid unwrapping and displacement estimation —Processing shorter periods of no more than 3 months at a time will minimise DEM corrections going stale • Soil moisture effects are expected in the rainy summer period • Dry atmosphere effects will be significant, especially in the pits. Wet atmosphere can be highly turbulent in August and September • Ionospheric effects are not expected • The chance of unwrapping errors will increase in proportion to the magnitude displacement for an area
Wavelength	X-band will work well in the arid, vegetation free environment, providing high precision results
Resolution	High-resolution data of 3 m or less is required to capture and help identify small localised displacement areas
Known displacement	Many known areas of displacement exist, and most are displacing at magnitudes where prior knowledge is not required. A critical areas of interest from the site will help focus InSAR analysts when producing reports
Satellite options	• C-band options are not ideal; Sentinel-1 with 20-m spatial resolution is too coarse; Radarsat-2 has high-resolution modes, but its 24-day repeat is limiting • ALOS-2 L-band is not precise enough • Ascending and descending CSK or TSX are good options. Data from both ascending and descending directions is required to improve coverage, displacement sensitivity, and provide displacement direction information • Careful consideration of satellite incidence angles is required to optimise coverage and sensitivity to displacement on slopes

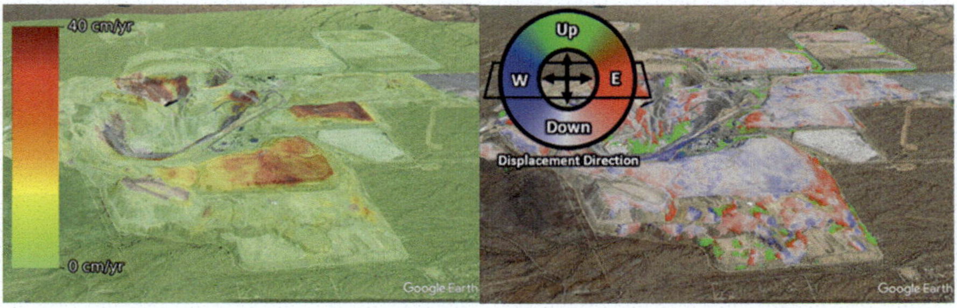

Fig. 2.17 Left: High magnitude displacement over a copper mine. Right: Data from both ascending and descending directions provided information regarding the direction of displacement in the up-down-east-west plane. Optical image credit: Google Earth, Landsat/Copernicus

required for success. These help to narrow down and inform data choices in terms of wavelength, resolution, and satellite look geometry.

These and other ground and satellite characteristics have been discussed to help laypeople to assess the applicability of InSAR for their site. Real-world case studies have been included to show examples of this decision process.

References

Bamler, R., and P. Hartl. 1998. Synthetic aperture radar interferometry. *Inverse Problems* 14: R1–R54. https://doi.org/10.1088/0266-5611/14/4/001.

Bekaert, D.P.S., A. Hooper, and T.J. Wright. 2015. A spatially variable power law tropospheric correction technique for InSAR data. *Journal of Geophysical Research - Solid Earth* 120: 1345–1356. https://doi.org/10.1002/2014JB011558.

Berardino, P., G. Fornaro, R. Lanari, and E. Sansosti. 2002. A new algorithm for surface deformation monitoring based on small baseline differential SAR interferograms. *IEEE Transactions Geoscience Remote Sensing* 40: 2375–2383. https://doi.org/10.1109/TGRS.2002.803792.

Cattabeni, M., A. Monti-Guarnieri, and F. Rocca. 1994. Estimation and improvement of coherence in SAR interferograms. In *Proceedings of IGARSS '94 - 1994 IEEE International Geoscience and Remote Sensing Symposium*, 720–722. Pasadena: IEEE.

Ferretti, A., C. Prati, and F. Rocca. 2001. Permanent scatterers in SAR interferometry. *IEEE Transactions Geoscience Remote Sensing* 39: 8–20. https://doi.org/10.1109/36.898661.

Ghiglia, D.C., and M.D. Pritt. 1998. *Two-Dimensional Phase Unwrapping: Theory, Algorithms, and Software*. New York: Wiley.

Hanssen, R.F. 2001. *Radar Interferometry: Data Interpretation and Error Analysis*. Dordrecht: Kluwer Academic.

Hooper, A. 2008. A multi-temporal InSAR method incorporating both persistent scatterer and small baseline approaches. *Geophysical Research Letters* 35: L16302. https://doi.org/10.1029/2008GL034654.

Hooper, A., D. Bekaert, K. Spaans, and M. Arıkan. 2012. Recent advances in SAR interferometry time series analysis for measuring crustal deformation. *Tectonophysics* 514–517: 1–13. https://doi.org/10.1016/j.tecto.2011.10.013.

Mora, O., J.J. Mallorqui, and A. Broquetas. 2003. Linear and nonlinear terrain deformation maps from a reduced set of interferometric SAR images. *IEEE Transactions Geoscience Remote Sensing* 41: 2243–2253. https://doi.org/10.1109/TGRS.2003.814657.

Pepe, Antonio, and Fabiana Calò. 2017. A review of interferometric synthetic aperture RADAR (InSAR) multi-track approaches for the retrieval of earth's surface displacements. *Applied Sciences* 7: 1264. https://doi.org/10.3390/app7121264.

Woodhouse, I.H. 2006. *Introduction to Microwave Remote Sensing*. Boca Raton: Taylor & Francis.

Chapter 3
Sentinel-1 InSAR Data for the Continuous Monitoring of Ground Deformation and Infrastructures at Regional Scale

Nicola Casagli, Silvia Bianchini, Andrea Ciampalini, Matteo Del Soldato, Pablo Ezquierro, Roberto Montalti, Monan Shan, Lorenzo Solari, and Federico Raspini

3.1 Introduction

Ground deformation affecting man-made linear infrastructures is a serious task in Italy and in many countries worldwide. The assessment and forecasting of the impact of a deformative event (*e.g.*, landslide, subsidence, consolidation) and the evaluation of the vulnerability to deformation of infrastructures of strategic interest (*e.g.*, metro systems, dams, pipelines, roads, bridges, overpasses and flyovers, highways, railways, and aqueduct) are important tasks for both Civil Protection Authorities, local authorities and public and private companies that manage transportation networks.

Potential damages of these features represent not only loss of financial investment and money but also loss of significant landmark in the world, producing high indirect impacts on the society. Moreover, long-lasting deformation can pose a serious threat also to human lives, when progressive damages lead to sudden and catastrophic failure of the infrastructure. The monitoring of deformation affecting

N. Casagli · S. Bianchini · M. Del Soldato · R. Montalti · M. Shan · F. Raspini (✉)
Earth Sciences Department, University of Firenze, Firenze, Italy
e-mail: nicola.casagli@unifi.it; silvia.bianchini@unifi.it; matteo.delsoldato@unifi.it; roberto.montalti@unifi.it; monan.shan@unifi.it; federico.raspini@unifi.it

A. Ciampalini
Earth Sciences Department, University of Pisa, Pisa, Italy
e-mail: andrea.ciampalini@unipi.it

P. Ezquierro
Geological Survey of Spain (IGME), Madrid, Spain
e-mail: p.ezquerro@igme.es

L. Solari
Centre Tecnològic de Telecomunicacions de Catalunya (CTTC), Castelldefels, Spain
e-mail: lorenzo.solari@cttc.cat

V. Singhroy (ed.), *Advances in Remote Sensing for Infrastructure Monitoring*, Springer Remote Sensing/Photogrammetry,
https://doi.org/10.1007/978-3-030-59109-0_3

linear structures is a challenging task that draws the attention of many geoscientists in recent times.

Several methods are used for measuring, mapping, and monitoring spatial extent and temporal evolution of regional and local deformation affecting man-made linear infrastructures. The conventional way of measuring deformation is based on geodetic surveying techniques such as levelling or GPS (Global Positioning System—*e.g.*, Kaloop et al. 2017). Conventional methods for the production of ground deformation maps rely on the materialization of a network of geodetic benchmarks, designed to cover the extension of the deforming area. Repeat surveys of benchmarks, referenced to a stable control point, allow the estimation of the extent and rate of deformation over time. These techniques, used for decades with standardized procedures, are robust and reliable, but, on the other hand, are time consuming and require a great deal of time and economic resources for timely updates. When dealing with long infrastructures located in mountainous areas, conventional surveying techniques are not suitable because of high costs and infeasibilities in terms of logistic. In other situations (*e.g.*, runway areas of airport) geodetic methods are not of practical help due to restriction of access within the area to be monitored. In order to deeply understand the deformation pattern along linear infrastructures such as airports, railways, or highways, characterized large linear extent and irregular distribution, a monitoring technique coupling wide coverage and high precision is required.

In this framework, EO (Earth Observation) techniques have successfully demonstrated in the last decades to be highly valuable in measuring ground motion in a wide range of application fields, including the analysis of several natural phenomena, such as earthquakes, volcanoes, landslides, and subsidence (Tralli et al. 2005). More recently, SAR technology and Interferometric Synthetic Aperture Radar (InSAR) have been applied in the analysis of deformations occurring on dams (Milillo et al. 2016), bridges (Fornaro et al. 2012), airport (Del Ventisette et al. 2015; Raspini et al. 2013) transportation network (Giannico et al. 2012; Ozden et al. 2016; Bakon et al. 2014; Polcari et al. 2019; Chang et al. 2016; North et al. 2017; Wasowski et al. 2017), for assessment of tunnelling effects (Bandini et al. 2015) and single-building stability (Bianchini et al. 2015; Ciampalini et al. 2014). These applications were possible, thanks to the capability to deliver very precise (few millimetres according to Colesanti et al. 2003) and spatially dense information (from hundreds to thousands measurement points/km^2) for wide area (thousands km^2). The potential of detecting differential deformation along the tracks makes it one of the most powerful and economical means for monitoring the safety and stability of the infrastructure on a weekly basis, thanks to the unprecedented revisiting time of the Sentinel-1 constellation (Torres et al. 2012).

In this work, we exploited C-band Sentinel-1 (SNT) and X-band Cosmo-SkyMed (CSK) InSAR data to detect anthropogenic-induced local subsidence that involve railways, highways, and infrastructures in the Tuscany Region, Central Italy. The Tuscany Region is the first region in the world to have a continuous ground deformation monitoring using Sentinel-1 constellation. Subsidence phenomena, as well as landslides and floods, are one of the main geohazards affecting the Tuscany Region. The monitoring of related ground deformations plays a key role in their management to avoid long-term effects on buildings and infrastructures. The total length of road network in Tuscany is 4.922 km. The length of the rail network is 1.400 km. Some sectors of these networks are affected by land subsidence, some other threatened by landslides. Considering that we can neither instrument all the landslides nor prevent all the risks posed by ground deformation and that it is not feasible to monitor with conventional techniques the whole road–rail system, a different strategy for risk mitigation urged to be conceived.

3.2 The Tuscany Region

Benefits deriving from the exploitation of satellite data for infrastructures monitoring are presented and discussed through the case study of the Tuscany Region (Central Italy). Located on the Tyrrhenian side of the Italian peninsula (Fig. 3.1), Tuscany, with an area of approximately 23.863 km^2, extends between the regions of Liguria, Emilia-Romagna, Marche, Umbria, and Lazio. Tuscan territory includes also a little islands archipelago composed by six main islands and other small rocks. Tuscany is administratively subdivided into 10 provinces with the regional capital, Firenze, located in the northern part of the Region.

The main reliefs rise in the northern and eastern parts of the region, and they belong to the Northern Apennine system, an arcuate orogenic belt that marks along the border of the region from NW to SE. The Northern Apennines is a NE-verging fold and thrust belt formed since Upper Cretaceous by the stacking of the Ligurian Units over the Tuscan—Umbrian Units (Bortolotti 1992; Vai and Martini 2001). The Ligurian Units are the most elevated in the nappe pile and consist of marine sedimentary rocks and ophiolite sequences. These units overlap, from west to east, the Umbro-Tuscan unit deposited in a passive margin marine environment and consisting in a thick sedimentary package of carbonate and silicoclastic deposits, that composes the backbone of the Northern Apennine ridge.

The present morphology (relief orientation and distribution) is controlled by the eastward migrating extension–compression pair which affects the Northern Apennines from the Oligocene (Elter et al. 1975; Bartolini et al. 2003). Being characterized by heterogeneous physical settings, with Apennines flysch ridges alternated with NW-SE trending intermontane basins, Tuscany Region represents a perfect scenario to test, tune, and refine this new monitoring approach.

Tuscany has been specifically selected due to its peculiar geological setting prone to ground instability phenomena. Its territory, mainly hilly (66.5%) with few plains where the main rivers flow (8.4%), is surrounded and crossed by major mountain chains (25.1%), and it results to be a very landslide-prone area (Lu et al. 2010). Landslide processes have pervasively shaped the Tuscan landscape (Rosi et al. 2018 mapped more than 90,000 landslide) and are a major issue for regional

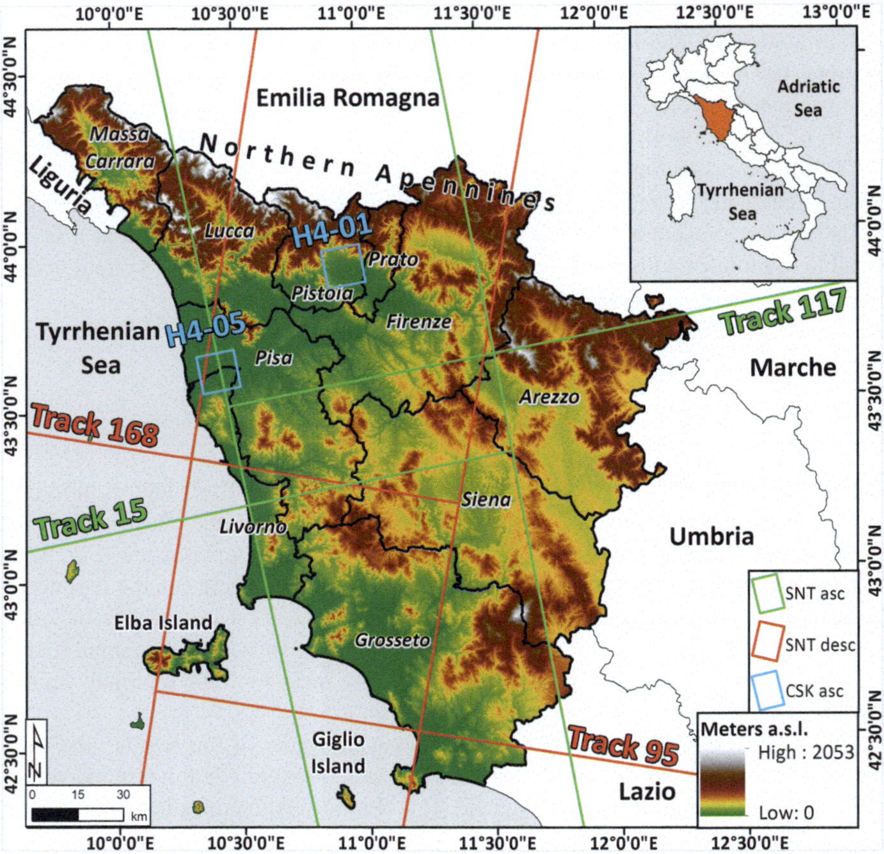

Fig. 3.1 Physical setting of the Tuscany Region shown through a digital elevation model. Tracks of Sentinel-1 and COSMO-SkyMed satellites used in this work are also included. The Sentinel-1 coverage of the Tuscany Region is achieved by two different frames in both ascending and descending geometries, distributed along two tracks along each orbit

authorities. In some situations, extensive protective measures have been planned and carried out (D'Amato Avanzi et al. 2006), some landslides have been instrumented (Rosi et al. 2013), some others slowed through remedial works (Farina et al. 2006), but it is not possible to reduce all the threats or to protect all the threatened areas.

Land subsidence has been also identified throughout the region, related both to groundwater exploitation (Del Soldato et al. 2018) and to compaction of soft sediments (Solari et al. 2016). Finally, extensive subsiding area has been mapped also in the geothermal district of Central Tuscany (Rosi et al. 2016). Land subsidence is a subtle threat in both urban and agricultural areas, where the continuous lowering over time can cause damage to buildings, sink houses, crack roads, and can cause severe impacts on the environment and economy.

3.3 The PSI (Persistent Scatterer Interferometry) Technique

The ground deformation maps for the Tuscany region have been generated by analysing two large stacks of C-band Sentinel-1 acquisitions by means of the SqueeSAR algorithm (Ferretti et al. 2011). This algorithm represents the evolution of PSInSAR, the first technique belonging to the PSI family specifically implemented for the processing of multi-temporal radar imagery (Ferretti et al. 2001). The novelty of PSInSAR algorithm is the capability of identifying a network of coherent radar targets, *i.e.*, point-like-targets (PS, Permanent Scatterers) within the radar image. Such pixels correspond to man-made objects, outcropping rocks, debris areas, or buildings that register a stable radar signal over the whole observation time period. The definition of these points allows reducing decorrelation phenomena and obtaining a high signal-to-noise ratio (Ferretti et al. 2001), thus discriminating phase contributions related to displacement from those due to atmosphere, topography, and noise.

Over urban fabric, where many stable reflectors can be identified, PS density can be very high (up to hundreds of PS per square km). One limitation of this technique is the impossibility of identifying "radar friendly" targets outside of urban and peri-urban areas and in highly vegetated zones.

The development of the SqueeSAR techniques (Ferretti et al. 2011) contributed to extend the field of application of PSInSAR to natural terrain, overcoming the main limitation of this technique. SqueeSAR estimates deformation rates not only from point-like-targets but also from partially coherent pixels, called Distributed Scatterers (DS). DS points correspond to homogeneous ground surfaces and not to single objects, *i.e.*, uncultivated areas, deserts, debris covered slopes, and scattered outcrops.

LOS (Line of Sight) deformation rate can be estimated with an accuracy theoretically lower than 0.1 mm/year, at least for very stable PS during a long time span (Colesanti et al. 2003). PSI analysis is designed to generate time series of ground deformations for individual reflectors. The accuracy of the single measurement in correspondence of each SAR acquisition ranges from 1 to 3 mm (Colesanti et al. 2003). Each measurement is referred temporally and spatially to a unique reference image and to a stable reference point. All the provided ground displacement measurements fulfil a set of basic requirements from the local authorities (a) regional coverage, (b) <1 cm accuracy, (c) continuous delivery of information, (d) timely delivery of updated products, (e) low cost (considering the number of measurement points for the spatial extent of Tuscany). The PSI data stack covers the whole Tuscany territory, with different spatial densities of measurement points on different types of land covers. For each measurement point, both the average displacement rate (in mm/year) and the time series of displacement (in mm), along the satellite line of sight, are provided.

3.4 Results at Regional and Local Scale

For the initial implementation, in October 2016, of the continuous monitoring of the Tuscany Region, the images archive of the ESA Sentinel-1 C-band images (centre frequency 5.4 GHz, wavelength 5.6 cm, and incidence angle ranging from 36 deg to 40 deg) was acquired and then processed by means of the SqueeSAR technique (Ferretti et al. 2011). At the beginning of the project, while Sentinel-1B "operationally qualified products" were yet not available, the project relied on Sentinel-1A images only, with an acquisition frequency of 12 days (Fig. 3.2). This first phase of the project ended with the provision of ground deformation maps with almost 2 million PS points, covering the whole Tuscan territory, data useful to understand and assess the stability evolution of the territory in the period 2014–2016.

In the second phase of the project, started after the provision of the baseline, the database of ground deformation measurements was updated by continuous processing of Sentinel-1 images. Starting from January 2017, both sensors (A and B) were included in the processing chain, reducing the time interval between two acquisitions to six days and granting the fully exploitation of the potential of the constellation. Once a new Sentinel-1 image is available, it is automatically downloaded and added to the existing archive. The new data stack is then entirely reprocessed to generate new ground deformation maps and updated displacement time series. The main driver of this work is the design of new paradigms for monitoring ground deformation at regional and/or national scale. Aimed at the transition from static snapshots to a dynamic streamline of displacement information, a series of subsequent updating is created every 12 days, maximizing the exploitation of Sentinel-1. A processing chain, optimized to reduce processing times, has been set up to provide in near real time an updated database of displacement data over 23,000 km^2. At the end of May 2019, a total of 71 updates have been created for the Tuscany Region, evenly distributed over the monitoring period (29 updates in 2017, 29 in 2018, and 13 at May 2019). In Fig. 3.3 velocity maps obtained from the processing of all the Sentinel-1 images up to May 2019 are reported. Each velocity map includes more than 900,000 points and provides a synoptic view of the regional displacement field. Moreover, through the analysis of displacement time series, it is possible to

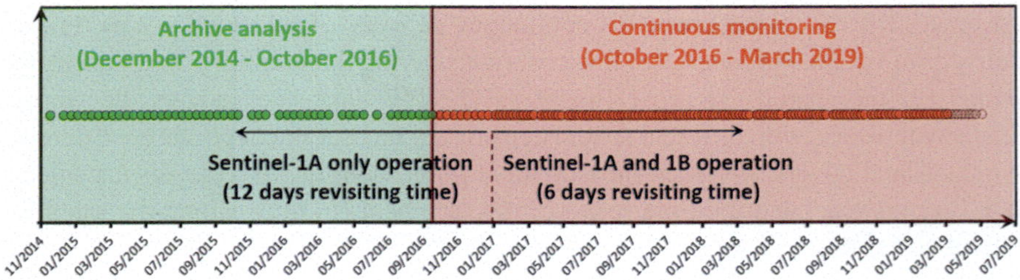

Fig. 3.2 Different phases of the monitoring project for the Tuscany region. Following the processing of images archive to set up the baseline, a series of analysis is performed every 12 days to provide a dynamic streamline of displacement

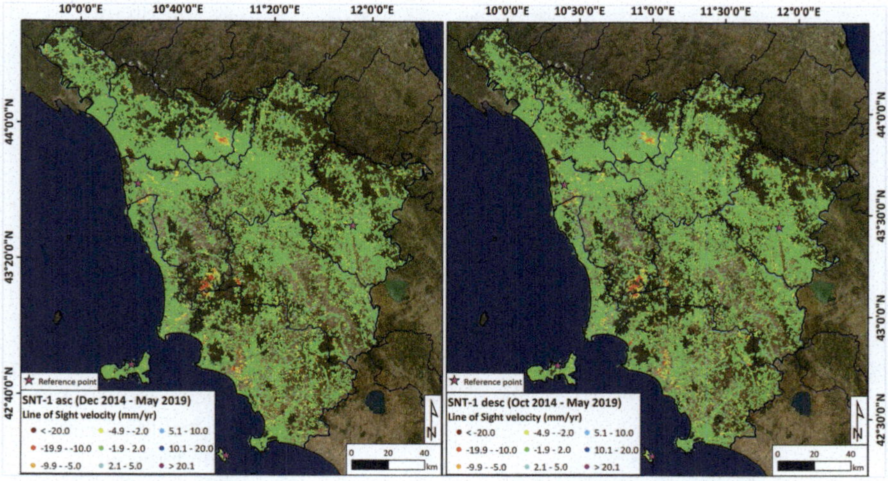

Fig. 3.3 Overview of the results obtained by SqueeSAR processing of C-band Sentinel-1 data in the two acquisition geometries (ascending on the left and descending on the right). Measurement points are represented by dots coloured according to their mean velocity along the line of sight (LOS) of the satellite, expressed in millimetres per year, with positive values of mean velocity representing displacements towards the satellite, and negative values representing displacements away from the satellite. Stability interval has been set according to the standard deviation of the mean velocity values

highlight "anomaly of movements", *i.e.*, changes in ground deformation temporal patterns (Raspini et al. 2018).

These maps include a wealth of quantitative information on a range of phenomena, including slow-moving landslides (with the possibility to highlight the acceleration of movements that typically precede collapse and to progressively refine the forecasting of the failure time), subsidence (related to aquifer depletion or load imposition), uplift, consequences of geothermal resources exploitation, and on the impact of these phenomena on buildings and linear infrastructures.

Figure 3.4 shows the infrastructures network for the Tuscany Region and the deformation rates extracted for the different roads and railways over the area of interest, using a buffer area of 200 m. Systematic monitoring of human infrastructure and of communication and transportation routes (highways, railways, roads, bridges, viaducts) using satellite data is an effective tool that can be used for maintenance planning activities as well as for infrastructure risk assessment. The information contained in Fig. 3.4 are particularly valuable, as they allow, scanning wide areas, to spot unstable zones, supporting the definition of priorities, starting from the situation deemed to be most urgent. While most of the roads show stable behaviour or very low deformation rates, significant deformation is revealed in the zone of Pistoia (central part of the Region) and in the area of the freight terminal "Amerigo Vespucci" close to the coastline.

To refine the analysis of on-going deformation patterns at local scale, these two areas became targets for detailed analysis with high-resolution COSMO-SkyMed

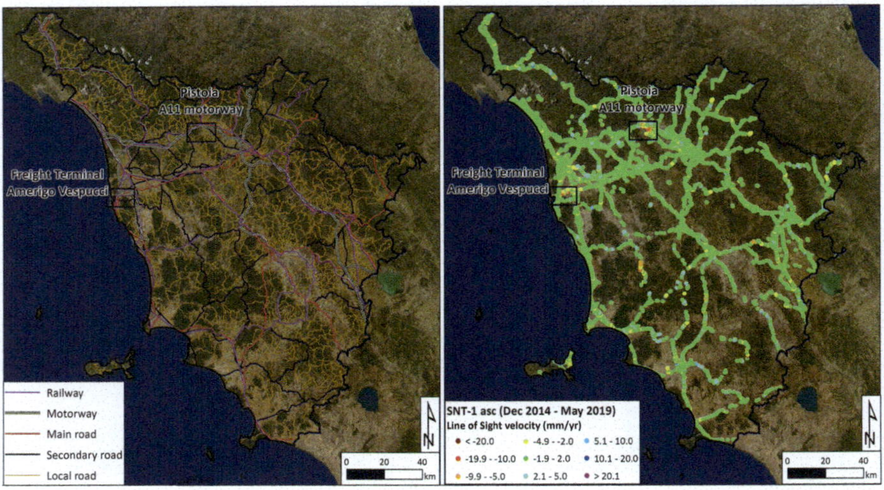

Fig. 3.4 Railways and road system for the Tuscany region, including motorways as well as primary, secondary, and local roads (left). Motion rates along infrastructures network on Tuscany Region (right)

images, to create a virtual constellation, where different satellite data sources are used in synergy to create a more effective and robust Earth observation system. In a virtual constellation different space segments overlap, in order to meet different requirements. The use of different sensors may help to mitigate intrinsic limitations of any one particular sensor, providing a common set of Earth observation data. For the target areas medium resolution, wide area (Sentinel-1) and high resolution-detailed scale (COSMO-SkyMed) data are merged and integrated to derive information to contribute to a quantitative analysis.

3.5 The Amerigo Vespucci Freight Terminal

The first area target of a detailed COSMO-SkyMed analysis is the recently built Amerigo Vespucci freight terminal (Guasticce terminal), at the border between Livorno and Pisa provinces, within the Arno coastal plain. This area, composed of several commercial sheds, was selected for its strategic position (about 6 km far from the Livorno commercial port) and for the proximity to the transportation network (both highways and railways). The Guasticce terminal is affected by subsidence since its construction, started in early 2000. The ground lowering is triggered by the load imposition, as freight terminal includes several warehouses with an extension varying from 10,000 m^2 to 17,000 m^2 and a height between 12 and 14 m above the ground. Deformation induced structural damages to several sheds, leading to long-term economic losses for the shipping companies (Ciampalini et al. 2019).

The upper part of the Arno coastal plain is filled by an alternation of 100 m of continental and coastal marine deposits cut by two 30-m deep incised-valley systems filled by a fining upward stratigraphic sequence. This sequence is overlaid by a highly compressible silty clay layer, 5–15 m thick, deposited within a coastal lagoon. This layer lies beneath a fluvio-deltaic succession, 10–15 m thick, made of clay and silt with lenses of sand (Aguzzi et al. 2007; Sarti et al. 2015).

The availability of ground deformation maps generated exploiting different sensors (C-band Sentinel-1 and X-band Cosmo-SkyMed) allows depicting the pattern of ground deformation which affects the Amerigo Vespucci freight terminal and surrounding infrastructure, among which the Scolmatore floodway (a 28-km artificial flood control channel of the river Arno), the A12 motorway, and the Firenze-Pisa-Livorno highway (Fig. 3.5).

Sentinel-1 data indicate that the whole area where the freight terminal is built is suffering for subsidence with a mean rate of −10 mm/year. The highest velocities (−20.7 mm/year) have been recorded in the northern sector (along the right bank of the Scolmatore channel).

Thanks to its finer resolution, COSMO-SkyMed (centre frequency 9.6 GHz, wavelength 3.1 cm, and incidence angle at mid-range of around 30 deg along ascending orbits) allows refining the definition of the subsidence pattern of this area. More than one thousand measurement points have been retrieved over the freight terminal. LOS deformation is measured for all the buildings. Ciampalini

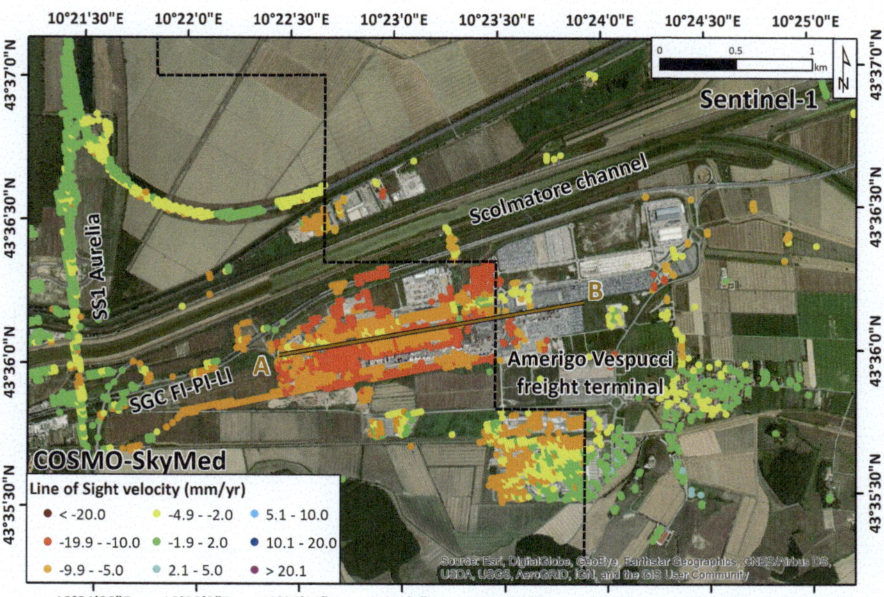

Fig. 3.5 LOS ground deformation maps for the area of the freight terminal Amerigo Vespucci derived with COSMO-SkyMed (2011–2014) and Sentinel-1 (2014–2019) images acquired along ascending orbits. The difference of PS density between COSMO-SkyMed (on the left) and Sentinel-1 (on the right) is marked by the dashed line

et al. 2019 highlighted that the rates are clearly time-dependent from the age of the buildings. Obtained results show that the oldest buildings are still affected by a deformation rate close to −5 to −10 mm/year, whereas recent buildings register rates around −40 mm/year. Time series of deformation suggest that the deformation rates decrease over time following a time-dependent trend that approximates the typical consolidation curve for compressible soils (Terzaghi and Peck 1967).

To reconstruct in detail the subsurface stratigraphy of the study area, a series of available on-site survey data was analysed, including boreholes, dilatometer tests (DMT), and piezocone penetration tests (CPT). These investigations have been performed between 1992 and 1998 for the determination of geotechnical parameters for the construction of the freight terminal. Six litho-geotechnical units have been recognized and described from top to bottom as follows (Fig. 3.6):

- Unit 1: massive brownish clayey silt locally passing to sandy silt. Unit 1 is present in the whole area of the freight terminal with an average thickness of few metres;
- Unit 2: present in the whole area of the freight terminal with a thickness decreasing from 22 m in the western part to 15 m in the eastern part. It is made of light grey silty clay with low silty fraction. The content in organic matter is high (peat, wood fragments);
- Unit 3: represented by thin lenses and/or layers usually interbedded to the units 2 or 4. This unit is made of silty fine sand to coarse sand;
- Unit 4: made of light grey clayey silt and locally of silty clay with a maximum thickness (42 m) in the central and eastern sectors of the freight terminal area. Occasionally sandy layers have been observed;
- Unit 5: consists of silty fine to coarse sand with occasional intercalations of few centimetres thick of sandy silt.
- Unit 6: thin unit made of gravel and sandy gravel.

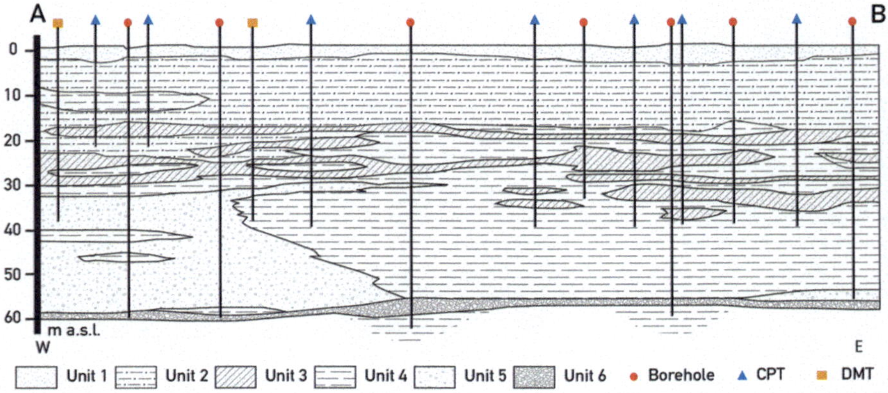

Fig. 3.6 Stratigraphic section AB reconstructed from the subsurface investigations. The trace of the stratigraphic sections is reported in brown in Fig. 3.5

The Unit 1 and Unit 2 are expected to be the soil layers with the biggest contribution to the total vertical displacements. Unit 2 is characterized by high thickness, poor geotechnical characteristics (*i.e.*, high natural water content, low friction angle, and specific weight), and high coefficient of consolidation. Despite the geotechnical features of the clay belonging to Unit 2 were improved by the ageing effect, they depreciate when clay is disturbed and remodelled after geotechnical investigations and/or when foundations are built.

For this area it is observed that the subsidence depends on the thickness of compressible deposits and on applied load (Fig. 3.7). Indeed, different settlement patterns observed along velocity profile of Fig. 3.7 result from the combination between the applied load (in terms of both magnitude and time of imposition) and the mechanic behaviour of the loaded soils over time, which is linked also to the fine soil thickness, hydraulic conductivity, and drainage conditions.

3.6 The Area of Pistoia

The second study area is the southern part of the city of Pistoia, a place with a long history of land subsidence related to overexploitation of several confined aquifers present at different depths (Colombo et al. 2003; Canuti et al. 2006). The city of Pistoia is located on the western margin of the Firenze–Prato–Pistoia basin, a flat area with an elongated shape in the NW–SE direction and approximately 35 km wide and 100 km long. This plain is an intermontane basin formed during the late extensional phase of the formation of the northern Apennines. The basin is bounded by a major structural border ideally connecting Pistoia to Florence, 50 km south-eastern, corresponding to a normal fault SW dipping. The activity of this fault since Lower Pliocene created an asymmetric semi-graben basin later filled by lacustrine and fluvial deposits, reaching a thickness of hundreds of metres in its north-eastern portion. In general, the bottom of the stratigraphical sequence is represented by lacustrine clay and silt with organic levels. On top of this clayey layer, the sequence

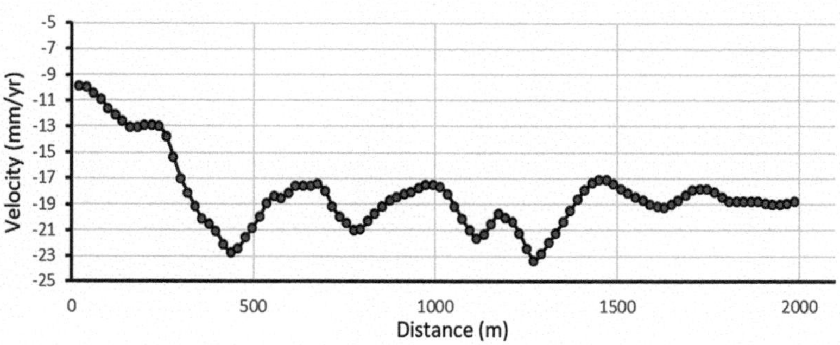

Fig. 3.7 COSMO-SkyMed ground deformation (2011–2014) along section AB at the Amerigo Vespucci freight terminal. The trace of the sections is reported in brown in Fig. 3.5

follows the paleogeographic evolution of the basin, with thick gravel and sand fan deltas deposits at the mouth of lateral streams and sand and silt level deposited by paleo-rivers wandering along the basin (Capecchi et al. 1975). Thus, the stratigraphical asset is quite complex and the thickness of deposits varies within the basin, with strong lateral variation that determine a highly variable hydrogeological context with multilayer aquifers with different hydraulic conductivity.

Again, the availability of multi-band ground deformation maps allows characterizing ground deformation affecting the southern sector of Pistoia, which hosts several strategic infrastructures, among which the A11 motorway, the railway connecting the city to Firenze and to northern Tuscany, the Pistoia by-pass, and the hospital.

A general agreement between ground deformation maps provided by Sentinel-1 (2014–2019) and COSMO-SkyMed (2015–2018) is observed (Fig. 3.8). The difference in the magnitude in the deformation pattern is likely related to the different time intervals covered by the two satellite systems, to their different imaging characteristics and to the different processing approaches.

Sentinel-1 data indicate that the whole area straddling the A11 motorway, which stretches east-west within the plain, is affected by land subsidence with a mean value of −8 mm/year and peaks of velocity between −20 and −30 mm/year in the area of Bottegone and southern than the Pistoia hospital.

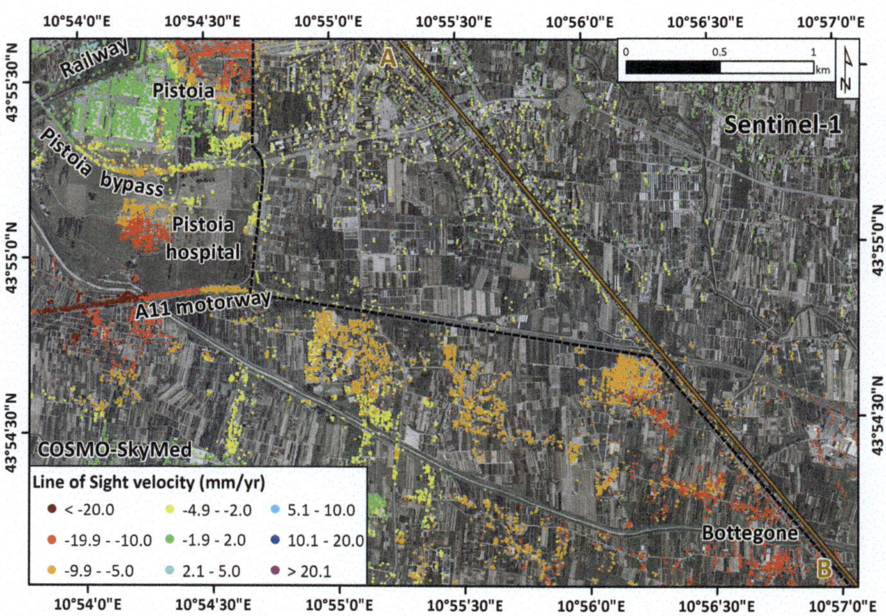

Fig. 3.8 LOS ground deformation maps for the area south of Pistoia derived with COSMO-SkyMed (2015–2018) and Sentinel-1 (2014–2019) images acquired along ascending orbits. The difference of PS density between COSMO-SkyMed (on the left) and Sentinel-1 (on the right) is marked by the dashed line

COSMO-SkyMed (incidence angle at mid-range of around 27° in ascending orbits), ensuring an almost spatially continuous coverage of information on surface deformation and related hazards, confirmed the deformation pattern and increased its characterization. Such density of points (*e.g.*, 620 measurements points for the 60,000 m^2 wide hospital) makes COSMO-SkyMed InSAR ideally suited to map the extension of threatened areas, improving confidence on spatial pattern of the examined phenomenon. Groundwater overexploitation and related compaction of silt and clay terrains are the main reasons for the observed subsidence, with additional local contribution related to load imposition, such as in the case of the Hospital of Pistoia (built between 2010 and 2012), whose time series show a decrease of deformation over time.

The stratigraphical asset below the area of south-eastern sector of Pistoia reflects its position within the Firenze–Prato–Pistoia basin. Basement rocks outcrop few kilometres northern and western than the city centre with the thickness of the sedimentary sequence increasing from 30 m in the north-western part of the city to more than 150 m in its south-eastern portion. In the northern part, thick gravel and pebbles levels are dominant, while moving southward within the centre of the basin, silt and clay terrains become prevalent.

For this area it is observed that the thicker the cumulate compressible deposits the higher the subsidence rates. The correlation between greater subsidence values and thickness of the compressible sediments along the NW-SE oriented section AB is clear (Figs. 3.9 and 3.10).

Indeed, towards the central part of the plain area, where compressible sediments gain considerable thickness, ground deformation increases severely; moving northwest, the intense gradient in subsidence decrease is clearly related to the thinning of lacustrine deposits. Compaction of the sediment system, measured as subsidence rates, depends not only on the geometrical and geotechnical characteristics of the aquifer deposits, namely thickness and compressibility, but also on piezometric level variations. Variability of both water head level and thickness of compressible fine-grained deposits contributes to the different deformation patterns observed along velocity profile of Fig. 3.10. In particular, deformation between 500 m and

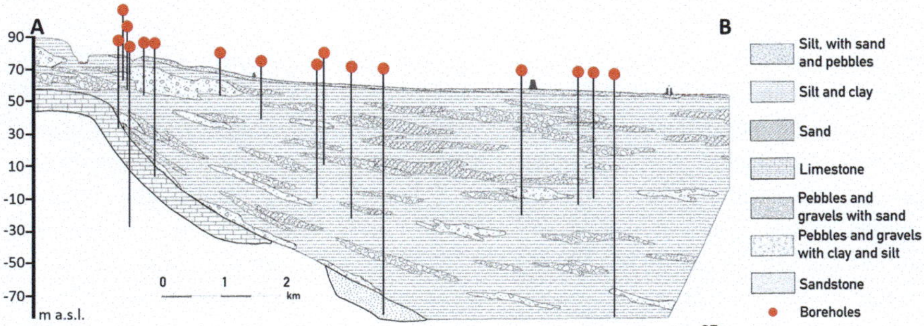

Fig. 3.9 Stratigraphic section AB reconstructed from the subsurface investigations. The trace of the stratigraphic sections is reported in brown in Fig. 3.8

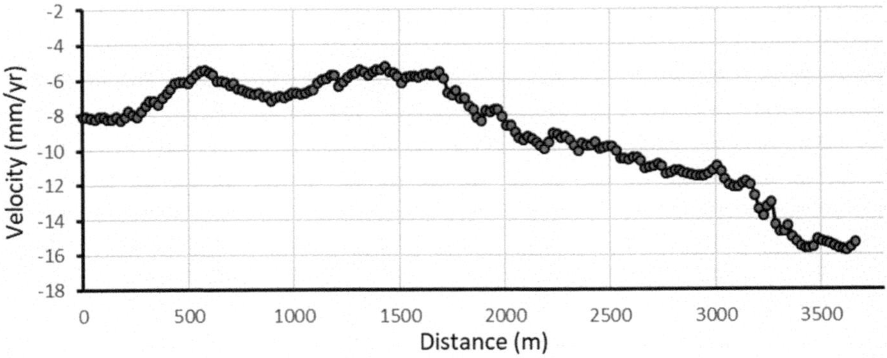

Fig. 3.10 COSMO-SkyMed ground deformation (2015–2018) along section AB south of the city of Pistoia. The trace of the sections is reported in brown in Fig. 3.8

1500 m is lower than expected because the presence of a thick, shallow lens of pebbles and gravels and the limited groundwater overexploitation (with respect to other sectors of the plain).

3.7 Discussion

In the last two decades, multi-temporal InSAR techniques have experienced major developments, gaining increasing attention within the scientific, stakeholders, and end users' communities. Despite the unique capability to observe very large areas, this technology has been typically applied to the analysis of relatively small sites or for post-event back-analysis, concentrating on the spatial investigation of ground movements, and possibly look retrospectively to capture precursory signals of an already set up motion.

The Tuscany Region adopted in 2016, first in the world, a monitoring operative service based on InSAR technology, which relies on the continuous processing, through advanced persistent scatterer interferometry techniques, of SAR images acquired by the Sentinel-1 satellites. Designed to acquire for decades to come, the Sentinel-1 satellites are ideally suited to provide a streamline of surface deformation measurements covering wide areas: for instance, Sentinel-1 images are used to map ground movements over Norway The European Space Agency (2019) (https://www.esa.int). Thanks to the systematic images acquisition, Sentinel-1 mission provides regular (every 6 days) and accurate updates of surface movement on a continuous basis. This constellation profitably contributes to the design and implementation of a monitoring system capable of monitoring and measuring ground deformation affecting linear infrastructures at an appropriate temporal scale.

The use of COSMO-SkyMed constellation, which proved its great potential for detailed analysis, represents a value-added to refine pattern of deformation and helps to overcome the necessary trade-offs that govern the design and implementation of a monitoring system at regional scale.

Despite the longer revisiting time (16 days with COSMO-SkyMed with respect to the 6 days with Sentinel-1), a much larger number of PSs can be detected using COSMO-SkyMed data. The finer ground resolution (3 m × 3 m, compared to 14 m × 5 m) allows detecting more radar targets and collecting a coherent backscattering signal from smaller spots of bare soil in vegetated terrains.

In order to increase the availability and to foster the use of the analyses carried out within the project, all PSI products (*i.e.*, ground deformation maps and displacement time series) have been published in the Regional Geoportal, managed by LaMMA Consortium (Environmental Modelling and Monitoring Laboratory for Sustainable Development). The interferometric measurements are freely accessible and downloadable by everybody. Geologists, geophysicists, geotechnical and structural engineers can access the PSI measurements to get quantitative information about a range of phenomena, including slow-moving landslides, subsidence, uplift, geothermal activities ground effects, and on the impact of these phenomena on buildings and infrastructures. The portal systematically publishes the updated catalogue of InSAR results, along with other geospatial information for open use (*e.g.*, the road and railway network database or the landslides inventory map). The whole community of operational users (both Civil Protection authorities, local authorities and stakeholders) can benefit from InSAR support.

Following the Tuscany example, also Valle d'Aosta started monitoring its territory (Solari et al. 2019). Outputs of this project may seed future European activities, such as the scheduled European-wide Copernicus Ground Motion Service (EGMS), whose implementation, governed by the European Commission in partnership with the European Environment Agency, may be enriched by the experience and protocols resulting for the current proposal.

3.8 Conclusion

Advances in the performance of satellite systems, development of sophisticated processing chains for SAR images, and increases in computational capabilities have led, in recent years, to the increase of applicability of SAR images for the mapping and monitoring of slow deformation affecting linear infrastructures. The effectiveness of satellite InSAR techniques is significant for wide regions, for which conventional analysis would not provide comparable potential, in terms of timely update, cost-efficiency, systematic coverage, density of measurements, accuracy, and precision.

C-band Sentinel-1 image has been exploited to implement a continuous monitoring of ground deformation for the Tuscany Region. Sentinel-1 results provide information that can be used to scan wide areas, to highlight unstable zones, to reconstruct the deformation history of the observed scene back to 2014, and to define a list of priority situations, where further analysis is required. These areas became target of a X-band Cosmo-SkyMed InSAR analysis to investigate localized deformation phenomena affecting railway, road network, and strategic infrastructures. InSAR-based analysis highlighted the large presence of anthropogenic-induced land subsidence, related to the load imposition (for the Amerigo Vespucci freight terminal) and to the depletion of aquifer (for the Pistoia plain), which could potentially damage railways and road infrastructures.

Acknowledgements The ground deformation monitoring system presented in this paper has been requested, founded, and supported by the Regional Government of Tuscany, under the agreement "Monitoring ground deformation in the Tuscany Region with satellite radar data". Authors are grateful to TRE ALTAMIRA for having processed Sentinel-1 data for the Tuscany Region. COSMO-SkyMed images for the area of the Amerigo Vespucci freight terminal have been processed in the framework of the PST-A project. COSMO-SkyMed images for the area of Pistoia have been processed with the CPT (Coherent Pixels Technique) approach.

References

Aguzzi, M., A. Amorosi, M.L. Colalongo, M. Ricci Lucchi, V. Rossi, G. Sarti, and S.C. Vaiani. 2007. Late quaternary climatic evolution of the Arno Coastal Plain (Western Tuscany, Italy) from subsurface data. *Sedimentary Geology* 202 (1-2): 211–229.

Bakon, M., D. Perissin, M. Lazecky, and J. Papco. 2014. Infrastructure non-linear deformation monitoring via satellite radar interferometry. *Procedia Technology* 16: 294–300.

Bandini, A., P. Berry, and D. Boldini. 2015. Tunnelling-induced landslides: The Val di Sambro Tunnel Case Study. *Engineering Geology* 196: 71–87.

Bartolini, C., N. D'Agostino, and F. Dramis. 2003. Topography, exhumation, and drainage network evolution of the Apennines. *Episodes* 26 (3): 212–216.

Bianchini, S., F. Pratesi, T. Nolesini, and N. Casagli. 2015. Building deformation assessment by means of persistent scatterer interferometry analysis on a landslide-affected area: The Volterra (Italy) Case Study. *Remote Sensing* 7 (4): 4678–4701.

Bortolotti, V. 1992. *The Tuscany–Emilian Apennine. Regional Geological Guidebook 4*. S.G.I. BEMA ed, 329.

Canuti, P., N. Casagli, P. Farina, A. Ferretti, F. Marks, and G. Menduni. 2006. Analisi dei fenomeni di subsidenza nel bacino del fiume arno mediante interferometria radar. *Giornale di Geologia Applicata* 4: 131–136.

Capecchi, F. 1975. Il Bacino Lacustre Di Firenze-Prato-Pistoia Geologia Del Sottosuolo E Ricostruzione Evolutiva.

Chang, L., R.P. Dollevoet, and R.F. Hanssen. 2016. Nationwide railway monitoring using satellite SAR interferometry. *IEEE Journal of Selected Topics in Applied Earth Observations and Remote Sensing* 10 (2): 596–604.

Ciampalini, A., F. Bardi, S. Bianchini, W. Frodella, C. Del Ventisette, S. Moretti, and N. Casagli. 2014. Analysis of building deformation in landslide area using multisensor PSInSAR technique. *International Journal of Applied Earth Observation and Geoinformation* 33: 166–180.

Ciampalini, A., L. Solari, R. Giannecchini, Y. Galanti, and S. Moretti. 2019. Evaluation of subsidence induced by long-lasting buildings load using InSAR technique and geotechnical data: The case study of a freight terminal (Tuscany, Italy). *International Journal of Applied Earth Observation and Geoinformation* 82: 101925.

Colesanti, C., A. Ferretti, R. Locatelli, F. Novali, and G. Savio 2003. Permanent scatterers: Precision assessment and multi-platform analysis. In International Geoscience Remote Sensing Symposium, Toulouse, France: IEEE.

Colombo, D., P. Farina, S. Moretti, G. Nico, and C. Prati 2003. Land subsidence in the Firenze-Prato-Pistoia Basin measured by means of spaceborne SAR interferometry. In Proceedings of the IGARSS 2003 IEEE International Geoscience and Remote Sensing Symposium, Toulouse, France, 21–25 July 2003; IEEE: Piscataway, NJ, 2003; pp. 2927–2929.

D'Amato Avanzi, G., D. Marchetti, and A. Puccinelli. 2006. Cultural heritage and geological hazards: The case of the Calomini Hermitage in Tuscany (Italy). *Landslides* 3 (4): 331–340.

Del Soldato, M., G. Farolfi, A. Rosi, F. Raspini, and N. Casagli. 2018. Subsidence evolution of the Firenze–Prato–Pistoia Plain (Central Italy) combining PSI and GNSS data. *Remote Sensing* 10 (7): 1146.

Del Ventisette, C., F. Raspini, A. Ciampalini, F. Di Traglia, M. Moscatelli, A. Pagliaroli, and S. Moretti. 2015. Use of PSInSAR data to map highly compressible soil layers. *Geologica Acta: An International Earth Science Journal* 13 (4): 309–323.

Elter, P., G. Giglia, M. Tongiorgi, and L. Trevisan. 1975. Tensional and compressional areas in recent (Tortonian to present) evolution of Northern Apennines. *Transactions-American Geophysical Union* 56 (3): 166–169.

Farina, P., D. Colombo, A. Fumagalli, F. Marks, and S. Moretti. 2006. Permanent scatterers for landslide investigations: Outcomes from the ESA-SLAM Project. *Engineering Geology* 88 (3–4): 200–217.

Ferretti, A., A. Fumagalli, F. Novali, C. Prati, F. Rocca, and A. Rucci. 2011. A new algorithm for processing interferometric data-stacks: SqueeSAR. *IEEE Transactions on Geoscience and Remote Sensing* 49 (9): 3460–3470.

Ferretti, A., C. Prati, and F. Rocca. 2001. Permanent scatterers in SAR interferometry. *IEEE Transactions on Geoscience and Remote Sensing* 39 (1): 8–20.

Fornaro, G., D. Reale, and S. Verde. 2012. Bridge thermal dilation monitoring with millimeter sensitivity via multidimensional SAR imaging. *IEEE Geoscience and Remote Sensing Letters* 10 (4): 677–681.

Giannico, C., A. Ferretti, L. Jurina, and M. Ricci 2012. Application of satellite radar interferometry for structural damage assessment and monitoring. In Life-cycle and Sustainability of Civil Infrastructure Systems: Proceedings of the Third International Symposium on Life-Cycle Civil Engineering (IALCCE'12), Vienna, Austria, October 3–6. Boca Raton, FL: CRC Press, p. 420.

Kaloop, M., E. Elbeltagi, J. Hu, and A. Elrefai. 2017. Recent advances of structures monitoring and evaluation using GPS-time series monitoring systems: A review. *ISPRS International Journal of Geo-Information* 6 (12): 382.

Lu, P., N. Casagli, F. Catani, and V. Tofani. 2010. Persistent scatterers interferometry hotspot and cluster analysis (PSI-HCA) for detection of extremely slow-moving landslides. *International Journal of Remote Sensing* 33 (2): 466–489.

Milillo, P., R. Bürgmann, P. Lundgren, J. Salzer, D. Perissin, E. Fielding, Filippo Biondi, and G. Milillo. 2016. Space geodetic monitoring of engineered structures: The ongoing destabilization of the Mosul dam, Iraq. *Scientific Reports* 6: 37408.

North, M., T. Farewell, S. Hallett, and A. Bertelle. 2017. Monitoring the response of roads and railways to seasonal soil movement with persistent scatterers interferometry over six UK sites. *Remote Sensing* 9 (9): 922.

Ozden, A., A. Faghri, M. Li, and K. Tabrizi. 2016. Evaluation of synthetic aperture radar satellite remote sensing for pavement and infrastructure monitoring. *Procedia Engineering* 145: 752–759.

Polcari, M., M. Moro, V. Romaniello, and S. Stramondo. 2019. Anthropogenic subsidence along railway and road infrastructures in Northern Italy highlighted by Cosmo-SkyMed Satellite Data. *Journal of Applied Remote Sensing* 13 (2): 024515.

Raspini, F., S. Bianchini, A. Ciampalini, M. Del Soldato, L. Solari, F. Novali, S. Del Conte, A. Rucci, A. Ferretti, and N. Casagli. 2018. Continuous, semi-automatic monitoring of ground deformation using Sentinel-1 satellites. *Scientific Reports* 8 (1): 7253. https://doi.org/10.1038/s41598-018-25369-w.

Raspini, F., C. Loupasakis, D. Rozos, and S. Moretti. 2013. Advanced interpretation of land subsidence by validating multi-interferometric SAR data: The case study of the Anthemountas Basin (Northern Greece). *Natural Hazards and Earth System Science* 13 (10): 2425–2440.

Rosi, A., V. Tofani, A. Agostini, L. Tanteri, C.T. Stefanelli, F. Catani, and N. Casagli. 2016. Subsidence mapping at regional scale using persistent scatters interferometry (psi): The case of Tuscany Region (Italy). *International Journal of Applied Earth Observation and Geoinformation* 52: 328–337.

Rosi, A., V. Tofani, L. Tanteri, C.T. Stefanelli, A. Agostini, F. Catani, and N. Casagli. 2018. The New Landslide Inventory of Tuscany (Italy) updated with PS-InSAR: Geomorphological features and landslide distribution. *Landslides* 15 (1): 5–19.

Rosi, A., P. Vannocci, V. Tofani, G. Gigli, and N. Casagli. 2013. Landslide characterization using satellite interferometry (PSI), geotechnical investigations and numerical modelling: The case study of Ricasoli Village (Italy). *International Journal of Geosciences* 4: 904–918.

Sarti, G., V. Rossi, and S. Giacomelli. 2015. The Upper Pleistocene "Isola di Coltano sands" (Arno coastal plain, Tuscany, Italy): Review of stratigraphic data and tectonic implications for the southern margin of the Viareggio Basin. *Atti Della Società Toscana di Scienze Naturali* 122: 75–84.

Solari, L., A. Ciampalini, F. Raspini, S. Bianchini, and S. Moretti. 2016. PSInSAR analysis in the Pisa Urban Area (Italy): A case study of subsidence related to stratigraphical factors and urbanization. *Remote Sensor* 8 (2): 120.

Solari, L., M. Del Soldato, R. Montalti, S. Bianchini, F. Raspini, P. Thuegaz, Davidi Bertolo, Veronica Tofani, and N. Casagli. 2019. A Sentinel-1 based hot-spot analysis: landslide mapping in North-Western Italy. *International Journal of Remote Sensing* 40 (20): 7898–7921.

Terzaghi, K., and R.B. Peck. 1967. *Soil Mechanics in Engineering Practice*. Hoboken, NJ: Wiley.

The European Space Agency. 2019. https://www.esa.int/Applications/Observing_the_Earth/Copernicus/Sentinel-1/Copernicus_Sentinel-1_maps_Norway_in_motion. Accessed 10 Jan 2020.

Torres, R., P. Snoeij, D. Geudtner, D. Bibby, M. Davidson, E. Attema, Pierre Potin, et al. 2012. GMES Sentinel-1 Mission. *Remote Sensing of Environment* 120: 9–24.

Tralli, D.M., R.G. Blom, V. Zlotnicki, A. Donnellan, and D.L. Evans. 2005. Satellite remote sensing of earthquake, volcano, flood, landslide and coastal inundation hazards. *ISPRS Journal of Photogrammetry and Remote Sensing* 59 (4): 185–198.

Vai, G.B., and I.P. Martini. 2001. *Anatomy of an Orogen: The Apennines and Adjacent Mediterranean Basins*, 632. Dordrecht: Kluwer Academic Publishers.

Wasowski, J., F. Bovenga, R. Nutricato, D.O. Nitti, and M.T. Chiaradia. 2017. High resolution satellite multi-temporal interferometry for monitoring infrastructure instability hazards. *Innovative Infrastructure Solutions* 2 (1): 27.

Chapter 4
Protecting Canada's Railway Network Using Remote Sensing Technologies

David Huntley, Peter Bobrowsky, Drew Rotheram-Clarke, Roger MacLeod, Robert Cocking, Jamel Joseph, Jessica Holmes, Shane Donohue, Jonathan Chambers, Philip Meldrum, Paul Wilkinson, Michael Hendry, and Renato Macciotta

4.1 Introduction

In many parts of the world, transportation infrastructure, public safety, the environment, and natural resources are increasingly at risk from landslides triggered by extreme weather events, earthquakes, volcanoes, changes in sea level, storms, floods, drought, wildfires and anthropogenic activities. A key role for both the Geological Survey of Canada (GSC) and British Geological Survey (BGS) is to provide fundamental knowledge to ensure resilience to multi-geohazard risks.

Vital railway infrastructure and operations are at risk from landslides across much of Canada. A particularly vulnerable section of the Canadian National (CN) and Canadian Pacific (CPR) railway corridors runs through the Thompson River valley between Ashcroft and Spences Bridge in southern British Columbia (Fig. 4.1;

D. Huntley (✉) · D. Rotheram-Clarke · R. Cocking · J. Joseph
Geological Survey of Canada, Vancouver, BC, Canada
e-mail: david.huntley@canada.ca

P. Bobrowsky · R. MacLeod
Geological Survey of Canada, Sidney, BC, Canada

J. Holmes
School of Natural and Built Environment, Queen's University Belfast, Belfast, UK

British Geological Survey, Keyworth, UK

S. Donohue
School of Civil Engineering, University College Dublin, Dublin, Ireland

J. Chambers · P. Meldrum · P. Wilkinson
British Geological Survey, Keyworth, UK

M. Hendry · R. Macciotta
Department of Civil and Environmental Engineering, University of Alberta,
Edmonton, AB, Canada

V. Singhroy (ed.), *Advances in Remote Sensing for Infrastructure Monitoring*,
Springer Remote Sensing/Photogrammetry,
https://doi.org/10.1007/978-3-030-59109-0_4

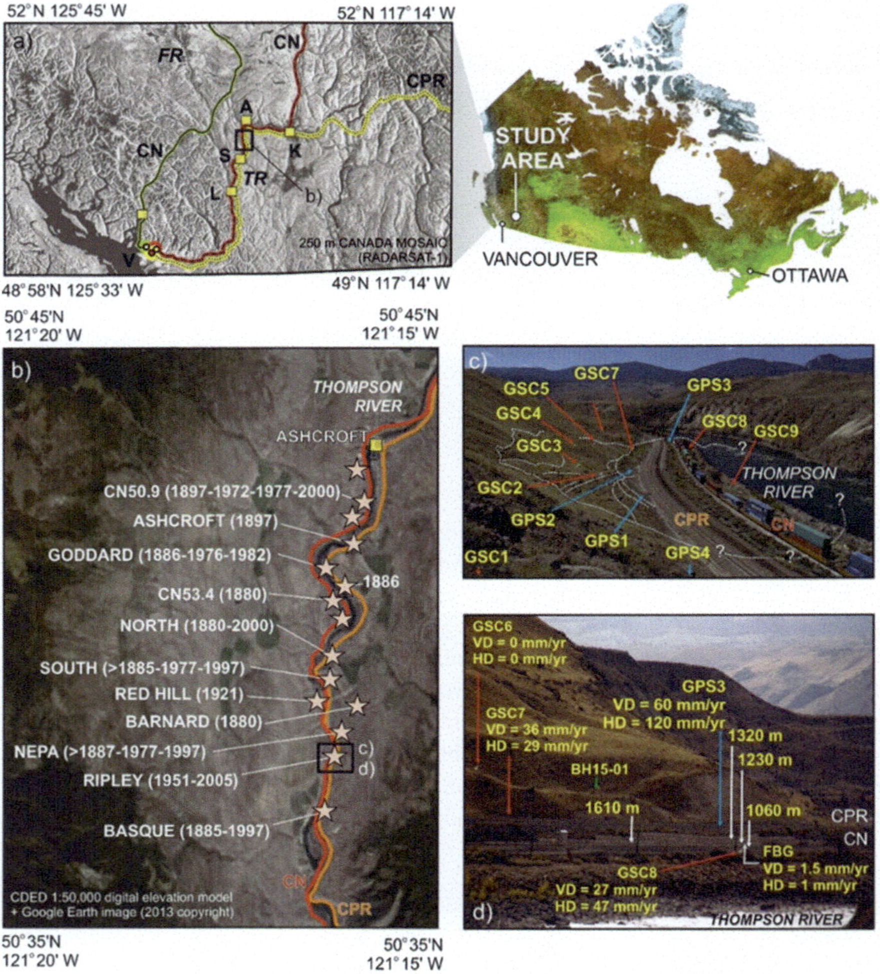

Fig. 4.1 The study area. (**a**) Rail transportation corridors in southwestern British Columbia with the location of the Thompson River valley area of interest: **A**—Ashcroft; **K**—Kamloops; **L**—Lytton; **S**—Spences Bridge; **V**—Vancouver; **FR**—Fraser River; **TR**—Thompson River. (**b**) Landslides of the Thompson River valley, with the location of Ashcroft, the railway transportation corridor and Ripley Landslide test site. (**c**) Overview of the test site highlighting the location of GNSS monitoring stations (GPS1-4), InSAR corner reflectors (GSC1-9), and the monitored retaining wall dividing the CN and CPR tracks—view to the south. (**d**) The south flank of landslide with sagging retaining wall separating CN and CPR tracks with displacement vectors for FBG, GNSS and InSAR stations; strain points detected by BOTDR (white arrows and distance in metres from datum); **VD**—vertical displacement; **HD**—horizontal displacement detected by FBG (grey arrow with displacement values expressed as mm/year); and location of BH15-01 (green arrow)

Bobrowsky et al. 2014, 2016). The economic importance of the Thompson River valley corridor, along with the need to understand and manage the safety risk related to the landslides that threaten this route, make this area a priority for the Government of Canada. Since 2013, the GSC, along with international and national partners, has pioneered innovative research and monitoring of landslides in the Thompson River valley. This research and development provides government agencies, university partners and the national railway companies with vital information to understand geohazard risk, predict landslide movement, improve the safety, security and resilience of Canada's transportation infrastructure, and reduce such risks to the economy, environment, natural resources and public safety.

This overview paper shows how conventional and emerging fixed and automated (i.e. remote) technologies provide the fundamental geoscience required to monitor landslides along a strategically important section of the national railway network. Such baseline knowledge will help build more robust risk tolerance, remediation and mitigation strategies to maintain the resilience and accessibility of critical transportation infrastructure, while also protecting the natural environment, community stakeholders and the national economy.

4.2 Canada's Landslide Laboratory

Landslides in the Thompson River valley (Fig. 4.1b) serve as a field-based laboratory to test and compare the reliability and effectiveness of different static, dynamic and real-time monitoring technologies. Research and development objectives are to: (1) understand slope instabilities affecting rail transport in the Thompson River valley; (2) generate information to help forecast landslide movement; (3) improve the safety and security of Canada's transportation infrastructure; and (4) reduce risks to the national economy, physical environment, natural resources and public safety. Collaboration with other government agencies, universities, industry stakeholders and international partners has been key to the successful delivery of the research activities.

The focus of attention is Ripley Landslide, approximately 220 m wide (N–S) by 150 m-long (E–W), with an estimated volume of 400,000 m^3 (Fig. 4.1c). The landslide has been active since at least 1951, but displacement across the slide body increased after 2005 when a rail siding was constructed across its middle portion (Bunce and Chadwick 2012). During construction, embankments were extended upslope and a lock-block retaining wall, separating the CN and CPR tracks, was installed (Fig. 4.1d). Pronounced sagging of the retaining wall and bulging of lock blocks has occurred since 2005 (Huntley et al. 2016). To accommodate continual lateral and vertical displacement across the landslide, both rail companies periodically add ballast, in addition to lifting and realigning their tracks.

Remote Monitoring of Landslide and Infrastructure

Satellite Interferometric Synthetic Aperture Radar (InSAR) was identified to provide the best opportunity to determine the extent of landslide activity in the Thompson River valley (Bobrowsky et al. 2014; Journault et al. 2018). Processed images obtained by the Canadian Space Agency (CSA) RADARSAT-2 since 2013 have measured land deformation over a significant portion of the railway transportation corridor (Fig. 4.2).

Ongoing, intensive and extensive monitoring at Ripley Landslide (Fig. 4.1c, d) provides the opportunity to compare InSAR interpreted displacements to those measured remotely with a differential global positioning systems (dGPS), ShapeAccelArrays™ installed within boreholes, Fibre Bragg grating (FBG) and Brillouin optical time domain reflectometry (BOTDR) (Bunce and Chadwick 2012; Macciotta et al. 2014; Hendry et al. 2015; Huntley et al. 2016, 2017a).

Numerous conventional and experimental continuous monitoring technologies (Fig. 4.1c, d) provide insight into the activity, deformation mechanisms and potential acceleration triggers of Ripley Landslide. Each of these techniques record increased landslide activity in winter when river and groundwater levels are lowest (Macciotta et al. 2014; Hendry et al. 2015; Schafer et al. 2015; Journault et al. 2018).

Four permanent dGPS monuments installed across the landslide in 2008 (Fig. 4.1c) record cumulative annual displacement on the order of 100 mm/year to 200 mm/year, which peaks in winter (Bunce and Chadwick 2012; Macciotta et al. 2014; Hendry et al. 2015).

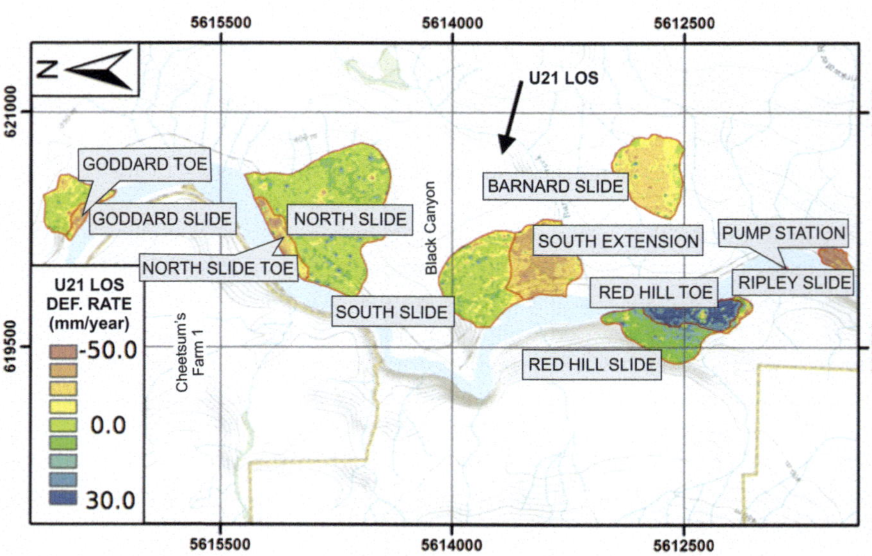

Fig. 4.2 Displacement intensity maps for the six zones of concentrated displacement along with the line of sight (LOS) of U21, azimuth 280°, incidence angle 45° (source: Journault et al. 2018)

InSAR results from 2013 to 2014 indicate similar magnitudes and spatial-temporal patterns of displacement (Fig. 4.3; Huntley et al. 2017a; Journault et al. 2018). Ground movement concentrated within the centre of the sliding mass averages 39 mm/year, with the fastest displacements detected upslope from the railway tracks and on the southern flank. Average and maximum line-of-sight displacement

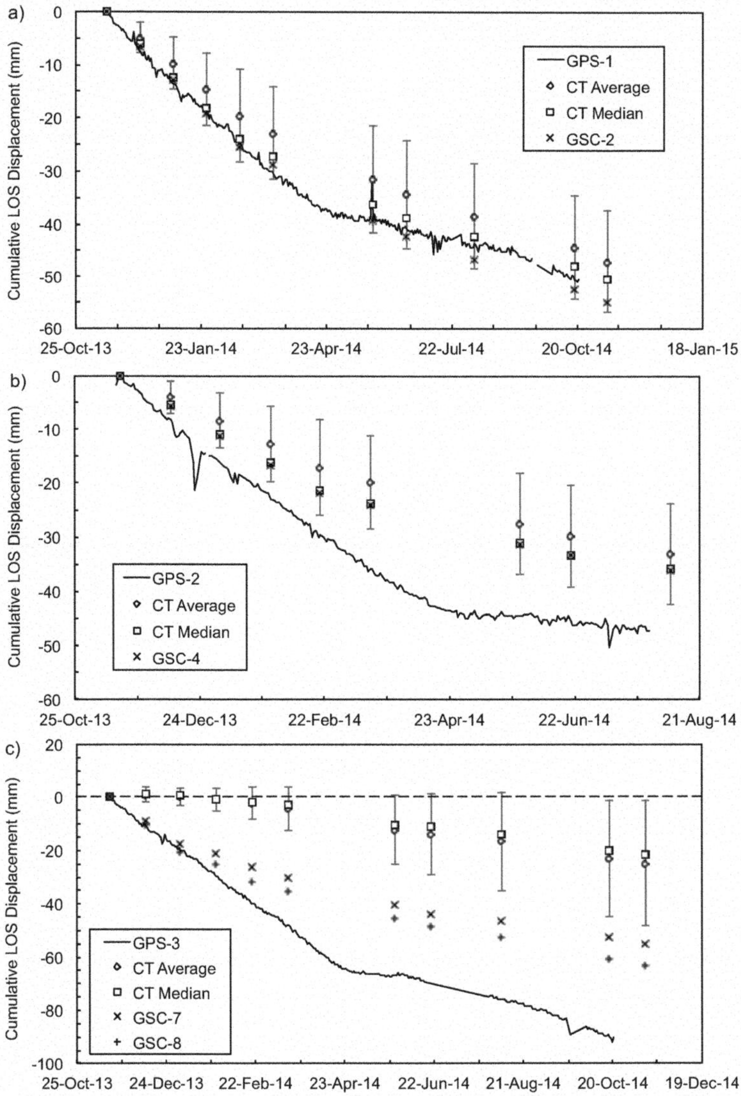

Fig. 4.3 Displacement data from October 2013 until December 2014, for GPS 1 (**a**) GPS 2 (**b**) and GPS 3 (**c**) compared with displacements of nearby corner reflectors and coherent targets (CT), with average, median, the range within a standard deviation shown as error bars (source: Journault et al. 2018)

rates (equivalent to the downslope direction for the west-facing test site) of trihedral corner reflectors (Fig. 4.1c) and other coherent targets (e.g. buildings, large boulders) are 49 mm/year, and 77 mm/year, respectively; with greater displacement from November to March (Huntley et al. 2017a; Journault et al. 2018).

Fibre Bragg grating (FBG) and Brillouin optical time domain reflectometry (BOTDR) monitoring of the retaining wall from 2013 to 2015 detected ~2 mm of accumulated strain in the wall, including the displacement of individual blocks at its southern end (Fig. 4.1d), with peak activity occurring in the fall and winter months (Huntley et al. 2016, 2017b).

Subsurface borehole monitoring combining ShapeAccelArray (SAA) inclinometry with piezometer head levels indicates that the main slide body is failing along sub-horizontal, weak, basal shear surfaces in highly plastic clay beds (Macciotta et al. 2014; Hendry et al. 2015; Schafer et al. 2015). The central and northern parts of the slide are translating sub-horizontally (2.1° to 2.5°) whereas the southern portion near the lock-block retaining wall has a steeper (28°) slide surface.

Previous studies suggest that landslides in the Ashcroft area are controlled, at least in part, by pre-sheared clay layers developed in a zone of elevated pore pressure at the transition from generally coarse-grained valley fills to underlying glaciolacustrine units with low permeability (e.g. Porter et al. 2002; Clague and Evans 2003).

A wide range of geophysical techniques have been tested at Ripley Landslide as non-invasive means to characterize subsurface landslide properties. Such techniques have provided a better understanding of the nature and behaviour of the earth materials comprising the landslide (e.g. Best et al. 2018; Huntley et al. 2017c, d, 2019a, b). Electromagnetic resistivity tomography (ERT) surveys conducted on land (November 2013) and along Thompson River (November 2014) provide the most useful information on the distribution of clay beds and other earth materials, in addition to the distribution and quantity of groundwater within various surficial deposits and bedrock (Huntley et al. 2017d, 2019c).

Geophysical mapping of geologic materials, including stratigraphic layers and geological structures (e.g. joints, faults, shear planes and tension cracks) have provided critical information on how and where groundwater flows through the landslide and surrounding bedrock. Regional and local groundwater conditions contribute to high pore pressures and slope instability in the Thompson River valley. Elevated soil moisture has been implicated as a driver of slope failure along the transportation corridor since the first slope stability study in the area (Stanton 1898). Since clay impedes groundwater flow, pore pressure increases along the surface of clay layers contributing to reduced material strength and reduced landslide stability (Porter et al. 2002; Clague and Evans 2003).

4.3 Methods: Monitoring with InSAR, UAVs, Bathymetric Surveys, GNSS and ERT

InSAR Change Detection

Ground motion measured by space-borne InSAR produced results with precision comparable to dGPS measurements (Fig. 4.3), but have the advantage of monitoring displacement over large areas (e.g. Huntley et al. 2017c; Journault et al. 2018). Remote sensing is effective in the Thompson River valley as a first approach for determining the risks posed from ground hazards and whether site investigation. Additional testing is required before RADARSAT-2 can be broadly used for change detection across Canada (Fig. 4.4a). Datasets must be compared along with other satellite platforms, low-cost remote change detection monitoring techniques employing unmanned aerial vehicles (UAVs), geophysical investigations (e.g. electrical resistivity tomography, ERT), and bathymetric surveys.

PCI Geomatica (2018) was used to create a series of interferograms from a stack of U21 Ultrafine RADARSAT-2 images. The stack has an average incident angle of 45.4° and was acquired with a right-looking geometry from descending orbital passes east of the study site. Reliable InSAR targets were identified from the stack of SAR imagery by examining the degree of phase stability over time for every resolution cell within the study area. By considering only targets with persistent scatterer characteristics and correcting for phase change contributions caused by orbital position errors, topographic and atmospheric effects a time series of deformation for each target was recovered with possible measurement accuracies of several mm (cf. Henschel et al. 2015). The trihedral corner reflectors (Fig. 4.1c) installed on-site served as highly coherent artificial InSAR targets (Fig. 4.4b, c). Submillimeter accuracy of corner reflector time series' has been demonstrated experimentally (Ferretti et al. 2007).

UAV Change Detection

UAVs allow flexible, inexpensive acquisition of low-altitude aerial imagery. Various off-the-shelf photogrammetric software packages enable the rapid production of high-quality digital surface models (DSM) from such images. Beginning in September 2016, repeat UAV surveys of Ripley Landslide have aimed to capture changes in landslide morphology. Overflights in May and October 2018 were undertaken using the GSC-owned DJI Phantom 4. This small UAV has a weight of 1.4 kg and includes a fixed payload consisting of a 12.4 MB RGB camera (Fig. 4.5a). Plan-view and oblique colour aerial photographs (e.g. Fig. 4.5b) captured using the DJI Phantom 4 were merged in using Structure from Motion software, and DSMs have been generated for these three surveys at the cm-level (e.g. Fig. 4.5c, d). The UAV surveying was conducted using the following autonomous surveying applications:

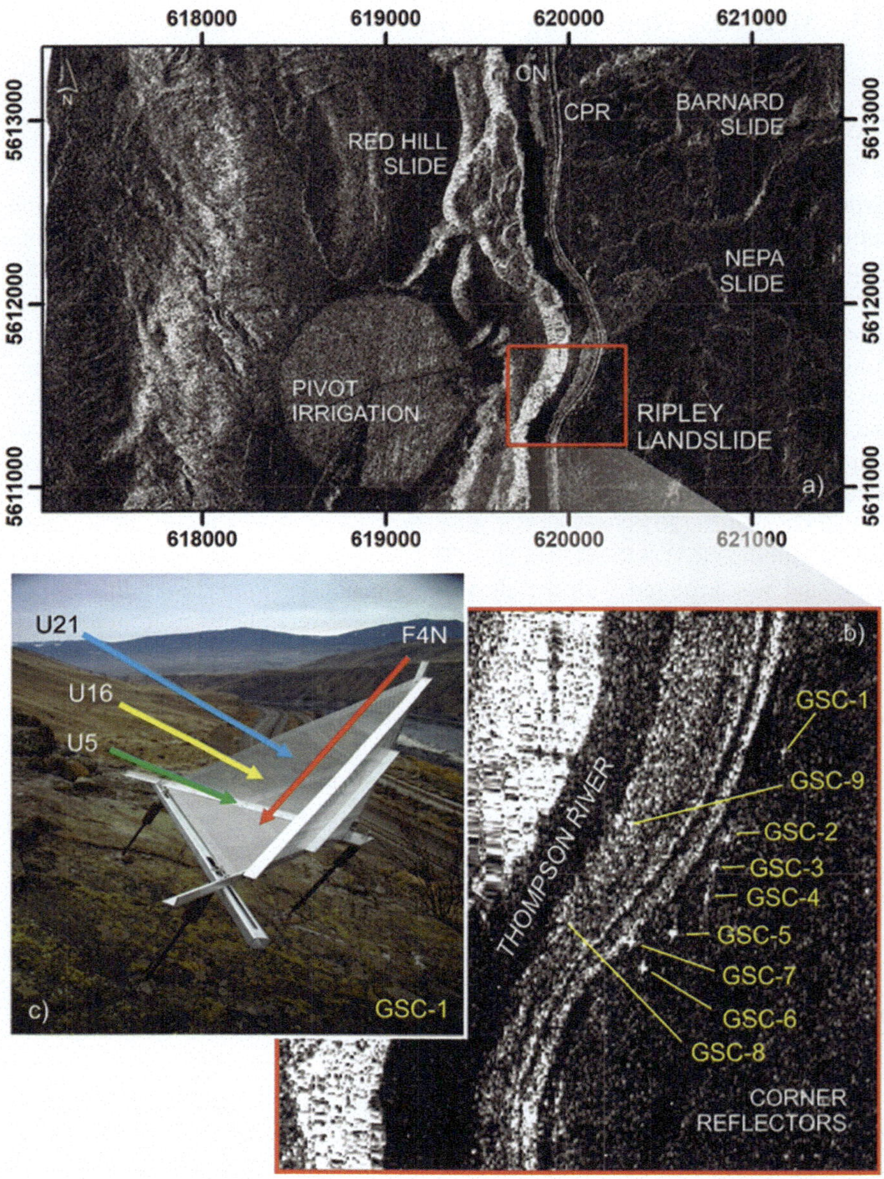

Fig. 4.4 (**a**) Regional orthorectified RADARSAT-2 SAR amplitude image, showing Ripley Landslide in relation to other slow-moving landslides and major anthropogenic features in the Thompson River valley; (**b**) orthorectified amplitude image of Ripley Landslide, showing trihedral corner reflectors (GSC-1, master stable reflector on bedrock); (**c**) stable trihedral corner reflector (GSC-1) showing diagrammatic Ultrafine beam paths for three descending nodes along an azimuthal line-of-sight of 280° (U5 34°, U16 42°, and U21 45°), interferograms were generated from a stack of U21 images; corner reflectors are not oriented to utilize the ascending node along an azimuthal line-of-sight of 088° (F4N 44°)

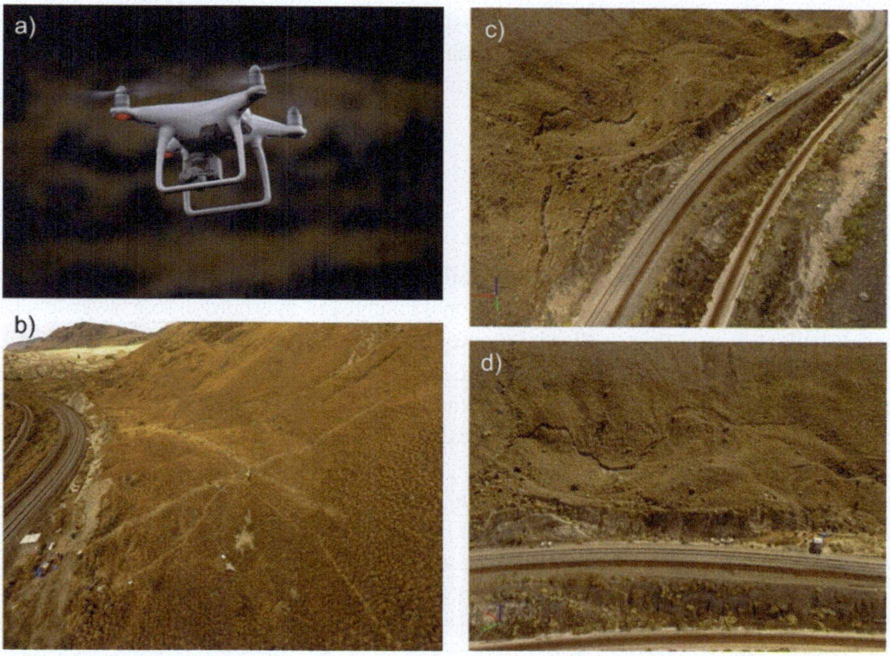

Fig. 4.5 UAV used to remotely map Ripley Landslide: (**a**) DJI Phantom 4; (**b**) oblique aerial photograph showing intersecting PRIME arrays, view north; (**c**) DSM generated from UAV overflight, view to the south, showing the twinned CPR tracks, single CN track and the southern portion of separating lock-block retaining wall; (**d**) DSM generated from UAV overflight, view to the east, showing the twinned CPR tracks, main slide body, tension cracks and headscarp

Pix4D Capture (2016, 2017) and Map Pilot (2018); the SfM modelling was undertaken using Pix4D Mapper.

A displacement map for September 2016 to October 2018 was created using a plug-in available for the remote sensing software called ENVI 5.2, called Cosi-Corr (Co-registration of Optically sensed Images and Correlation) developed by the California Institute of Technology (Caltech). The co-relation was conducted on hill-shaded UAV DSMs of the landslide created on Sept 2016 to Sept 2017 and Sept 2017 to Oct 2018. Hill-shading was applied at 1.5 times exaggeration and 310° azimuth and 45° sun altitude. Hill-shaded images were masked using ArcMap 10.5 prior to correlation; however, the choice of masked areas was derived from comparing an initial correlation attempt with the areas of vegetation and recent track ballast work. This was to reduce the areas indicating substantial changes that were not related to the movement of the slide.

The horizontal displacement was calculated by first using Cosi-Corr to derive E/W (x) and N/S (y) displacement maps. Elevation changes (z) were calculated in ArcMap using map algebra on the SfM derived DEM. Then, 3D displacement maps were calculated by squaring each image (x, y, z) to convert all measurements to positive, then added, and finally squared. This produced a single raster containing

3D displacement values that were all positive (larger values = more displacement). Horizontal displacement *vectors* were derived using Cosi-Corr and applied using the E/W and N/S images. Colouring of the images was done in ArcMap 10.5.

Bathymetric Change Detection

For landslides in the Thompson River valley, downslope displacements across their main bodies are partly triggered by seasonal changes in overburden (e.g. river level, track ballast, heavy trains), fluvial erosion (e.g. scour pools) and deposition (e.g., rock falls and gravel bars), and dynamic groundwater conditions (as suggested by piezometer data, Schafer et al. 2015) in the slide toes. The risk to CN and CPR railway infrastructure crossing Ripley Landslide generated interest for detailed bathymetric surveys of Thompson River.

In November 2013, high-resolution single-beam sound navigation ranging (SONAR) and ground-penetrating (GPR) bathymetric surveys over the submerged toe of Ripley Landslide were undertaken with the aim of providing details of surface morphology and subsurface composition, respectively, below water level (Fig. 4.6a, b). The single-beam acoustic bathymetric survey was completed for a 10 km reach of Thompson River from Ashcroft to the south of Ripley Landslide in 2017 (Fig. 4.6c, d). Extreme weather conditions and very low river levels over the winter of 2016–2017 produced operational safety concerns, limiting the initial phase of the survey to the thalweg. The river levels were tied into benchmark survey points and a LiDAR digital elevation model provided by CN. The GSC collaborated with Fisheries and Oceans Canada's (DFO) Canadian Hydrographic Services (CHS) between October 2017 and November 2018 with the aim of surveying reaches of Thompson River with a conventional Norbit multibeam SONAR navigation and ranging survey system. An emphasis was placed on the mapping of the river bottom adjacent to Ripley Landslide and South Slide (Huntley et al. 2018; Fig. 4.7a–d).

GNSS Change Detection

Permanent dGPS monuments provide a baseline measure of landslide activity for the assessment of other monitoring techniques at Ripley Landslide (Macciotta et al. 2014; Hendry et al. 2015). These displacement records capture full (i.e. three-dimensional) displacement vectors with high temporal resolution (Fig. 4.3). Thus, localized high-resolution spatial data are complementary to remote sensing techniques that provide spatially expansive records of motion with generally lower accuracy (e.g. UAV photogrammetric surveys; Bobrowsky et al. 2018) or are restricted to LOS components (e.g. InSAR; Journault et al. 2018).

The innovative Geocube (Ophelia) high-resolution (millimetric) differential GNSS has been deployed at Ripley Landslide (cf. Benoit et al. 2015). Geocubes are

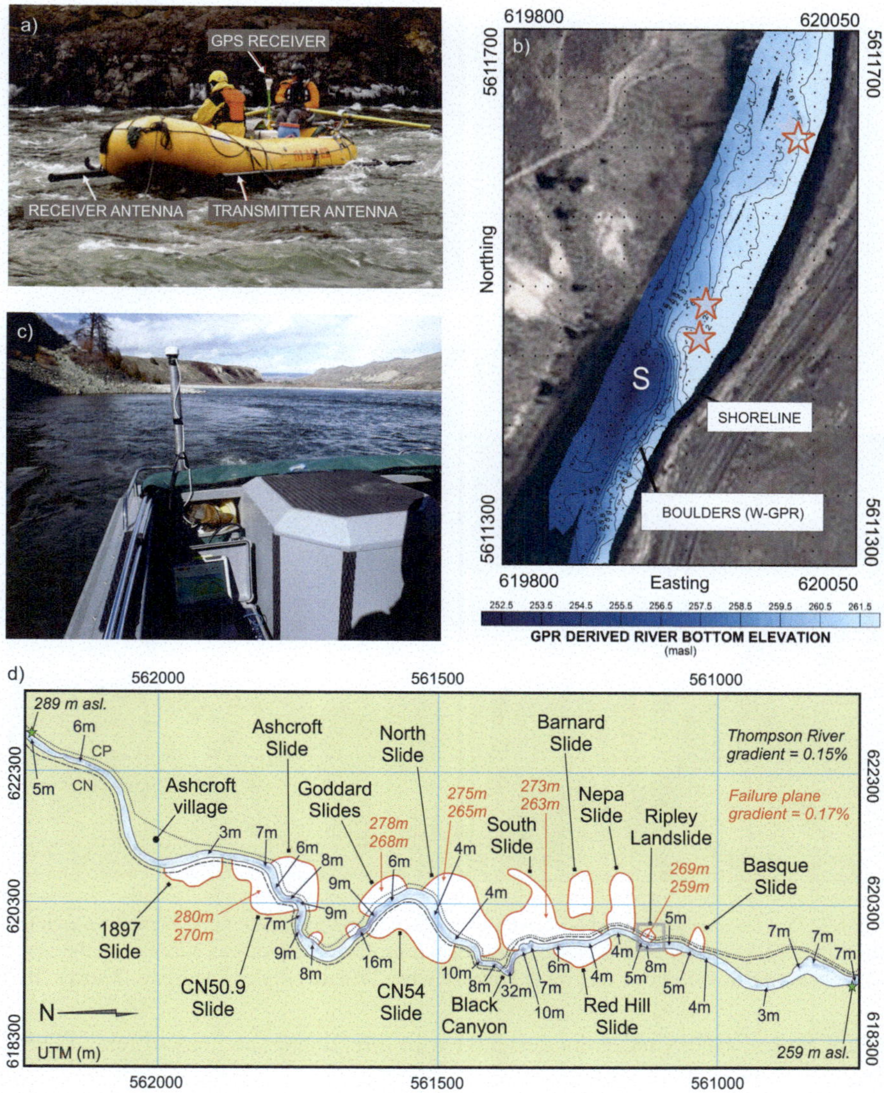

Fig. 4.6 Provisional bathymetric surveys of Thompson River: (**a**) GPR survey river at Ripley Landslide using rubber-hulled raft; (**b**) bathymetry of Thompson River at Ripley Landslide derived from GPR data, showing the location of boulders on channel floor (Huntley et al. 2017d); (**c**) BioSonics® DTX hydro-acoustic (single-beam SONAR) echo sounder mounted to stern of jet boat; (**d**) distribution of major landslides in the Thompson River valley study reach from Ashcroft to Basque Ranch, with location and depths of scour pools (northern and southern limits of the figure correspond to limits of the March 2017 single-beam survey)

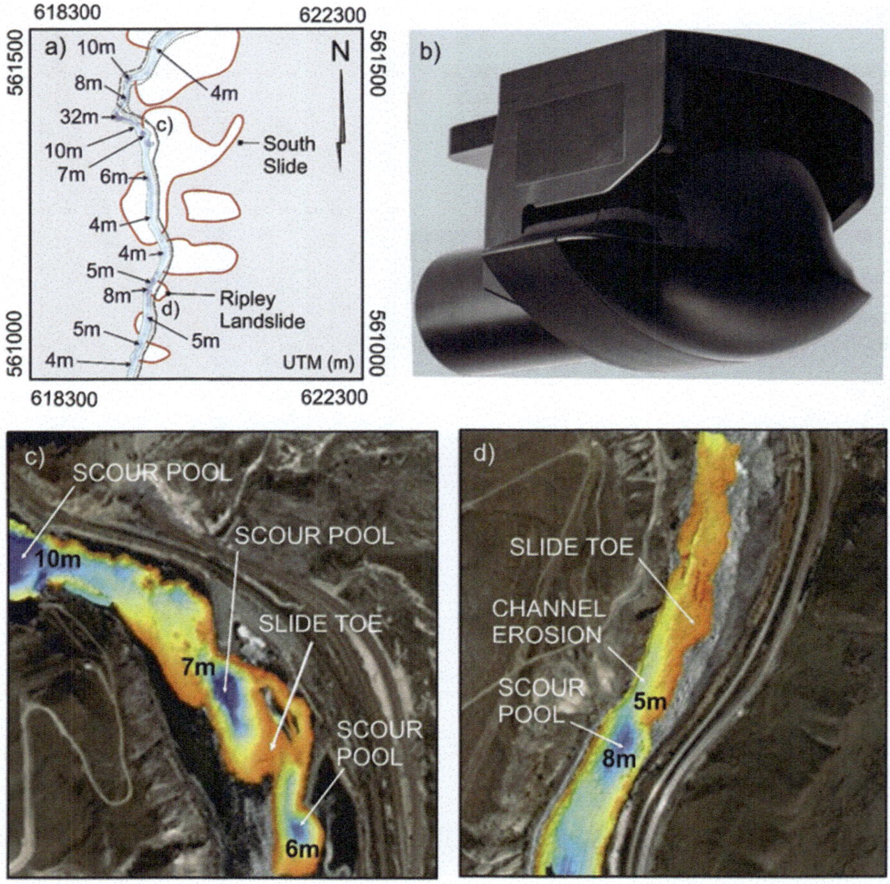

Fig. 4.7 Provisional multibeam bathymetry results: (**a**) location of scour pools in relation to South Slide and Ripley Landslide; (**b**) Norbit multibeam sensor, this system was attached by the metal beam to the hull of the jet boat; (**c**) South Slide, October 2017; (**d**) Ripley Landslide, October 2017

small, rugged GNSS receivers designed with directional antennas to relay GNSS data to Geocoordinator units. The Geocoordinator stores GNSS data from multiple Geocubes and was connected to a modem and an omnidirectional antenna. This allowed access to a local 3G network and provided internet access to the GNSS data (Fig. 4.8a, b).

For Ripley Landslide, one Geocube was installed on stable terrain; the remaining eight were positioned across the slide body to capture spatial variation in displacement. The stable location (i.e. reference location) selected was adjacent to stations GSC1 and GPS 4: a bedrock outcrop that confined the landslide in the northeast (Figs. 4.1c and 4.8c). There was excellent line-of-sight visibility across the

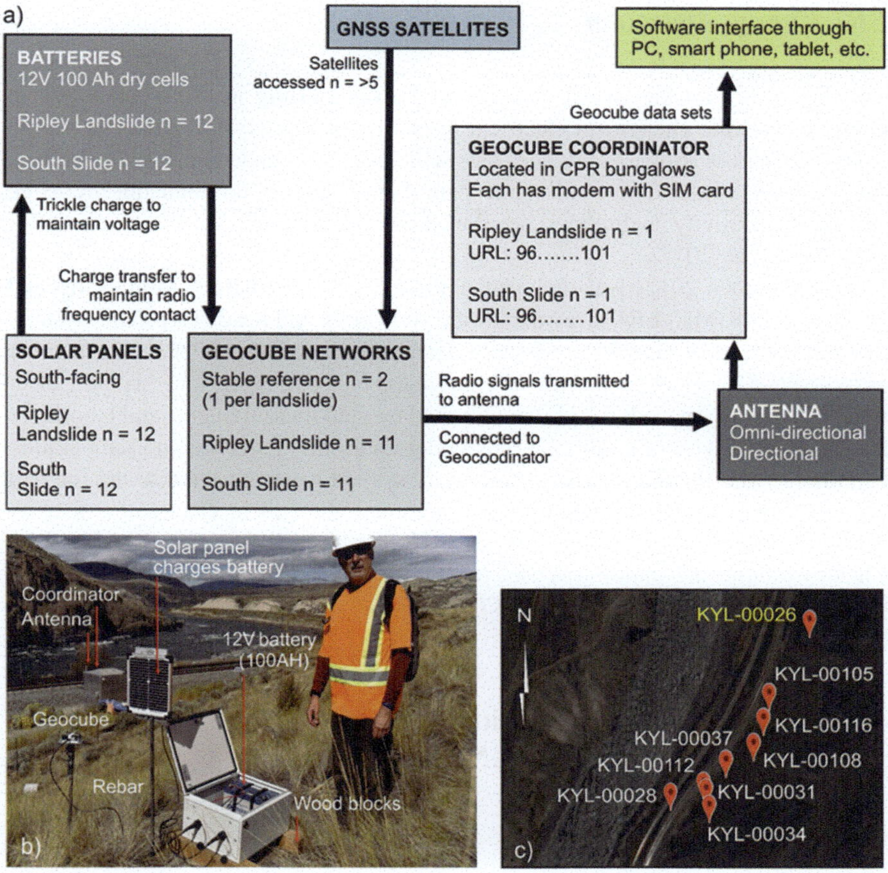

Fig. 4.8 Geocubes installation: (**a**) schematic of Geocube system; (**b**) components of the Geocube system; (**c**) Geocube network installed at Ripley Landslide (yellow text labels are stable reference positions)

landslide from this vantage point. Railway infrastructure (e.g. rolling stock, tracks and lock-block retaining wall) were anthropogenic obstacles between the coordinator, reference site and Geocubes.

A disadvantage of employing ground-based instrumentation is that units often experience damage attributed to local wildlife (e.g. bears, deer, cattle, rats, snakes). Within 2 months of initial installation at Ripley Landslide and South Slide, Geocube cables were damaged on a number of units. Wildlife cameras established the culprit to be a herd of deer grazing on the rangeland. After some experimentation, the most cost-effective and animal-proof approach was to cover exposed cable with a combination of tightly-woven fine wire mesh, 2 mm-thick spiral-wound plastic tubing and rubberized adhesive electrical tape.

ERT Change Detection

Collaboration with the British Geological Survey starting in 2017 has aimed to monitor real-time changes in electrical resistivity as a proxy for ground moisture conditions across Ripley Landslide. This will allow us to understand hydrogeological processes, and monitor soil water percolating through the landslide by using electrical resistivity techniques developed in the UK (cf. Merritt et al. 2014; Uhlemann et al. 2017).

In November 2017, we deployed a Proactive Infrastructure Monitoring and Evaluation (PRIME) ERT array across the slide body and head scarp. This system, providing near-real time 4-D resistivity data, consisted of two intersecting Wenner arrays (NE-SW, 91 m-long with 45 electrodes; NW-SE, 54 m-long with 27 electrodes, all evenly spaced; Fig. 4.9), calibrated by field measurements and laboratory testing of resistivity of earth materials under controlled moisture and temperature conditions. The PRIME electrical resistivity system was connected to the internet for real-time, continuous monitoring of seasonal changes in electrical resistivity across the headscarp and main slide body (Holmes et al. 2018). Weather data from a nearby monitoring station was used to quantify the degree of surface water caused by extreme rain events (Sattler et al. 2018).

4.4 Results: Change Detection with InSAR, UAVs, Bathymetric Surveys, GNSS and ERT

InSAR Change Detection

A network of interferograms was created for SAR acquisitions between fall 2018 and spring 2019 (Table 4.1). Several examples were chosen to illustrate the phase change over three discrete intervals, as well as the contributions of various effects such as seasonal changes in coherence (Fig. 4.10). Wrapped phase measurements start to become visible in the longer interval interferograms, which present an integer ambiguity challenge since phase values are only known *modulo* 2π.

Highly accurate cumulative deformation and rate measurements cannot be determined from the analysis of single interferograms. It is difficult to distinguish between reliable and unreliable targets and the statistical redundancy of a multiply interconnected network cannot be leveraged. Qualitative assessment and rough estimates of deformation are still possible.

When corrected for all other phase contributions, one fringe cycle ($-\pi$, π) corresponds to $\lambda/2$ or 2.8 cm of the line-of-sight deformation with λ equal to the RADARSAT-2 wavelength of 56 mm. The change in phase over the sliding area ranges from nil to approximately 1.6 radians in the 48 days between 27 September 2018 and 14 November 2018.

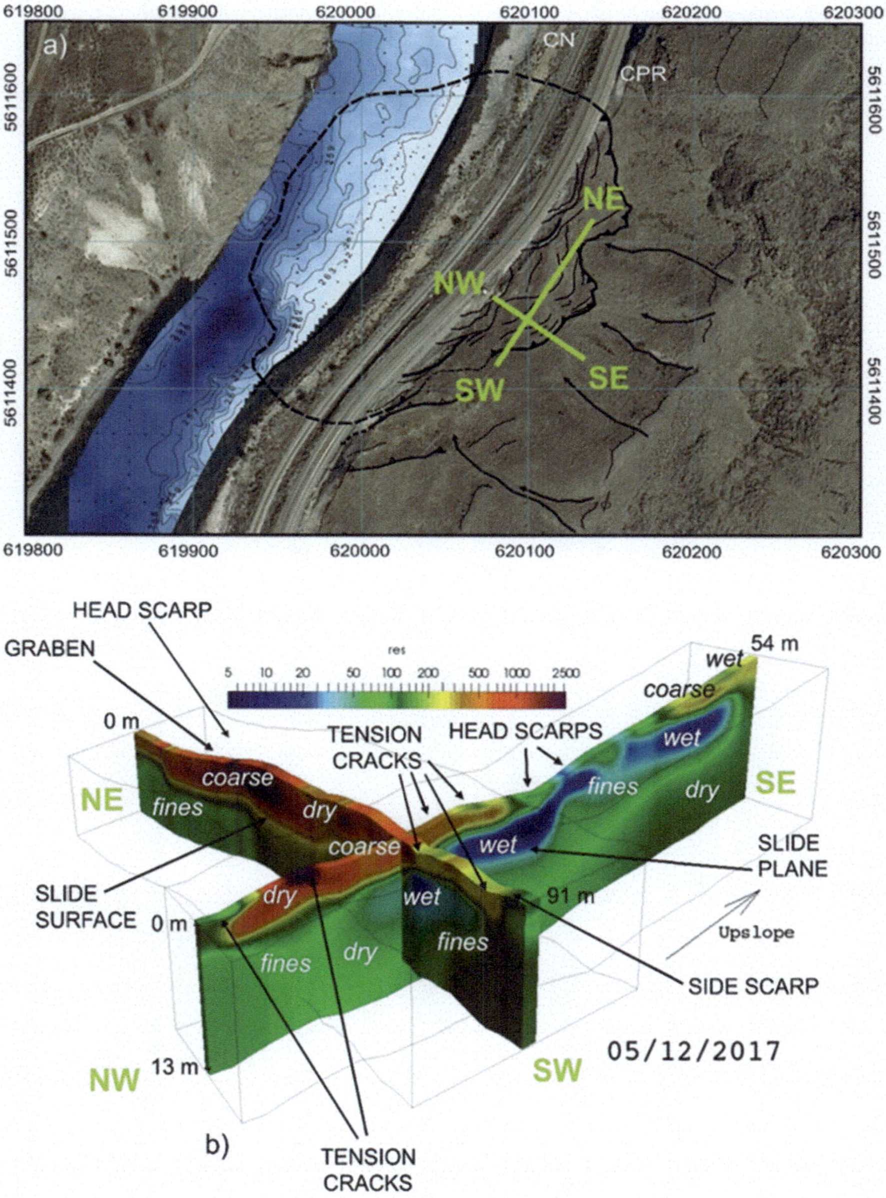

Fig. 4.9 Location of PRIME arrays (X-X′ and Y′Y′) on Ripley Landslide (green lines); also showing the location of other instrumentation in the paper. The ERT results show resistive poorly-sorted, well- to rapid-drained, coarse-grained sand, silt, clay, gravel, cobbles and boulders in the upper 3–5 m. Below this, to a depth of 15 m, lies conductive, moderately well- to poorly-drained, moderate- to well-sorted fine-grained silt, clay, sand, cobbles, boulders, diamicton (modified from Huntley et al. 2019c)

Table 4.1 SAR acquisition intervals for interferograms of Ripley Landslide (see Fig. 4.10)

Start Date	End Date	Interval
2018/09/27	2018/11/14	48 days
2018/11/14	2019/03/14	120 days
2019/03/14	2019/05/25	72 days

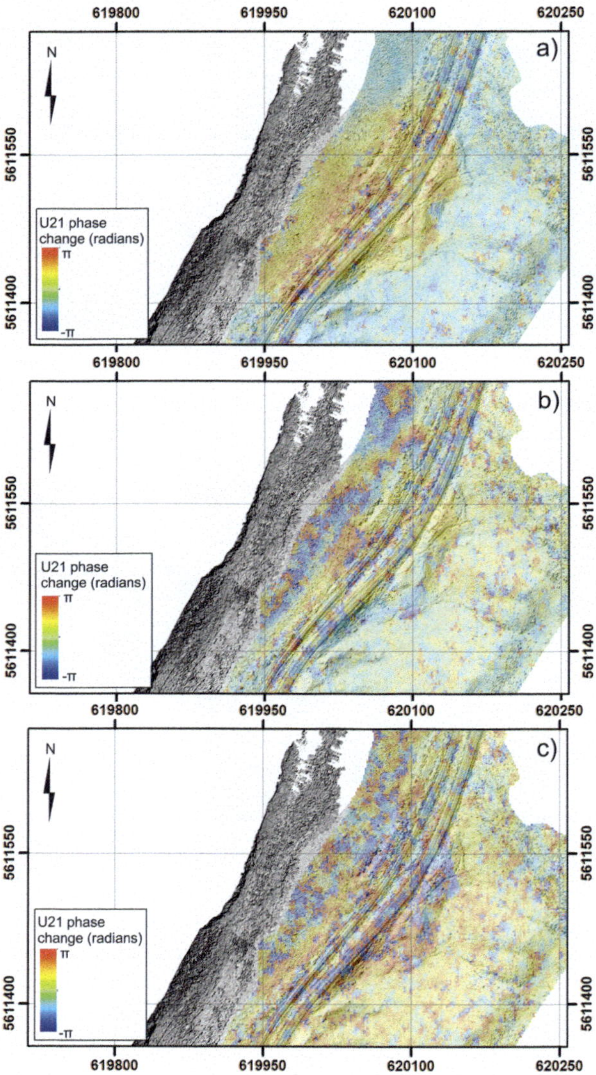

Fig. 4.10 Ripley Landslide interferograms for: (**a**) September–November 2018; (**b**) November 2018–March 2019; (**c**) March 2019–May 2019

Between 14 November 2018 and 14 March 2019 at least one full fringe cycle is visible; however, there is a notable increase in phase noise. This may be associated with both regular temporal decorrelation effects, in addition to seasonal decreases in coherence caused by changes in the surface due to precipitation. The 72 day period between 14 March 2019 and 25 May 2019 also shows a notable change in phase, which is spatially consistent with the slide extents. Spatial coherence of this interferogram is slightly lower than the 48 day pair as expected, and phase values for this period are also wrapped (Fig. 4.10).

UAV Change Detection

Individually, each DSM captures the surface condition of the landslide, including vegetation growth (e.g. grasses, shrubs and trees) that only partly obscure the bare earth. These conditions are sufficient for surface change detection mapping using successive DSMs. Vertical displacement modelling using the SfM show mostly changes in vegetation.

Metre-scale features (e.g. boulders, corner reflectors) are identifiable, and horizontal and vertical displacements of, respectively, 20–30 cm and 15 cm have been measured throughout the slide area on the UAV aerial photography (Fig. 4.11). Zones within the landslide show an average movement of 230 mm between September 2016 and September 2017; areas not expected to show movement (those outside the slide) show an average difference of 1.64 cm, and up to 4.4 cm (within the RMS error of the GNSS and 3D point cloud processing). The measured flow direction of the landslide material is consistently downslope, as would be expected. The absolute position of the rail tracks has moved 12 cm to 27 cm. These datasets now form the basis for the ongoing instrumentation of the slide.

Bathymetric Change Detection

High-resolution single-beam and multibeam acoustic and GPR bathymetric surveys provide details of surface morphology and subsurface composition, respectively, below water level. Bathymetric data acquired for the active Ripley Landslide between 2014 and 2018 establishes the river bed geometry and identifies river bottom erosion at the slide toe (Huntley et al. 2018).

Many boulder diffractions are visible in the GPR cross-section, and some boulder diffractions located on the riverbed are annotated (Fig. 4.6b). Secondary events or multiple reflections (multiples) are evident as the laterally continuous reflectors at depth. These are a function of multiple reflections off the river bottom and water/air interface (Huntley et al. 2017d). No obvious stratigraphy, sedimentary structures or sediment packages can be interpreted on the GPR cross-section due to the boulder diffractions in the modern fluvial sediments.

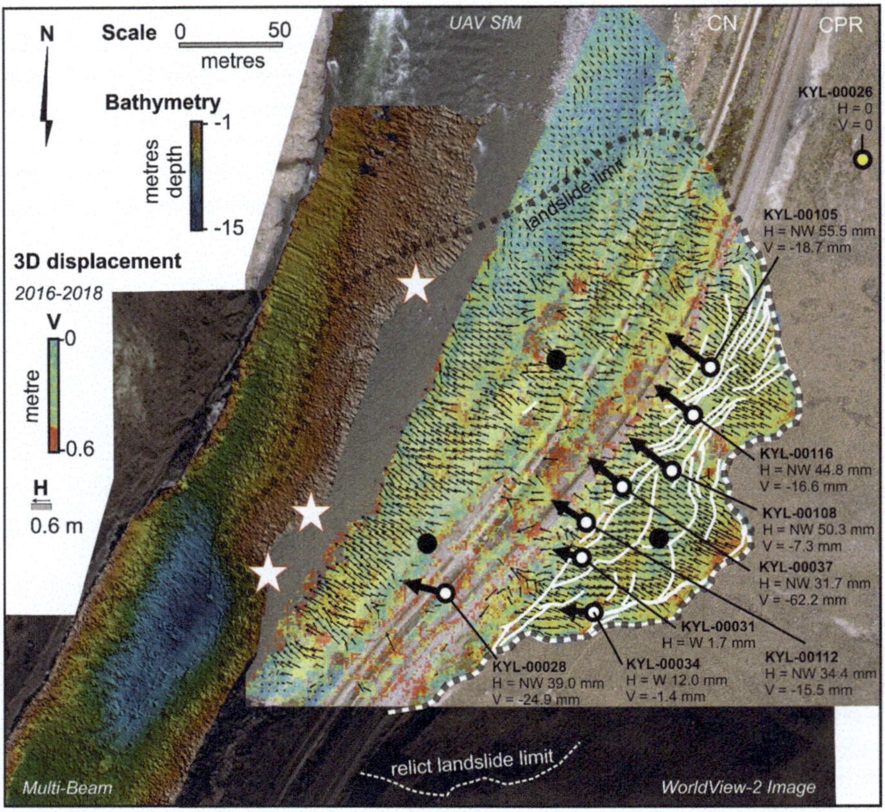

Fig. 4.11 Surface displacement data derived from UAV overflights in 2016 and 2018 and multibeam bathymetry data collected in 2018; plotted with Geocube displacement data (November–December 2018). Figure also merges multibeam bathymetry data of Thompson River with an outline of Ripley Landslide mapped from a WorldView-2 image of the study site. Stable Geocube—yellow dot; active Geocube—white dot; inactive Geocube—black dot

At the southern end of the GPR line, in the vicinity of the large resistive zones identified in waterborne electrical resistivity, radar velocities do not indicate competent bedrock, due to the apparently high water content reducing the velocities. The GPR data in this area indicate thick granular overburden with slower velocities and much diffraction, rather than igneous bedrock where higher radar velocities would be expected. Surface boulders are displayed in Fig. 4.6b. The largest boulders are located in the southern portion of the survey area where deeper bathymetry is recorded and shallow, high subsurface resistivity zones are identified.

Results of the 2017 and 2018 multibeam surveys reveal variations in river bed composition ranging between fines draping bedrock to coarse gravel and boulders overlying clay-rich valley fill. The resolution of the Norbit system is sufficient to identify boulders, suggesting this system is adequate for repeat surveys to map changes in river bottom profiles.

Provisional results (Fig. 4.11) show multibeam bathymetry layered in ArcGIS, on a World View satellite image, merged with other change detection datasets. The boulder-rich lag captured by the GPR survey is evident. Scour features at landslide toe are also revealed in more detail (compare with Fig. 4.6b). Clay-rich sediments appear to be exposed at the base of scour pool, with a depth of 15 m. Shallow waters are revealed with rapids adjacent to stable terrain, separated by deep (typically 5–10 m below river level) scour pools adjacent to the toes of all landslides in the Thompson River valley (Fig. 4.6d). This pool-riffle channel pattern reflects the intersection of the modern river channel with bedrock and deep deposits of glacial, boulder-rich clay preserved in paleochannels segments along the studied reach, and suggests that fluvial erosion plays a key role in the location of major slope instabilities. The channel bottom morphology was likely a significant pre-condition for slope failure in the late nineteenth Century. Understanding the erosional regime at these stable slide toes will help us understand whether they could possibly reactivate in the near future.

The discovery of scour pools incising landslide toe slopes are important for two reasons. Firstly, erosion within the pools removes support at the toes of the landslides, likely contributing to their instability. InSAR, UAV photogrammetry, and GNSS markers confirm parts of the landslides close to the water edge have significant movement, and these active parts are typically in proximity to deep pools (Figs. 4.6b, d and 4.11). Secondly, the scour pools can expose fractured and porous sediments at the river bed, improving hydraulic connectivity between the river and groundwater in the slide mass. Both of these apparent relationships will be investigated in more detail in the coming years using multibeam technologies. These remote bathymetric mapping techniques are thus important for expanding characterization of landslides toes where they are extrapolated to extend under the Thompson River, and in developing a complete understanding of form and function.

GNSS Change Detection

Here, we present displacement data from November to December 2018. Eight Geocubes installed across Ripley Landslide were active to some degree or other, and recording 3D displacement during this time interval (three units were not active due to wildlife damage or low battery charge). In Fig. 4.12, location data (x, y, z coordinates) over the 8 month-long observation trial were imported into an ArcGIS Geodatabase. Geocube tracks were plotted as points, colour-shaded from light to dark according to date (per month) from exported individual Shapefiles.

Immediately apparent is the generally NW displacement trend for all working Geocubes, with each unit charting a helical drift over the months of observation (Fig. 4.12). The exception is the fixed unit, KYL-00026, which shows no horizontal and vertical displacement over the 8 month observation period. These trends are consistent with the displacement vectors derived from the change detection analysis of UAV imagery using Cosi-Corr software (Fig. 4.11).

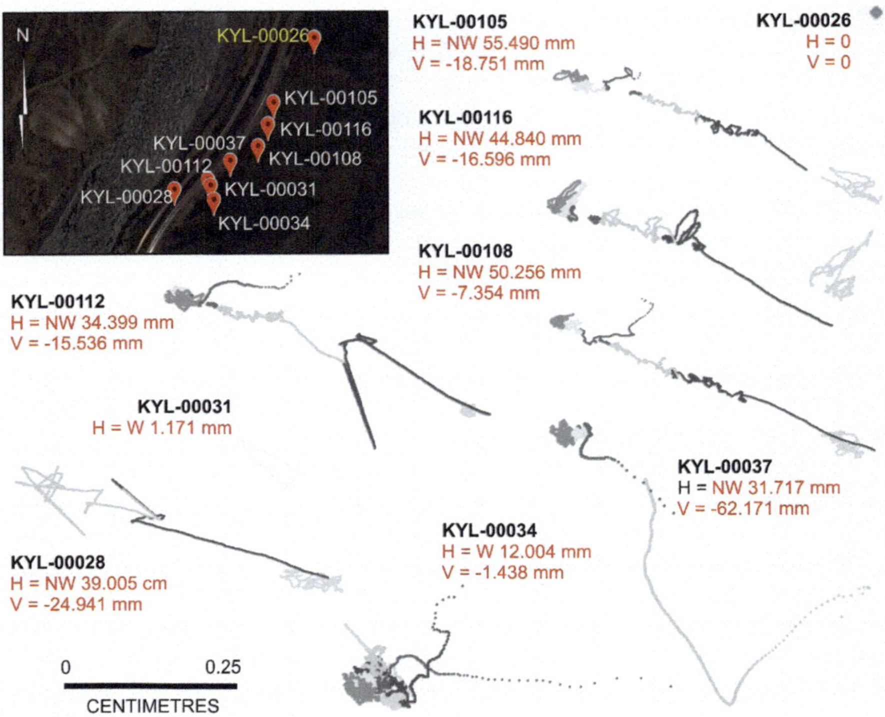

Fig. 4.12 Displacement trends of Ripley Landslide Geocube network. Google Earth image showing active units from December 2018 and provisional results for the November 2018 to June 2019 observation period

Geocubes at the north end of the landslide capture movement at varying rates from November 2018 to June 2019, suggesting the displacement of multiple slide blocks with shallow dipping slide planes. The most northerly moved horizontally NW 55.49 mm while dropping 18.75 mm in elevation (KYL-00105). A second block, forming a graben, moved NW 44.84 mm with a 16.60 mm drop in elevation (KYL-00116). The third block moved horizontally NW 50.26 mm, dropping 7.35 mm (KYL-00108) over the 8 month trial.

In the centre and south, generally slower horizontal, but greater vertical displacements were recorded over the 8 month observation period. KYL-00031 positioned below the southern head scarp, moved horizontally NW 1.17 mm prior to failure during the winter. KYL-00037 captured 31.72 mm horizontal movement to the NW with a drop of 62.17 mm in a central block of the slide main body (the greatest vertical downward displacement across the slide body). Similar to KYL-00116 in the northern sector of the slide body, KYL-00112 is located in a graben toward the southern margin of the slide. While experiencing only 34.40 mm of horizontal movement NW, the Geocube dropped 15.54 mm in elevation. On the lock-block retaining wall, KYL-00028 recorded 39.01 mm of horizontal movement NW and experienced 24.94 mm vertical downward displacement over the observation period (Fig. 4.12). Above the southern head scarp, KYL-00034 captured 12.00 mm of horizontal movement to W, while dropping in height by 1.44 mm. These small displacement values may be an indication of developing instability upslope of the active headscarp.

ERT Change Detection

The ERT models reveal a stratigraphy of coarse colluvial sediments overlying massive silt-rich diamicton with tension cracks >0.5 m wide and 1 m deep, and highly fissured laminated silts and clays in which the landslide failure plane lies.

Large decreases in surface resistivity (>50%) through spring (March to May) are due to an increase in moisture content from snowmelt and intense, short-duration precipitation events. Temperatures are consistently above 0 °C by this time, and despite a negative weekly effective rainfall during this season, the additional moisture resulting from snowmelt is sufficient to increase the moisture content of the slope. This observation demonstrates the need for subsurface imaging in such locations where both temperature and precipitation control groundwater hydrology since this hydrogeological regime could not be predicted from weather data alone. In addition, the spatial variations in resistivity changes are also revealed, which again highlights the need for subsurface investigation. The propagation of the wetting front along the failure plane is readily apparent in Fig. 4.13, indicating that the headscarp acts as a major conduit for the flow of groundwater. Long-term monitoring of the slope will provide an insight into the seasonal variations in subsurface moisture, and combining this with near-real time displacement data will enable a long-term goal of developing moisture thresholds for failure to be realized.

4.5 Discussion: Evaluating InSAR, UAVs, Bathymetric Surveys, GNSS and ERT

InSAR Change Detection

Beginning in 2019 and continuing through 2020–2025, RADARSAT-2 results will be compared against data acquired from EUROSPACE platforms (e.g. SENTINEL) with different spatial resolutions, view-angle capacities, and orbital paths. Dedicated processing systems and specific processing algorithms will together reduce processing time and likely improve the quality InSAR results. The application of advanced

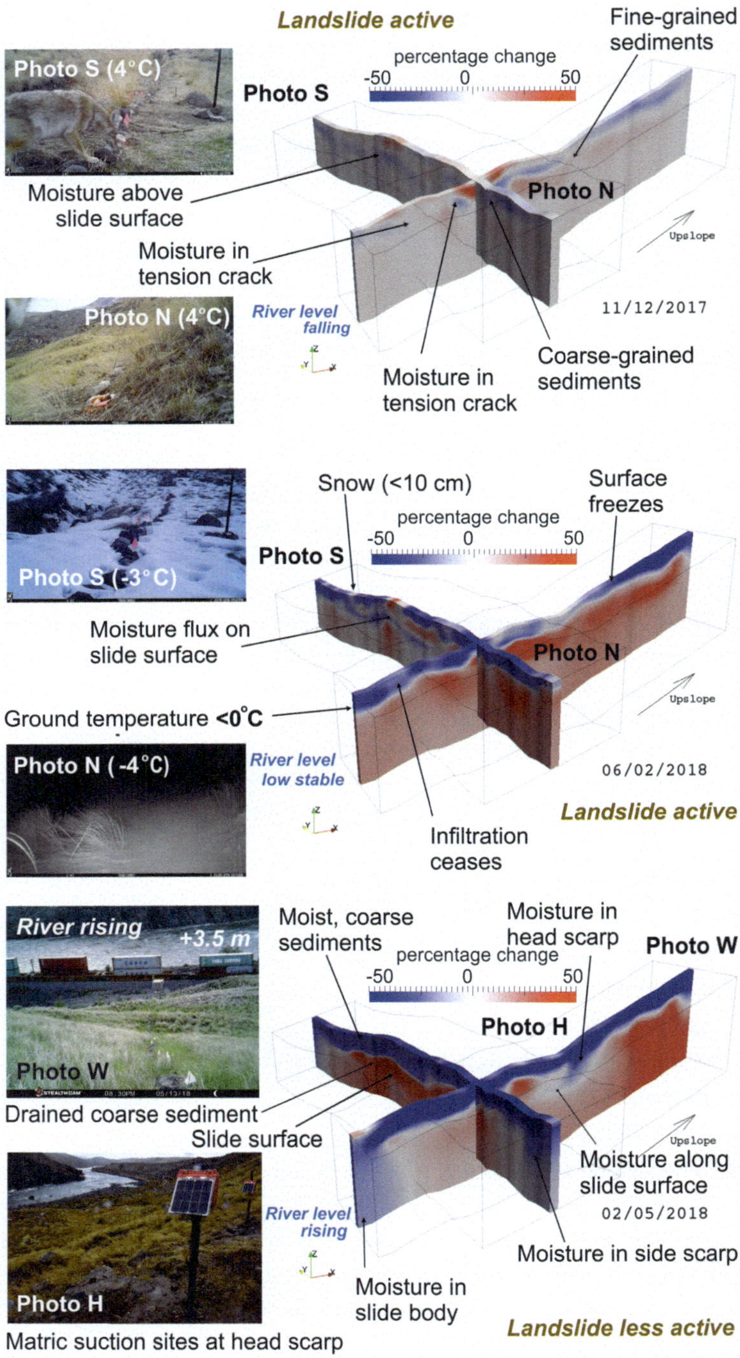

Fig. 4.13 Percentage (pc) changes in resistivity of PRIME array from 11 December 2017, to 02 May 2018; with ground conditions confirmed by wildlife cameras positioned at the ends of each ERT array (after: Holmes et al. 2018; Sattler et al. 2018; Huntley et al. 2019b)

processing techniques that consider both persistent scatterer targets and distributed targets (e.g. corner reflectors) will greatly improve the spatial density of displacement records.

The main expected outcomes of this activity will be: (1) analysis of the vertical and horizontal deformation components, combining ascending and descending acquisition geometries of satellite data; (2) analysis of the historical datasets and integration with GNSS and UAV data collected in situ; (3) improving knowledge of the behaviour and characteristics of the studied landslides; and (4) definition of a sequence of thresholds representative of increasing states of activity of the studied landslides.

UAV Change Detection

Preliminary UAV results are proof of concept (Bobrowsky et al. 2018). UAVs allow flexible, inexpensive acquisition of low-altitude aerial imagery. Various off-the-shelf photogrammetric (SfM) software packages enable the rapid production of high-quality digital surface models (DSM) from such images. Flying-height and battery restrictions limit the size of such surveys to areas on the order of magnitude of the largest landslides in the Thompson River valley. UAV photogrammetric surveys offer an excellent opportunity to generate high-quality repeat DSMs that can be compared to characterize the surface change of active landslides along the transportation corridor.

A *caveat*, however: wildfires and extremely poor air conditions during the 2017 and 2018 field seasons prevented fulfilment of the flight and survey objectives. With changing climate and weather conditions, poor air quality may limit future sortie times and ranges. In addition, federal aviation regulations are strictly enforced near railway property, requiring personnel with a flying licence and controlled traverses avoiding the railway right-of-way.

Additional UAV surveys to be conducted from 2020–2025 will extend surface change characterization of Ripley Landslide and others in the valley (e.g. South Slide and other active landsides). Comparison of these subsequent surveys will allow exceptionally precise quantification of points and rates of movement. These additional surveys will enable comparison of landslide motion over multiple periods, thus providing a further means to evaluate creep acceleration measured by CPR dGPS monuments and the GSC GNSS markers (Geocubes).

Bathymetric Change Detection

Similar to the other ground and aerial-based surveys conducted at the site, mapping the river bottom has introduced another set of challenges as a result of the harsh operating environment. Single-beam inertial measurement units (IMU) are too slow to accurately record the rapid changes in attitude, e.g. the scour hole in front of Ripley Landslide. Some scour pools (e.g. Black Canyon, >45 m) are too deep for strong single-beam acoustic signal returns. The multibeam system provides complete coverage, but only for the slide areas in deeper water (pools); additionally, surveying is not possible in rapids and in shallow waters (i.e. riffles). In the fast current and almost whitewater conditions, the jet boat is pitched, rolled and yawed more than typical watercraft.

Future survey passes should use a higher level of IMU (e.g. Applanix WaveMaster II) to compensate for this movement. Shallow water requires a large number of passes. In turbulent water, the SONAR system gets confused by air bubbles caused by the current. This produces a considerable noise requiring post-survey manual clean-up of data. Surveying ended early in November 2018 when the Norbit system hit a large boulder, shearing off the mounting column. The Norbit mounting beam must be re-designed before another survey takes place, and should be attached to a larger (5 m-long) jet boat.

GSC and CHS are currently processing data collected for all bathymetric surveys (single- and multibeam datasets). The GPR-, single-beam and multibeam-derived bathymetry match very well. Multibeam datasets capture key river bed features, including the thalweg channel and scour pools at the toe of Ripley Landslide and South Slide, and a >50 m deep scour pool in Black Canyon.

GNSS Change Detection

GNSS spatial data are used to help understand the behaviour and drivers of slope instabilities affecting rail transport in the Thompson River valley. Detailed examination of Geocube records provides insight on the rates and spatial pattern of creep, as well as the timing, and possibly precursors of changes in creep behaviour. To this end, the Geocube system is helping to characterize three-dimensional ground displacement patterns across Ripley Landslide (and at South Slide). This improved characterization of slope activity can then be combined with more spatially detailed line-on-sight displacements measured by InSAR and UAV photogrammetry. In addition, Geocube data is helping to evaluate the utility of precipitation, ambient temperature and soil moisture in predicting landslide activity based on comparison of three-dimensional displacement measurements with PRIME results and meteorological records.

The resulting knowledge will help to characterize landslide hazard, and ultimately reduce landslide risk. The locations of highest creep rates as indicated by the Geocube networks will help identify where track damage can be expected if the creep is continuous. Spatial variability in landslide motion will help inform the mechanism of the monitored landslides, thus improving landslide characterization and mitigation efforts. For example, determining whether failure involves complex interactions between structurally separate blocks that could preferentially move and damage railway infrastructure.

Comparing displacement trends with proxy records of possible landslide drivers—including temperature, precipitation, river level and irrigation will help establish landslide warning threshold based on environmental conditions. Such comparisons will include a detailed analysis of Geocube displacement histories spanning acceleration periods indicated by the CPR dGPS monuments in an effort to identify possible environmental triggers. If gradual acceleration is found to precede failure events, future Geocube-measured accelerated creep can be used to forecast impending failures.

The 'molecular' or 'signature' tracks (Fig. 4.12) are a reflection of precision and accuracy of satellite data used and their orbital paths; daily and seasonal atmospheric conditions; which satellites are visible to antennas; and developing instrumental failure in some instances. Filtering and data smoothing may remove some of this temporal drift. Rodriguez et al. (2018) used a 15 day moving average to get useable data from the Geocubes. As the network continues to collect data over the coming months it is expected that the displacement paths will become better defined. KYL-00031 and KYL-00034 failed in late November; while KYL-00105 and KYL-00116 behave erratically; and KYL-00028 functions intermittently. The poor performance is partly because Geocubes act as a network (i.e. relaying data from one unit to the next until their signals get to the Geocoordinator. If the network loses a critical nexus (e.g. trains and vegetation blocking signals, or wildlife damaging units), the whole network can slowly come down.

Lastly, the valley and landslide orientation limited sunlight to the solar panels during the winter (less than 2 h per day in mid-winter). Given the data uncertainties, limited interpretation of the Geocube plots is possible here. However, the horizontal and vertical displacement values are similar to those measured by the dGPS monuments, InSAR and UAV change detection methods (Figs. 4.3, 4.10, and 4.11).

As the Geocube network on Ripley Landslide (and South Slide) stabilizes through 2019, data collected (remotely and on-site, respectively) will be processed and presented to graphically show: (a) Displacement trends (coloured by month, ArcGIS shapefiles); (b) 3D displacement (mm) -vs- Date (month, year); (c) Surface angle of movement (angle, degrees) -vs- Date (month, year); (d) Precipitation (mm) and 3D displacement (mm) -vs- Date (month, year); and, (e) Temperature (°C) and 3D displacement (mm) -vs- Date (month, year).

ERT Change Detection

The PRIME system provides new insight into the hydrogeological structure of the slope (Fig. 4.13). Daily monitoring of electrical resistivity since November 2017 reveals complex hydrogeological pathways in the slide body, with differences in electrical resistivity of earth materials reflecting a combination of hydrogeological characteristics, responses to variations in ground temperature and precipitation, and the subsurface distribution of solute-bearing soil water. These new hydrogeological and geophysical datasets: (1) enhance our understanding of the composition, internal structures and groundwater flow paths through the landslide—critical information to develop moisture thresholds for slope failure; and, (2) provide important context for interpreting other multi-year slope monitoring efforts underway in the valley.

Relationships between suction and resistivity derived from field data demonstrate the importance of changing moisture content (and therefore changing suction and pore pressures) for resistivity (Holmes et al. 2018; Sattler et al. 2018). However, given the wide range of geological units present on the site, understanding the hydrogeological properties of each unit will be pivotal in providing an improved understanding of slope stability in the Thompson River valley. As such, future work will focus on the laboratory testing of samples from different units of the Ripley landslide, building upon relationships identified in the field to establish relationships between moisture content, suction and resistivity. These relationships will then be used to better constrain the inversion of the ERT data. This new understanding will guide the interpretation of multi-year monitoring datasets (e.g. InSAR, UAV and GNSS change detection) and future efforts to track landslide activity in the Thompson River valley.

4.6 Summary: Protecting Canada's Railway Network Using Remote Sensing Techniques

Resilient railway transportation networks require sustainable, cost-effective management of service operations to meet future socio-economic needs and ensure the protection of the natural environment. Where transportation corridors traverse unstable terrain, critical rail infrastructure is at risk of damage and presents potential local and national economic, social and environmental challenges. Remote monitoring unstable slopes and infrastructure at risk is a cost-effective hazard management practice that also provides important geoscience information to help develop appropriate mitigation measures.

In the Thompson River valley, remote change detection monitoring at Ripley Landslide reveals that peak displacement rates are observed through winter to spring, indicating that low river and groundwater levels for significant movement. Between late fall (November) and early spring (March), snowfall blankets the slope, the ground freezes (<0 °C) to an estimated depth <2 m. When air temperatures are >0 °C, rain and melting snow result in an increase in the moisture content of subsurface clay-rich units (Huntley et al. 2019b).

The greatest displacement rates indicated by InSAR, UAV photogrammetry, GNSS and ERT occur during winter and spring when transitional ground conditions allow snowmelt and rainfall to penetrate deep into the still-frozen (or thawing) slide body by way of tension cracks, planar fractures and bedding surfaces. During summer months, Thompson River levels are high and support the submerged portions of the toe slope. Groundwater is at its maximum level within the slide body. It is during this period that InSAR and other monitoring techniques indicate minimum rates of surface displacement (cf. Journault et al. 2018; Huntley et al. 2019c).

Railway infrastructure and operations are expected to face unique challenges in design, monitoring, adaptation, mitigation, reclamation and restoration in a scenario of future extreme weather events and climate change. Seasonal and spatial variations in precipitation, temperature, river levels and groundwater recharge are perceived as the dominant controls on landslide activity in the Thompson River valley. An understanding of the geographic distribution and temporal range of earth materials and geological hazards, and their potential responses to climate change is essential for a resilient and accessible transportation network, but also to protect the natural environment, local communities, land-use practices and the national economy.

Acknowledgements The research was undertaken as part of the International Consortium on Landslides (ICL) International Programme on Landslides (IPL) Project 202. Activities at this site have led to the World Centre of Excellence on landslide disaster reduction recognition for both the Geological Survey of Canada and the University of Alberta (iplhq.org). The research was funded by Transport Canada and Natural Resources Canada. Fieldwork would not be possible without the support of Trevor Evans (Canadian National Railway, Kamloops, BC) and Danny Wong (Canadian Pacific Railway, Calgary, Alberta). Vincent Decker, Nicolas Svacina and Kevin Murnaghan (Canadian Centre for Mapping and Earth Observation, Ottawa, Canada) helped setting up licences for PCI and Gamma to process RADARSAT-2 imagery. Geocube monitoring is supported by Remi Usquin and Frédéric Verluise (Ophelia Sensors, Paris, France). The contribution of Jessica Holmes was funded by the Department for the Economy (DfE), Northern Ireland. The manuscript was edited by Catherine Sidwell (PGeo) and critically peer-reviewed by an Anonymous Reader.

References

Benoit, L., P. Briole, O. Martin, C. Thom, J.-P. Malet, and P. Ulrich. 2015. Monitoring Landslide Displacements with Geocube Wireless Network of Low-Cost GPS. *Engineering Geology* 195: 111–121.

Best, M., P. Bobrowsky, D. Huntley, R. Macciotta, and M. Hendry. 2018. Integrating Terrestrial and Waterborne ERT Surveys at the Ripley Landslide Near Ashcroft, British Columbia, Canada. In *Environmental and Engineering Geophysical Society, Proceedings Volume of 31st SAGEEP*, Nashville, Tennessee.

Bobrowsky, P., D. Huntley, M. Hendry, R. Macciotta, M. Schafer, J. Journault Q. Zhang, X. Zhang, and Z. Lv. 2016. Assessing multi-sensor technologies and methods in monitoring landslide movement in Canada. In *35th International Geological Congress, Abstracts and Proceedings Volume*, Cape Town, South Africa.

Bobrowsky, P., R. MacLeod, D. Huntley, O. Niemann, M. Hendry, and R. Macciotta. 2018. Ensuring Resource Safety: Monitoring Critical Infrastructure with UAV Technology. In *Resources for Future Generations*, Conference Abstracts Volume, 1 p. Vancouver, Canada.

Bobrowsky, P.T., W. Sladen, D. Huntley, Q. Zhang, C. Bunce, T. Edwards, M. Hendry, D. Martin, and E. Choi. 2014. Multi-Parameter Monitoring of a Slow Moving Landslide: Ripley Slide, British Columbia, Canada. In *Engineering Geology for Society and Territory – Volume 2 Landslide Processes*, eds. G. Lollino, D. Giordan, G. Battista Crosta, J. Corominas, R. Azzam, J. Wasowski, and N. Sciarra, 155–159. IAEG (International Association of Engineering

Geology and the Environment) Congress, Turin, Italy, Springer Publishing (Contribution #20140007).

Bunce, C., and I. Chadwick. 2012. GPS monitoring of a landslide for railways. In *Landslides and engineered slopes - protecting society through improved understanding*, ed. E. Eberhardt et al., 1373–1379. London: CRC.

Clague, J.J., and S.G. Evans. 2003. Geologic Framework of Large Historic Landslides in Thompson River Valley, British Columbia. *Environmental & Engineering Geoscience* 9: 201–212.

Ferretti, A., G. Savio, R. Barzaghi, A. Borghi, S. Musazzi, F. Novali, C. Prati, and F. Rocca. 2007. Submillimeter Accuracy of InSAR Time Series: Experimental Validation. *IEEE Transactions on Geoscience and Remote Sensing* 45 (5): 1142–1148.

Hendry, M., R. Macciotta, and D. Martin. 2015. Effect of Thompson River Elevation on Velocity and Instability of Ripley Slide. *Canadian Geotechnical Journal, Volume* 52 (3): 257–267.

Henschel, M.D., J. Dudley, B. Lehrbass, S. Sato, and B.-M. Stöckel. 2015. Monitoring Slope Movement from Space with Robust Accuracy Assessment. *Southern African Institute of Mining and Metallurgy, Slope Stability*: 151–159.

Holmes, J., J. Chambers, S. Donohue, D. Huntley, P. Bobrowsky, P. Meldrum S. Uhlemann, P. Wilkinson, and R. Swift. 2018. The Use of Near Surface Geophysical Methods for Assessing the Condition of Transport Infrastructure. Civil Engineering Research Association, Special Issue on Structural Integrity of Civil Engineering Infrastructure. *Journal of Structural Integrity and Maintenance*, 6 p.

Huntley, D., P. Bobrowsky, Q. Zhang, X. Zhang, Z. Lv, M. Hendry, R. Macciotta, et al. 2016. Application of Optical Fibre Sensing Real-Time Monitoring Technology at the Ripley Landslide, Near Ashcroft, British Columbia, Canada. In *Canadian Geotechnical Society, Proceedings Volume of GeoVancouver2016 Annual Meeting*, 13 p.

Huntley, D., P. Bobrowsky, F. Charbonneau, J. Journault, and M. Hendry. 2017a. Innovative Landslide Change Detection Monitoring: Application of Space-Borne InSAR Techniques in the Thompson River valley, British Columbia, Canada. In *Landslide Research and Risk Reduction for Advancing Culture and Living with Natural Hazards*, vol. 3, 13 p. 4th World Landslide Forum (ICL-IPL), Ljubljana, Slovenia, Springer Nature.

Huntley, D., P. Bobrowsky, Q. Zhang, X. Zhang, and Z. Lv. 2017b. Fibre Bragg Grating and Brillouin Optical Time Domain Reflectometry Monitoring Manual for the Ripley Landslide, Near Ashcroft, British Columbia. *Geological Survey of Canada*, Open File 8258, 66 p.

Huntley, D., P. Bobrowsky, and M. Best. 2017c. Combining Terrestrial and Waterborne Geophysical Surveys to Investigate the Internal Composition and Structure of a Very Slow-Moving Landslide Near Ashcroft, British Columbia, Canada. In *Landslide Research and Risk Reduction for Advancing Culture and Living with Natural Hazards*, vol. 2, 15 p. 4th World Landslide Forum (ICL-IPL), Ljubljana, Slovenia, Springer Nature.

Huntley, D., P. Bobrowsky, N. Parry, P. Bauman, C. Candy, and M. Best 2017d. Ripley Landslide: The Geophysical Structure of a Slow-Moving Landslide Near Ashcroft, British Columbia, Canada. In *Geological Survey of Canada* Open File 8062, 59 p.

Huntley, D., P. Bobrowsky, R. MacLeod, and N. Roberts. 2018. New Insights Into Form and Function of Very Slow-Moving Landslides from Bathymetric Surveys of Thompson River, British Columbia. In *Geological Society of America Annual Meeting*, Session T53, Abstract Volume, 1 p.

Huntley, D., P. Bobrowsky, M. Hendry, R. Macciotta, and M. Best. 2019a. Multi-Technique Geophysical Investigation of a Very Slow-Moving Landslide Near Ashcroft, British Columbia, Canada. *Journal of Environmental and Engineering Geophysics* 24 (1): 85–108.

Huntley, D., P. Bobrowsky, M. Hendry, R. Macciotta, D. Elwood, K. Sattler, M. Best, et al. 2019b. Application of Multi-Dimensional Electrical Resistivity Tomography Datasets to Investigate a Very Slow-Moving Landslide Near Ashcroft, British Columbia, Canada. *Landslides*. https://doi.org/10.1007/s10346-019-01147-1.

Huntley, D., P. Bobrowsky, K. Sattler, D. Elwood, J. Holmes, J. Chambers, P. Meldrum, et al. 2019c. PRIME Installation in Canada: Protecting National Railway Infrastructure by

Monitoring Moisture in an Active Landslide Near Ashcroft, British Columbia. In *32nd SAGEEP, Environmental and Engineering Geophysical Society*. Proceedings Volume of the Annual Meeting, Portland, Oregon, USA; Geological Survey of Canada Open File 8548, 1 p.

Journault, J., R. Macciotta, M. Hendry, F. Charbonneau, D. Huntley, and P. Bobrowsky. 2018. Measuring Displacements of the Thompson River Valley Landslides, South of Ashcroft, B.C., Canada, Using Satellite InSAR. *Landslides* 15 (4): 621–636. https://doi.org/10.1007/s10346-017-0900-1.

Macciotta, R., M. Hendry, D. Martin, D. Elwood, H. Lan, D. Huntley, P. Bobrowsky, et al. 2014. Monitoring of the Ripley Slide in the Thompson River Valley, B.C. In *Geohazards 6 Symposium, Proceedings Volume*, Kingston, Ontario, Canada.

Merritt, A., J. Chambers, W. Murphy, P. Wilkinson, L. West, D. Gunn, P. Meldrum, M. Kirkham, and N. Dixon. 2014. 3D Ground Model Development for an Active Landslide in Lias Mudrocks Using Geophysical, Remote Sensing and Geotechnical Methods. *Landslides* 11 (4): 537–550.

Porter, M.J., K.W. Savigny, T.R. Keegan, C.M. Bunce, and C. MacKay. 2002. Controls on Stability of the Thompson River Landslides. In *Proceedings, Canadian Geotechnical Conference*, Niagara Falls, Ontario, pp. 1393–1400.

Rodriguez, J., M.T. Hendry, and R. Macciotta. 2018. Use of Low-Cost GPS Landslide Monitoring System for Obtaining High-Frequency Displacement Measurements at Multiple Measurement Points. In *Proceedings of the 71st Canadian Geotechnical Conference*, Geohazards 7, Edmonton, Canada, 8 p.

Sattler, K., D. Elwood, M. Hendry, R. Macciotta, D. Huntley, P. Bobrowsky, and P. Meldrum. 2018. Real-Time Monitoring of Soil Water Content and Suction in Slow-Moving Landslide. In *GeoEdmonton 2018*, Proceedings Paper, 8 p.

Schafer, M., R. Macciotta, M. Hendry, D. Martin, P. Bobrowsky, D. Huntley, C. Bunce, and C. Edwards. 2015. Instrumenting and Monitoring a Slow Moving Landslide. *GeoQuebec 2015* Paper, 7 p.

Stanton, R.B. 1898. The Great Land-Slides on the Canadian Pacific Railway in British Columbia. *Proceedings Civil Engineers* 132 (2): 1–48.

Uhlemann, S., J. Chambers, P. Wilkinson, H. Maurer, A. Merritt, P. Meldrum, O. Kuras, D. Gunn, and T. Dijkstra. 2017. Four-Dimensional Imaging of Moisture Dynamics During Landslide Reactivation. *Journal of Geophysical Research (Earth Surface)* 122: 398–418.

Chapter 5
Landslide Activity Assessment of a Subtropical Area by Integrating InSAR, Landslide Inventory, Airborne LiDAR, and UAV Investigations: A Case Study in Northern Taiwan

Ching-Fang Lee, Vernon Singhroy, Shih-Yuan Lin, Wei-Kai Huang, and Junhua Li

5.1 Introduction

Landslide and sediment-related hazards are a common, frequent natural phenomena that causes significant damage to public infrastructure as well as to the citizens in Taiwan during typhoon and monsoon season (June–November). The report published by World Bank indicated that Taiwan is simultaneously exposed to multiple natural hazards such as landslides, floods, and earthquakes; it also ranked Taiwan in the first level of high mortality risk (The World Bank 2005). The statistical report published by UNISDR pointed out the total losses achieved US$1791 billion from weather-related disasters around the world in the past ten years (2005–2015). This included storms, floods, and other geohazards. The direct economic damage is equivalent to 67% of losses attributed to all natural hazards (Center for Research on the Epidemiology of Disasters 2015). To address the issues, spaceborne remote sensing techniques have been developed and widely applied over the landslide-prone areas (Oliveira et al. 2014; Calvello et al. 2017;

C.-F. Lee (✉) · W.-K. Huang
Disaster Prevention Technology Research Center, Sinotech Engineering Consultants, Inc., Taipei, Taiwan, China
e-mail: cflee@sinotech.org.tw

V. Singhroy
EOSPATIAL, Ottawa, ON, Canada

S.-Y. Lin
Department of Land Economics, National Chengchi University, Taipei, Taiwan, China

J. Li
Canada Centre for Remote Sensing, Ottawa, ON, Canada

V. Singhroy (ed.), *Advances in Remote Sensing for Infrastructure Monitoring*, Springer Remote Sensing/Photogrammetry,
https://doi.org/10.1007/978-3-030-59109-0_5

Mateos et al. 2017). Also, synthetic aperture radar (SAR) sensors equipped with various microwave wavelengths were adopted to assess the potential risk and activity state of the slow-moving landslides at a regional scale (Vecchiotti et al. 2017; Schlögel et al. 2017; Bozzano et al. 2017; Ciampalini et al. 2015; Herrera et al. 2013). Together with multiple interferometric SAR (InSAR) approaches, such as D-InSAR, PS-InSAR, and SqueeSAR, time-series ground motion caused by landslides and sediment-related hazards can be successfully extracted (Singhroy and Molch 2004; Singhroy et al. 2012; Singhroy and Charbonneau 2014; Raspini et al. 2017; García-Davalillo et al. 2014; Martire et al. 2016; Bovenga et al. 2017). Concerning operating procedures related to disaster prevention and rescue, one can identify the landslide areas which frequently slide by updating landslide inventories based on InSAR interferograms, automatic interpretation, and statistical analysis. The historical set of SAR data acquired for a given period provides a wide-range of ground movement points, allowing for the assessment of the latest landslide activity in pre-existing landslide areas (Ventisette et al. 2014; Rosi et al. 2017; Bonì et al. 2017). Utilizing a two-dimensional hazard matrix based on a combination of landslide state and ground displacement from InSAR, some researchers have updated landslide activity maps at a basin scale, even though a slight deformation on the ground may be triggered by rainfall events. The reduction of the revisit period of an InSAR sensor, ranging from the 24 days (RADARSAT-2) to 6 days (Sentinel-1A&B, COSMO SkyMed 2), also helps to improve the temporal resolution of the analysis results. However, some landslide-prone areas that have no disaster history, but are characterized by slow movement, cannot be definitively extracted from the dense vegetation region. To deal with this problem, high-resolution SAR imagery is adopted to enhance the small displacement of terrain after the occurrence of geohazards. Considering global climate change and extreme rainfall, Taiwan faces more disaster impacts on the influence of both shallow landslides and debris flows under the RCP8.5 scenario (Gariano and Guzzetti 2016). Thus the dynamic assessment related to environmental geohazards is required to be improved in Taiwan since the urgent demand for public safety around the mountainous region is still very high.

This study aims to develop an assessment procedure with forecasting of high landslide susceptibility for the subtropical mountainous area. This paper intends to be an investigation of the time-series InSAR technique on regional landslide morphologic change. Ground displacement maps derived from high-resolution InSAR satellite images, historical landslide inventories, and LiDAR derivatives are integrated and validated for early warning and long-term monitoring purposes. Different types of sediment-related hazards, such as shallow landslides, deep-seated landslides, debris flows, and deposition on the riverbed show movement trending on both the hillslope and basin. In particular, an additional contribution to the identification of landslide-prone areas with a disaster record is useful for hazard management. This may assist the local government in finding the possible landslide hotspots in advance of the next rush rainfall event. Finally, this study provides a starting point to evaluate landslide precursors for slow creep which may transform into fast catastrophic failure.

5.2 Study Area

Geography and Geological Setting

The study area is located in Xindian and Wulai Districts, New Taipei City, Taiwan (Fig. 5.1). Nanshi River is the mainstream which comes from southern Mt. Taman (2130 m, m.a.s.l) and crosses the whole study area (where the basin area is 332 km^2). Both Tonghou and Pingguang creeks are where the tributaries converge at Wulai and Xindian area, respectively. Around this area, most hill and mountain terrain vary from 200 m at the northern part to 2130 m at the south. The total length and mean slope (elevation difference of the basin/river length) of Nanshi River are approximately 45 km and 3.6%. The valley landform is mostly in the upstream reach of the Nanshi basin instead of a terrace. Meander and lateritic terraces distribute over the downstream region. The whole study area is classified at the classical youth stage of valley terrain, so the natural landscape as waterfall, cliff, and gorge is observable. The slope gradient classification of the study area is categorized as 5 (40–55%) to 7 (>100%) grade.

For the geological setting, Wulai District is the part of the slate mountain block of the Hsuehshan Range. The lithological units of the study area are comprised principally of Kangkou, Mushan, Shihti, Shuichangliu, Szeleng, Taliao, Tatungshan,

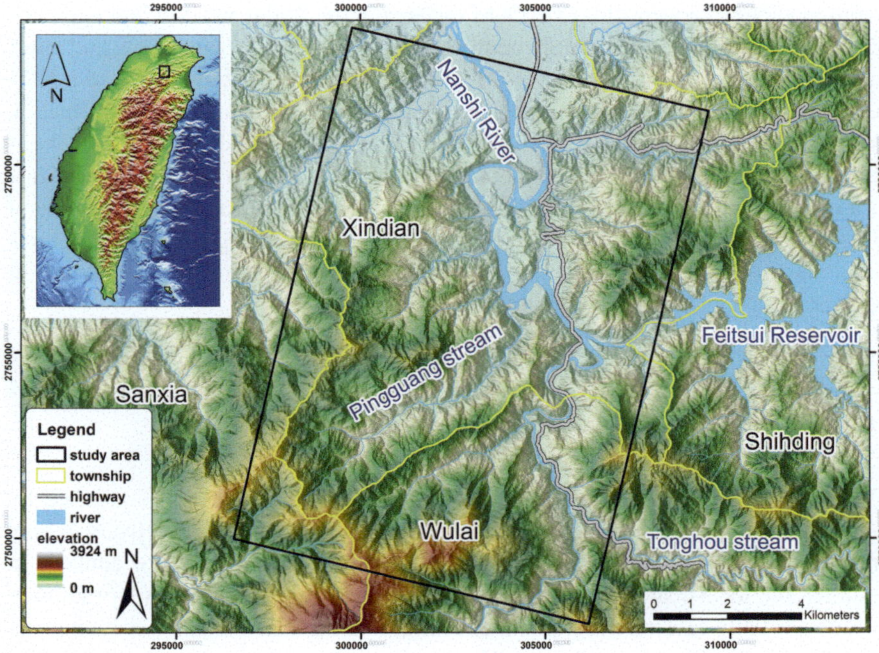

Fig. 5.1 Topography of the study area

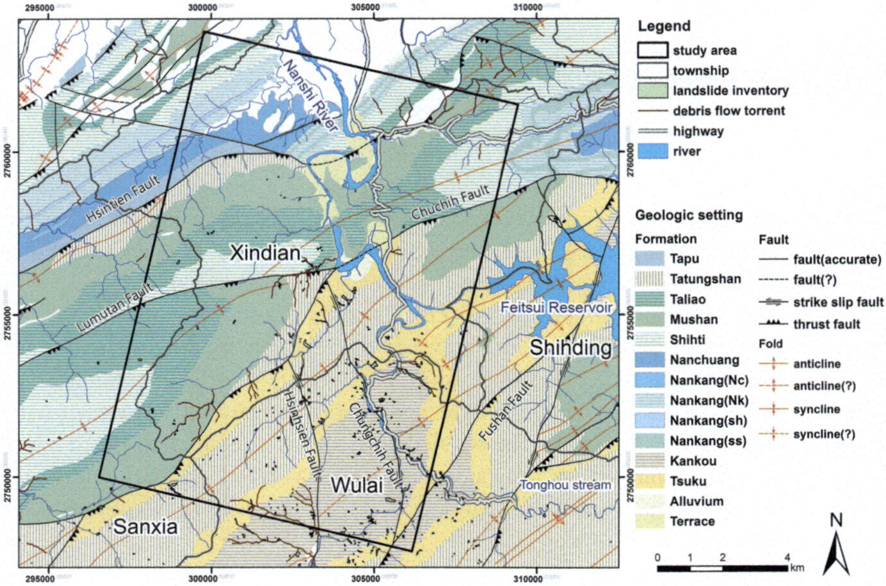

Fig. 5.2 Geological map of the study area

and Tsuku formation depending on the coverage area (Fig. 5.2). The formation mentioned above covers almost 90% of the area in the Wulai and Xindian Districts. The lithologic features are mainly composed of thick argillite, sandstone, and alternations of sandstone and shale. Several thrust and strike-slip faults oriented NNW-SSE (Hsinhsien and Chungchi faults) and NEE-SWW (Xindian, Lumutan, Chuchih, and Fushan faults), which pass through the Nanshi Basin, are shown in the geological map. Notably, the dip angle of Chuchih fault is almost close to the vertical plane, as a result of which, one can perceive the offsetting phenomena of stratum at lateral river banks at Wulai village. Due to the evolution of complex geological structures (anticline), hot-spring spots are found in many regions around Wulai, where it becomes a fashionable resort in northern Taiwan.

Hydrometeorological Conditions

The study area lies in the subtropical monsoon zone. The weather is affected by occasional monsoons and heavy rainfalls in summer and winter, respectively. It may snow periodically in the southern mountain, based on the water vapor content. In summer, the local-scale convection induced by the heating effect of solar radiation is active, usually causing a regional thunderstorm in the afternoon. The annual mean rainfall is 3000–4000 mm (Central Weather Bureau), with the majority of rainfall concentrated in summer and autumn (June–October), especially for typhoon events. When it comes to the seasonal temperature in Nanshi Basin, it is influenced significantly by elevation at the site. When the altitude is lower than 500 m, the mean temperatures in summer and winter are 28 °C and 13 °C, respectively. While the mean temperature in summer and winter are 20 °C and 6 °C for altitudes over 1000 m. The intense rainfall event is the primary triggering factor which combines with river erosion, infiltration, and surface washout on the sediment-related hazards.

The Events of Typhoon Soudelor and Dujua in 2015

On 8th of August 2015, a severe Typhoon (Soudelor) brought torrent rainfall to New Taipei City and Ilan County in northern Taiwan. The mean rainfall intensity exceeded 80 mm/h within 3 h, and the accumulated rainfall reached 769 mm in 24 h causing regional landslides, rockfalls, debris flows, and debris floods. The regional landslides started particularly at 5:00 AM on 8th August and caused some road closures on Route 9A (Xinwu Highway). Several villages located in remote mountainous areas became isolated and inundated by flash flooding. The post rainfall analysis found that the rainfall of 3-h (253 mm), 6-h (443 mm), and 12-h (670 mm) for Typhoon Soudelor exceed the 200-year return period (248, 367, and 633 mm). It was reported that the high-intensity rainfall within a short timespan was a leading cause in inducing environmental sediment-related disasters. The maximum flooding height even reached 1 m and 3 m at the downstream reach in Wulai and Xindian Districts, respectively. Soon later, Typhoon Dujua brought more serious damage around northern Taiwan on 28th of September 2015 (the total rainfall was 387 mm in 24 h). It triggered new geo-disasters and the expansion of pre-existing landslides again (Fig. 5.1). Shallow landslides occurred mostly in the region of slope gradient with 30–40° and dip slope. To sum up, the total landslide area was approximately 87.1 ha, and ten debris flows were triggered.

5.3 Material and Methodology

The interpretation methodology and material used in this study are summarized and presented in this section. The materials applied include event-based landslide inventories LiDAR shaded reliefs, InSAR datasets, and an active matrix assessment. Figure 5.3 shows the methodological flowchart used for each processing step. The detailed illustration is delineated below.

Event-Based Landslide Inventory

An event-based landslide inventory is an essential set of data for landslide susceptibility analysis. Depending upon which specific event is taken as a reference, the image selection must be close to the start time of the event being studied; therefore,

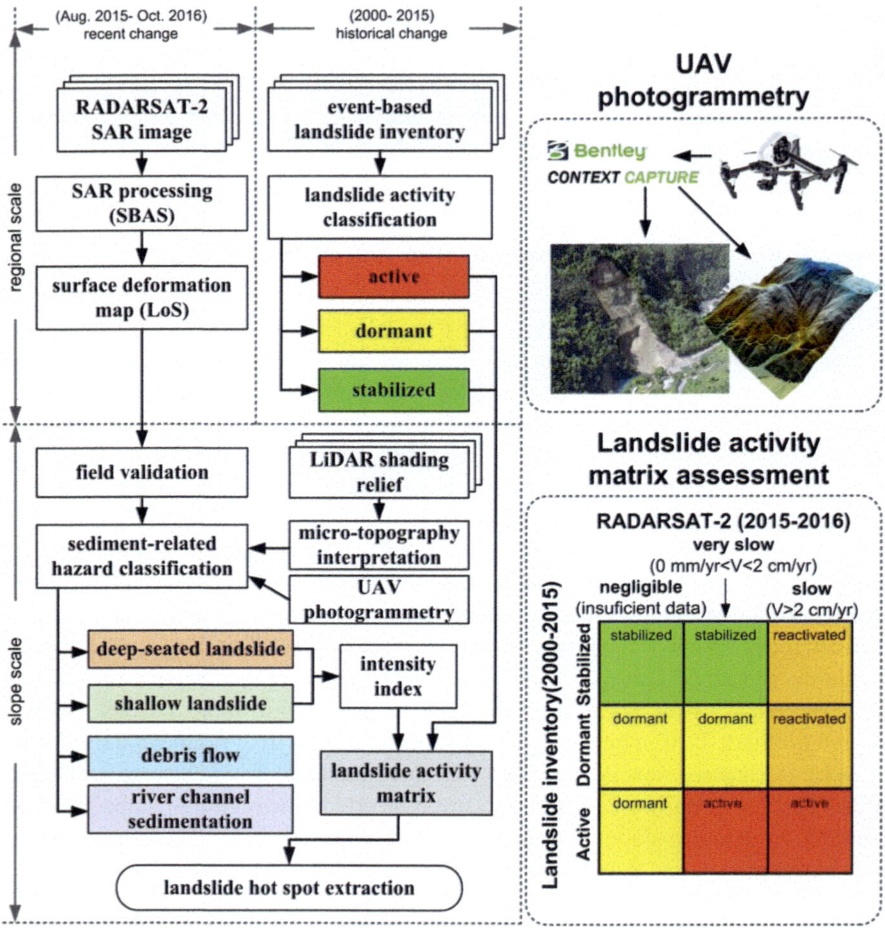

Fig. 5.3 Methodological flowchart used for each data processing step and corresponding description of the study

the precise spatial distribution of wide-ranging landslides can be obtained. This study collected seven event-based landslide inventories triggered by the typhoon, which represents curial influence around northern Taiwan (data producer: Central Geological Survey, Sinotech, and NCDR). The inventory dataset comprises Typhoon Xangsane (2000), Nari (2002), Aere (2004), Talim (2005), Krosa (2007), Jangmi (2008), and Soudelor (2015) which were mapped from 2000 to 2015. As shown in Fig. 5.4, it indicates that Typhoons Jangmi (2008) and Soudelor (2015) caused widespread landslide hazards. Most landslide areas frequently appeared near the riverbank, and the location seems to be related to dip slope. The status of activity of

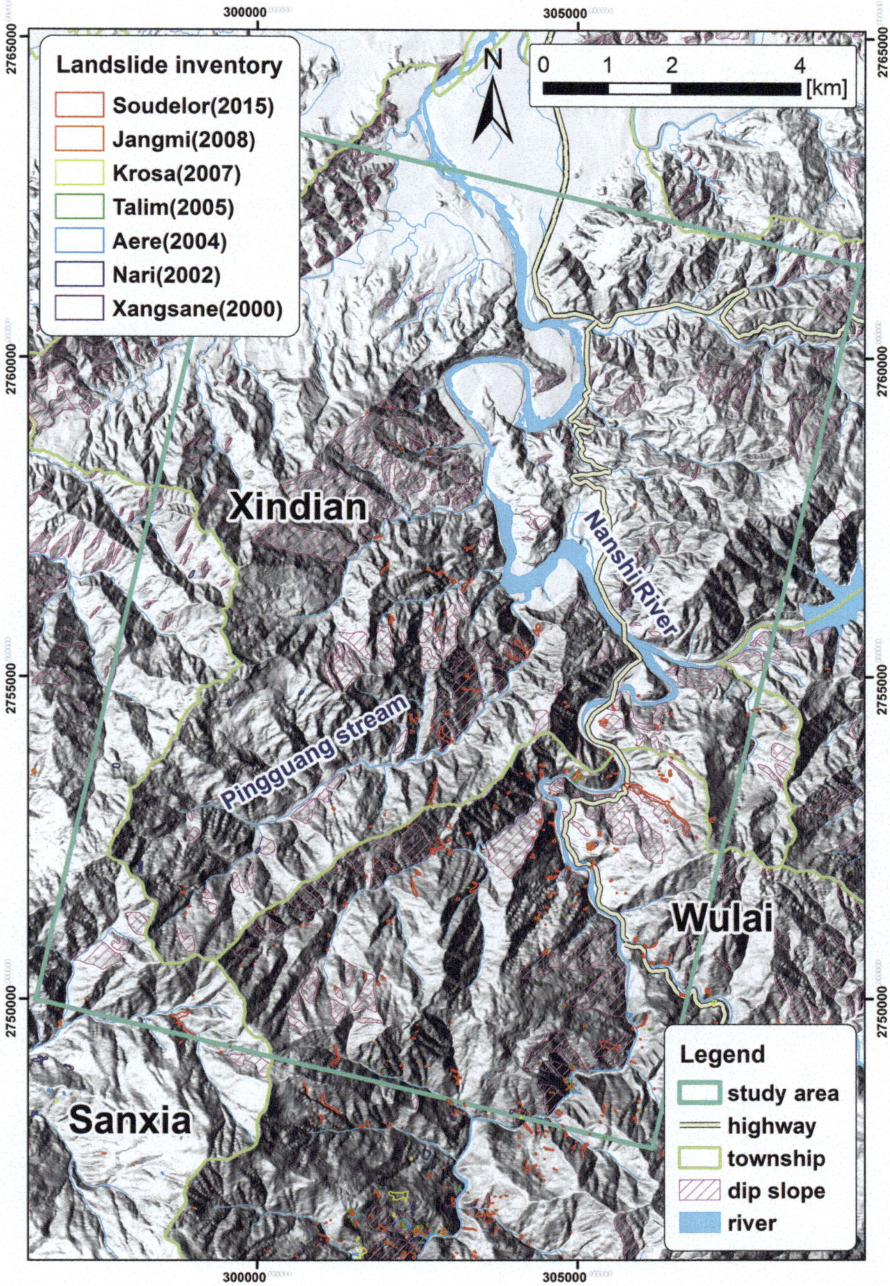

Fig. 5.4 Event-based landslide inventory in the study area (basemap: 5 m shading relief with 315° azimuth illumination)

landslides can be classified into three categories (The International Geotechnical Society 1993): (1) *Active*: the landslide is currently moving and landslide mass was remaining on the sliding plane; (2) *Dormant*: an inactive landslide within the last 12 months but can be reactivated or triggered by its original causes or other causes, including natural processes and artificial perturbation; (3) *Stabilized*: it also belongs to a state of an inactive landslide. This kind of landslide has been protected from its original causes by erosion and sediment control engineering. Using both InSAR data and landslide inventories to update the activity of landslide areas became a straightforward approach for an isolated community in Italy (Rosi et al. 2017; Bonì et al. 2017; Righini et al. 2012). While this principle focuses on the multi-stage landslide inventory, it is worth understanding the recent activity for the presentation of the landslide state. It offers a macroscopic phenomenon to assess the possible landslide zone regarding the adverse factor that may transform the stable state into instability.

Landslide Micro-Topography Interpretation

After the suffering of large-scale landslide disasters caused by Typhoon Morakot (8th of Aug. 2009), Central Geological Survey developed an airborne LiDAR scanning plan to investigate specific geohazards in the metropolitan area and mountain villages. This 6-years project covered the whole Taiwan Island and produced national-wide 1 m high-resolution DEMs. This work characterized topographic lineaments, scarps, and landslide mass in the Wulai and Xindian areas using LiDAR shaded relief derivatives. A powerful visualization technique called sky-view factor (SVF) relief is utilized here instead of the traditional analytical hill shading from multiple directions (Lee et al. 2017; Lo et al. 2017). In brief, SVF is able to suggest a geophysical parameter demonstrating the space of the sky visible from the given point on the ground surface (Kokalj and Hesse 2017; Zakšek et al. 2011). The value of SVF ranges from 0 (when located in a deep valley or sink, no sky could be visible) to 1 (total visibility in a hemisphere zone). Figure 5.5 illustrates the procedure used in the compilation of the SVF relief map, which involved compiling the slope map. Two parameters with considerable influence over SVF analysis computation are the horizontal search direction (HSD) and the maximum search radius (MSR). Following preliminary parameter analysis and using the size of the deep-seated landslide, we selected 16 directions and 20 pixels (1 pixel = 1 m) for the HSD and MSR to generate an SVF relief map. In such a setting, the landslide micro-topographic features, such as main (or minor) scarp, trench, landslide body, transverse ridge, and toe bulge are suitable for mapping, whatever the orientation and shape.

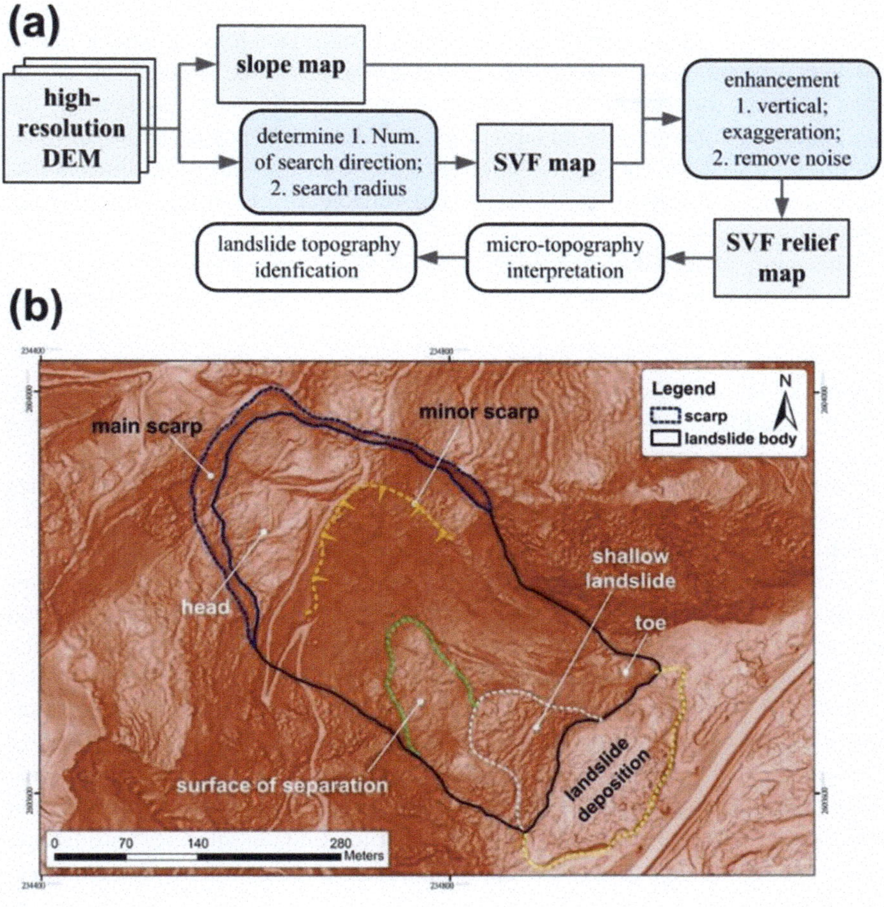

Fig. 5.5 SVF relief map compilation procedure (upper) and the result of landslide micro-topography interpretation (bottom)

RADARSAT-2 Images and InSAR Processing

Over the past ten years, satellite radar interferometry (InSAR) techniques are increasingly being used in monitoring slope stability and ground deformation (Singhroy and Molch 2004; Singhroy et al. 2012; Singhroy and Charbonneau 2014). An interferometric phase image (interferogram) represents the phase differences between the backscatter signals in two or more SAR images obtained from similar positions in space. On stable terrains, the phase differences can be converted into a digital elevation model. On unstable terrains, once the topographic phase is removed, the phase differences between two repeat-pass images are the result of changes in the line-of-sight distance (range) to the radar due to the displacement of the surface, as well as the changes in the atmospheric conditions between the scenes. With rapid improvements in image processing, the availability of coherent ground targets and the high frequency revisiting of radar satellites, InSAR monitoring techniques, capable of measuring millimeter displacement, are becoming widespread and reliable in observing slow-moving landslides.

This study analyzed fifteen RADARSAT-2 ultra-fine (3 m resolution) InSAR images from August 2015 to October 2016 in the Wulai and Xindian areas. The dataset spanned the period of Typhoons Dujua and Soudelorand and may record information of topographic change and secondary disaster caused by the typhoons. The InSAR processing method introduced here uses an improved SBAS (Small Baseline Subset) algorithm which minimizes both the effects of the atmosphere, and inaccuracies (or limited resolution) in the DEM, on the accuracy of the InSAR measurements. The method produces a nonlinear time-series of deformation over an extended period using interferograms with short periods, which are usually more coherent. In this approach, a high-pass filter with a Gaussian window was applied to remove the residual orbital and long wavelength atmospheric signals. A Singular Value Decomposition (SVD) inversion was applied to simultaneously solve the individual deformation rates and the residual topographic error. The final surface deformation information derived from the SBAS method is the line-of-sight (LoS) displacement between the scanning direction of a satellite and the object being detected. That is, the subsidence phenomenon can be identified as the distance between the satellite and the object being detected is increasing. On the contrary, the surface uplift can be reflected as the shortening of the distance between the object and the satellite. Through the interpretation, we learn the reactivation and recent activities of existing landslide hotspots. We can also provide an assessment of the source of soil delivery in catchment areas, as well as understand the impact of specific rainfall events on the current landslides as a basis for subsequent disaster prevention and management.

Landslide Activity Assessment

As listed in Table 5.1, the sliding velocity of a landslide can be roughly divided into seven levels, from extremely slow to extremely fast. The catalogue of event-type landslides provides the surface variation areas and the distribution of landslide places after heavy rainfall events, which are usually discontinuous in the domain of time, depending on the frequency of events. Therefore, it is only possible to know whether a particular area is a new landslide or an expansion following recent rainfall events. We can classify them according to the definition of activity introduced in the section "Event-Based Landslide Inventory." Time-series SBAS analysis can be used to analyze large-scale surface migration information by satellite-based radar imag-

Table 5.1 The class of landslide sliding velocity corresponding activity (Cruden and Varnes 1996; Lee et al. 2016)

Class	Description	Sliding velocity	
1	Extreme slow	–	–
2	Very slow	16 mm/year	0.5×10^{-6} mm/s
3	Slow	1.6 m/year	50×10^{-6} mm/s
4	Moderate	13 m/month	5×10^{-3} mm/s
5	Rapid	1.8 m/h	0.5 mm/s
6	Very rapid	3 m/min	50 mm/s
7	Extreme rapid	5 m/s	5×10^{3} mm/s

ing of a fixed return period, which can complement slope stability and movement characteristics between heavy rain events. We divide the surface displacement rate into three grades in this study, including slow (velocity > 2 cm/year), very slow (velocity = 0–2 cm/year), and non-slip sign Negligible (no PS point) by using a risk matrix (Fig. 5.3) combined with multiple-period landslide inventories and RADARSAT-2 analysis. Based on the intersection of a two-dimensional matrix, the landslide activity can be divided into active, reactivated, dormant, and stabilized levels to update the existing state of the landslide in question and clarify the potential instability features of its slope.

5.4 Results and Discussions

This section discusses the main analysis results and field validation which are related to the integrated application of InSAR, landslide inventory, airborne LiDAR, and UAV investigation, as follows.

Slope Movement Analysis of Time-Series SBAS

The ground deformation map derived from time-series SBAS analysis is shown in Fig. 5.6, in which the maps of existing deep-seated landslides, debris flows, and landslide inventories are superimposed. In SBAS analysis, artificial structures in this area are selected as reference points (fixed points) to eliminate the influence of the fluctuation of the earth. As typhoon Dujua is covered by RADARSAT-2 images applied in InSAR analysis, the current situation of the landslide and sand accumulation can be observed. As shown in Fig. 5.6, it is found that the main surface subsidence in this analysis occurs in the Xiluoan deep-seated landslide and in Laka village (in the Wulai area), while others are located in hilly areas on both sides of Pingguang Creek and downstream of the Nanshi River. It is revealed that, in the event-based landslide inventory after Typhoon Soudelor, there is a high recurrence in the activity

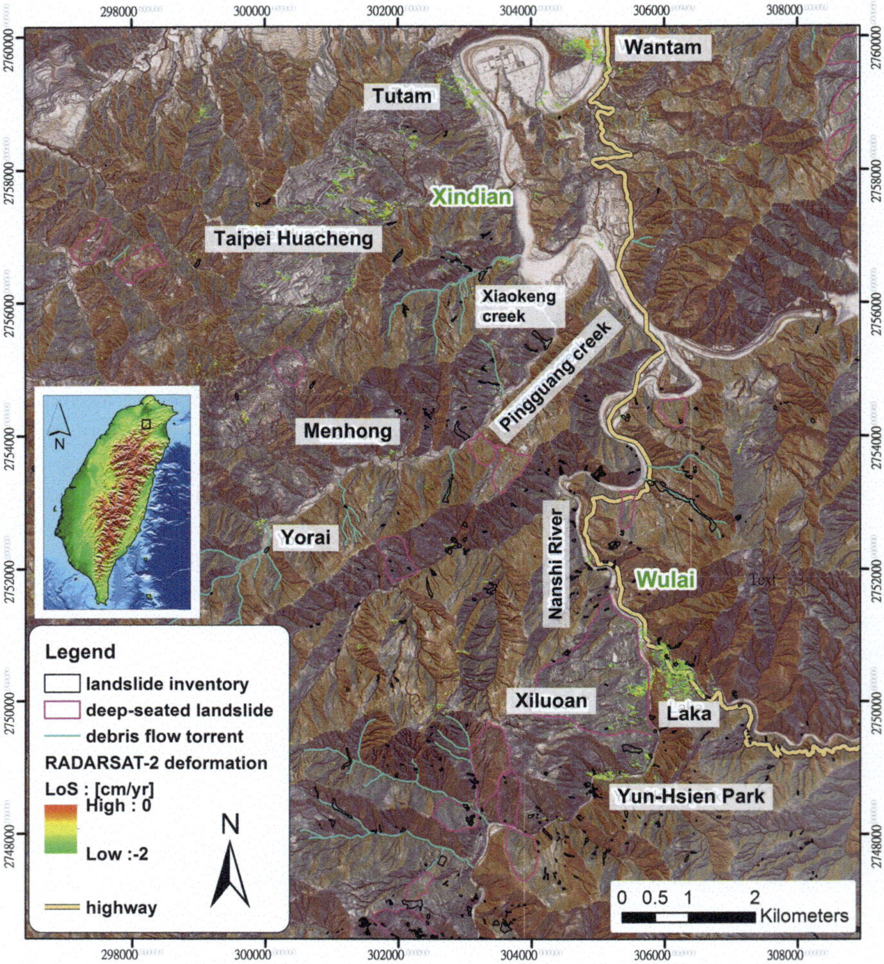

Fig. 5.6 Ground deformation (LoS) obtained from RADASAT-2 images onto the historic landslide inventory map (Aug 2015–Sep 2016)

of sediment-related hazards on both sides of the Wulai area stream. This means that the weathered surface soil in this region continues to slide along the slope after Typhoon Dujua, and even a shallow landslide caused by erosion of riverbanks and high sedimentation currents forms significant soil transportation. In order to understand the actual types of slope hazards identified through SBAS analysis, classification and validation were carried out in conjunction with the follow-up in situ survey, UAV 3D topographic modeling, and micro-topography interpretation.

Field Validation and Hazard Classification

In order to validate the InSAR results related to the movement of a mass of rock, earth, or debris flow, we investigated several landslide hotspots based on the ground deformation map from SABS analysis.

UAV (Unmanned Aerial Vehicle) photogrammetry is frequently used and provides a 3D topographic model to interpret landslide features from a bird's eye view. Although optical images taken by UAV cannot penetrate the surface vegetation as airborne LiDAR scanning does, it is still advantageous to obtain a detailed digital surface model (DSM) after disasters. In recent years, the DSM obtained by UAV 3D modeling can even use the tree canopy to evaluate the spatial distribution of old landslides under dense forests. In addition, UAV aerial photography has better spatial resolution than traditional aerial photography, and oblique photography is of great value to obtain the details of slope landform features.

Regarding the field validation, photogrammetric analysis and color orthoimage maps using aerial images have been adopted to explore the irregularity of forest or exposure area on hillslopes (software: Bentley Context Capture). The preliminary result of the validation is summarized in Table 5.2. We found that four types of geohazards can be identified, such as deep-seated landslides (DSL), shallow landslides (SL), debris flows (DF), and sediment deposition (SD) on the gully and stream bed. For example, shallow landslides, barren land, tilling trees, thin vegetation, and gully erosion are often observed in the area of dense signal response in the InSAR analysis results. Figure 5.7 is a site selected by ground surface displacement from SBAS analysis. It integrates micro-topography interpretation, UAV aerial photography, and ground survey instructions as follows.

As mentioned in the section "Slope Movement Analysis of Time-Series SBAS," after Typhoons Soudelor and Dujua hit within the study area, as shown in Fig. 5.8, the three most obvious surface displacements were located near the slope at the confluence of the Nanshi River and Tonghou Stream. Site 1 is the entrance of the Yun-Hsien Park, cable car, and hot-spring hotel area (as shown in Fig. 5.9). During the two typhoon events in 2015, the short but strong rainfall caused a shallow landslide on the side slope. After the disaster, a large amount of soil and sand accumu-

Table 5.2 Field validations of sediment-related hazard by field investigation and their topographic 548 feature

Type of sediment-related hazard		Common feature causing ground deformation
Shallow	Landslide	A group of tilling tree on the slope land; gully pattern with headward erosion; colluvium sliding on the slope surface
Deep-seated		A main steep scarp, sliding mass, and lateral boundary are obvious; tension cracks are perpendicular to sliding direction; toe bulge
Debris flow torrent		Source area with outcrop or debris slide; abundant gravel and soil deposition in the gully bed and downstream reach
Sedimentation on the stream channel		Sediment material consisted of fine grain deposits on the river bed or convex bank

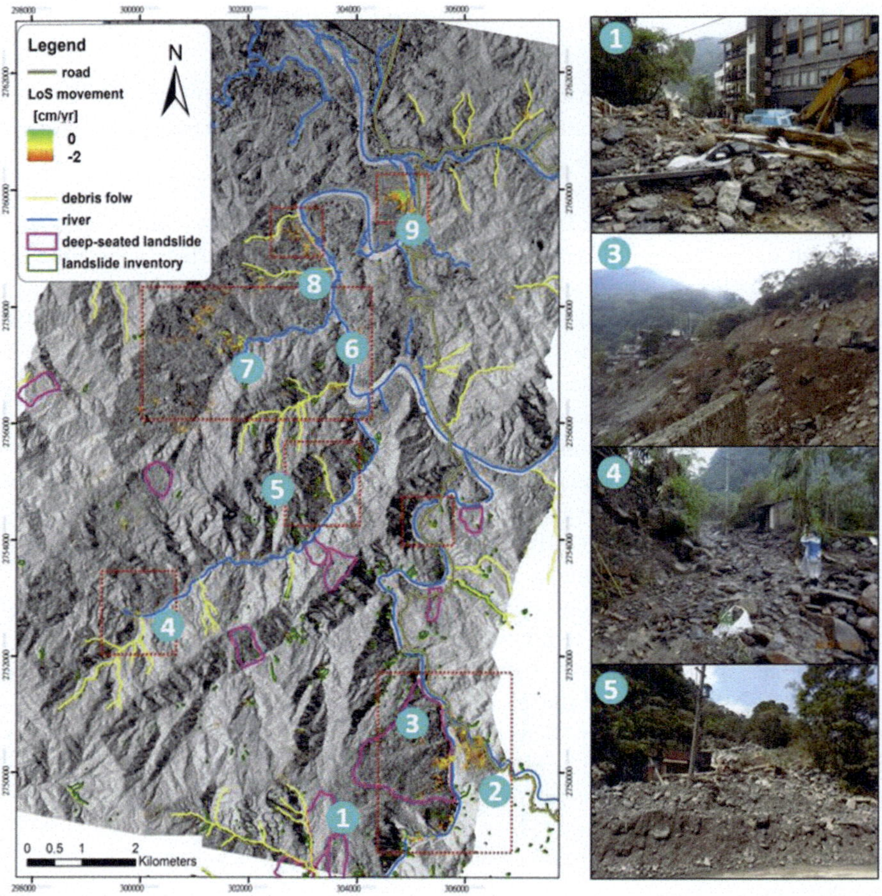

Fig. 5.7 The selected sites of field survey and corresponding disaster status during the analysis period

lated and buried the hotel lobby and blocked the two-way road traffic. The collapsed rocks from the slope also caused side bank erosion on the Nanshi River riverbed, which was consistently demonstrated in the SBAS results. The results of the InSAR analysis also reflect the current situation of debris flow and debris flood disasters caused by Typhoon Rhododendron in the upper reaches of Site 1 (Fig. 5.10). The spatial distribution of the erosion area in relation to road slope, highly agreed with the surface displacement area detected through InSAR analysis.

One ancient deep-seated landslide-prone area (called Laka DSL) was interpreted in the study area where buildings and resorts are located (Fig. 5.11a–e). The toe of Laka DSL, situated on the concave bank of Tonghou stream faces repeated scouring effects from the stream. Unstable colluvium around the wash zone may collapse and cause retrogression failure during intense rainfall. We noted that the displacement of toe failure also induced channel narrowing at the undercut slope (see the green

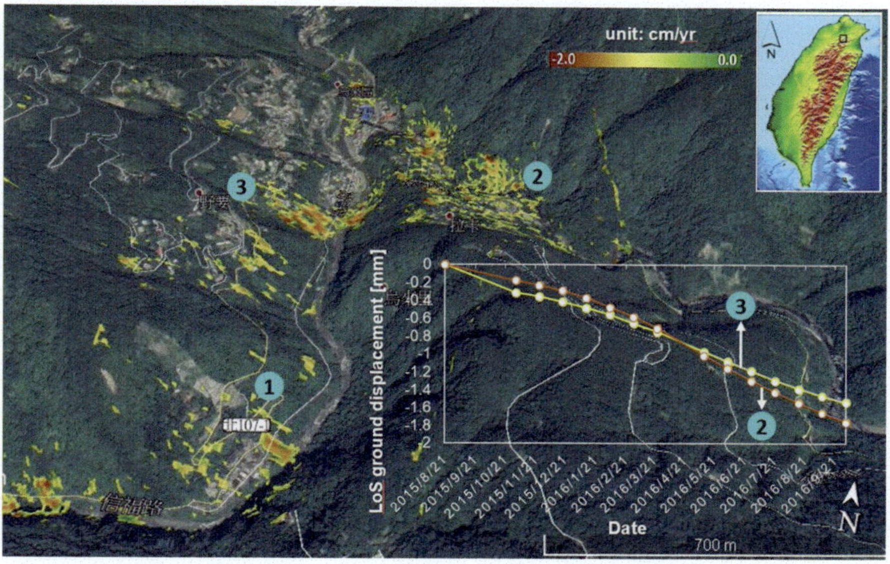

Fig. 5.8 The surface displacement map (LoS) and terrain movement velocity of SBAS for three different sites in Wulai area (basemap: Google Map)

Fig. 5.9 The surface displacement map (LoS) of SBAS and corresponding slope disaster in site 1 (basemap: UAV aerial image)

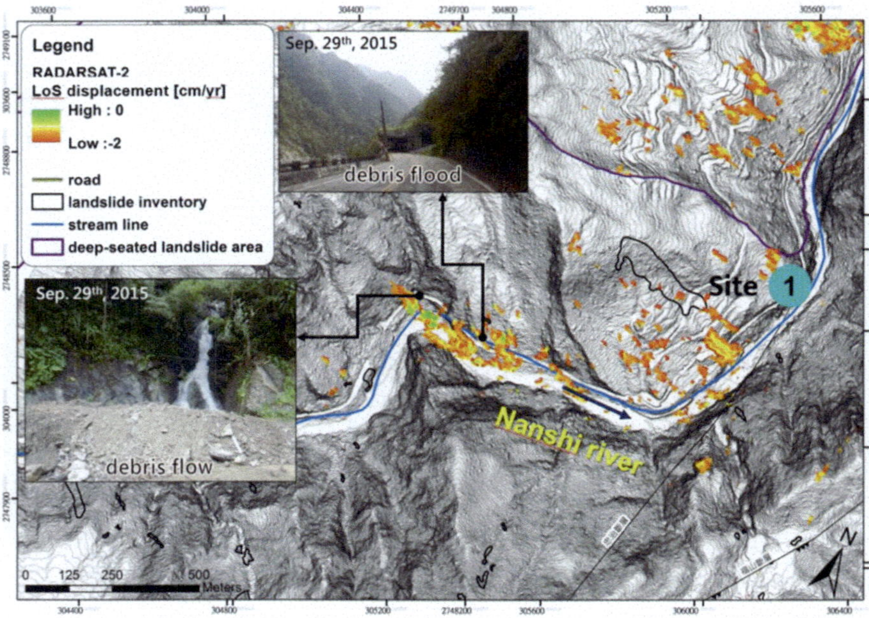

Fig. 5.10 The surface displacement map (LoS) of SBAS and corresponding slope disaster near site 1 (basemap: LiDAR shading relief)

frame in Fig. 5.11a). It was observed in the field that the spatial distribution of ground deformation point is consistent with the lateral boundary of landslide mass (Fig. 5.11d, e). These examples well illustrated the capability of the high-resolution SBAS technique in detecting slight topographic changes under the vegetation. The surface displacement of Laka DSL during Typhoons Soudelor and the Dujua was about 0.4 cm, and the cumulative deformation in the following year was 1.8 cm (until 2016/9/21; Fig. 5.8). Although the pavement rupture was repaired after the disaster, it was corroded by the high and steep concave bank of the Tonghou stream, where the main sliding body was located. It still impairs the stability of the Laka DSL toe.

Xiluoan area (site 3) is a well-known deep-seated landslide at Wulai area after Typhoon Soudelor (the area of sliding mass is approximately 30 ha). It is characterized by a deep sliding depth (~65 m) and marked as high sliding susceptibility (Fig. 5.12). From the micro-topography interpretation, we observe that within the scope of deep-seated landslides, minor scarps and slide mass can be subdivided into five parts. After Typhoon Soudelor, the surface displacement analyzed through InSAR mainly appeared at the slope toe, which was above the largest slide mass in the site, and extended to the Nanshi River. The average annual surface displacement obtained by SBAS analysis was about 1.5 cm (Fig. 5.8). Meanwhile, the official monitoring report indicated the constant sliding rate from GPS observation data was about 2 cm per year, especially in downslope (according to a report from Central

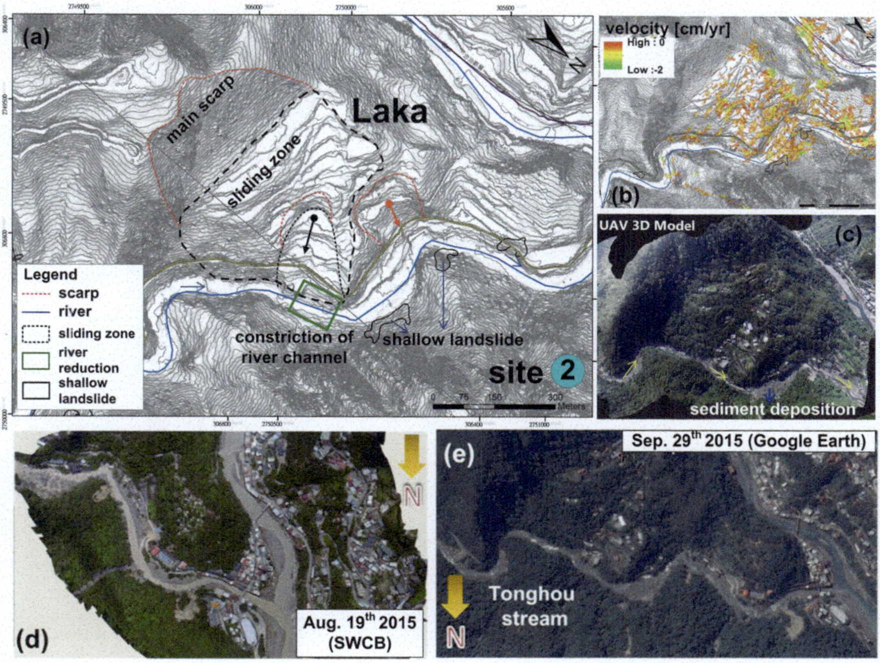

Fig. 5.11 Ancient deep-seated landslide mapping on high-resolution shading relief map in Laka village, Wulai area

Geological Survey). In this case, the similarity of creep motion between ground deformation of SBAS and GPS measurement suggests a higher sliding rate at the downslope of Xiluoan. The analysis results also confirm the fact that the Xiluoan slope top is relatively stable compared with the slope belly and toe, and there should be no significant slip in the near future.

Yorai area (site 4) is located in the upper reaches of Pingguan Creek and belongs to a potential debris flow torrent listed by Soil and Water Conservation Bureau (SWCB). In this area, two typhoon events in 2015 triggered different scale debris flow disasters (zones A and B indicated in Fig. 5.13). A large number of boulders (diameter of stone > 1 m) and mud flooded both sides of the stream confluence, forming alluvial fans on the ground. In addition to the upstream part of the Yorai area, the turbulent flow and boulders at the confluence also eroded the left bank of Pingguan Creek during heavy rains, resulting in the erosion of riverbank slopes as shown in the SBAS surface displacement map. In addition, by comparing the reconnaissance photos and LiDAR numerical topographic data after the disaster, the InSAR analysis results show a good spatial correspondence with the soil delivery area.

There is one other case that is important in confirming the capability of SBAS. This case is the recent shallow landslide site (Menhong shown in Fig. 5.14) near the midstream of Pingguang creek which demonstrates a reactivation of debris

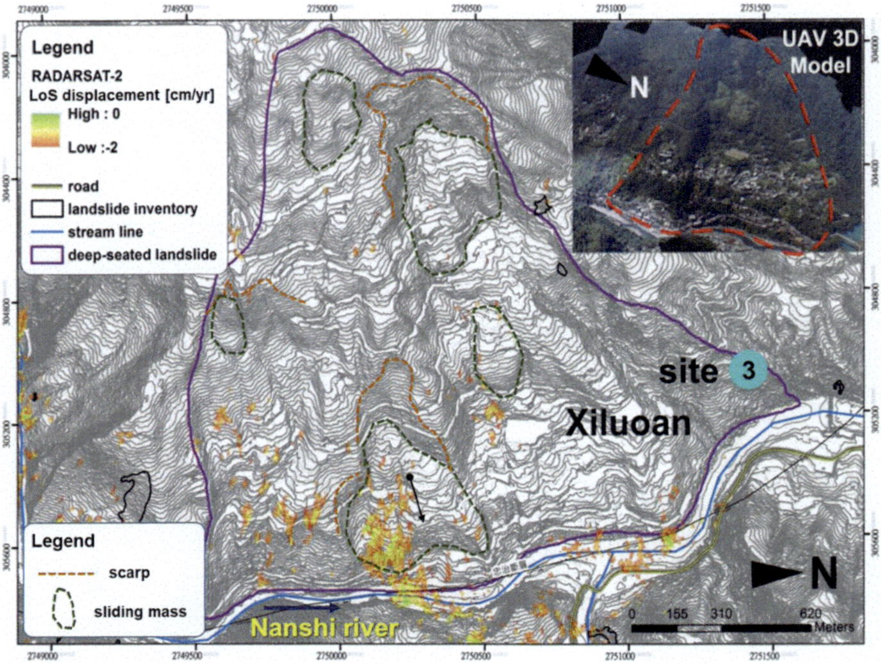

Fig. 5.12 Interpretation of Xiluoan deep-seated landslide on high-resolution shading relief and SBAS surface displacement (LoS) map in site 3, Xiluoan, Wulai area

flow that occurred within the past 20 years. According to resident interviews, this event seems to have been triggered by Typhoon Winnie (1997). As shown in Fig. 5.14, an aerial survey by UAV shows that sparse vegetation has been rehabilitated on the slope, confirming that the site is an old landslide. Through in-depth and in-situation exploration, it is found that there are tilling tree characteristics and tension crack extension lines of the surface in the collapsed area, which are suspected to be a lateral boundary of a potentially creeping landslide. The more the upslope is explored, the more mega-rocks larger than 3 m in diameter are visible, which are consistent with the early landslides that residents indicated. The hotspot, shown by SBAS analysis, is also the shallow landslide disaster caused by the typhoon in 2015.

Table 5.3 is the type of sediment-related hazards that can be identified by InSAR surface variability analysis after comprehensive site survey verifications (Fig. 5.7). It includes shallow landslides, deep-seated landslides, debris flows, and sediment deposition on riverside walls or stream beds. The preliminary evaluation results show that the good results from InSAR analysis is related to the radar image band we used. In terms of physical characteristics, although the penetration rate of C-band radar wave is limited in dense forest, the long-term intensive observation of 24-day return period, combined with the ultra-fine-descending image of 3 m high resolution, is indeed effective for regional slope soil and sand disaster monitoring after a heavy rainfall event. The hazards have good, quick screening effect, especially when the area of slope subsidence or collapse is more than 2500 m^2.

Fig. 5.13 Debris flow disaster on high-resolution shading relief and SBAS surface displacement (LoS) map in site 4, Yorai, Xindian area

Triggering Factors for Ground Deformation

The regional landslide disaster caused by Typhoon Soudelor in this study area resulted from concentrated rainfall. Figure 5.15a shows the map of the accumulated rainfall overlapping the catalogue of the Typhoon Soudelor landslides in the Nanshixi catchment area from August 6 to August 10. It is found that the rainfall center in the study area is located at the Fushan Rainfall Station (Fig. 5.15b), and the landslide occurrence locations are mostly concentrated in the accumulated rainfall areas with 500–600 mm. The rainfall pattern of this typhoon in the Nanshih River catchment area was recognized as the post-peak type. Its peak rainfall was concentrated at 5:00–07:00 on August 8, which agreed with the occurrence time of the disaster (5:00–8:00 on August 8). Most of the soil and sand cluster of disasters occurred at 1–2 h after the peak rainfall. The average rainfall intensity was 80 mm/h and the total rainfall was 789 mm. Compared with historical typhoons with similar paths, the cumulative rainfall of 1, 3, 6, and 12 h of Typhoon Soudelor is 95, 253, 442, and 655 mm, respectively, which is higher than the local historical records. Table 5.4 is the cumulative rainfall scale of Fushan Station in different recurrent periods, corresponding with the maximum cumulative rainfall of 3, 6, 12, and 24 h of Typhoon Soudelor, which shows that the maximum rainfall of 3, 6, and 12 h of Typhoon Soudelor is greater than 20 h. The maximum 24-h rainfall is less than the 25-year recurrent period. Further analysis of the relationship between the total rain-

Fig. 5.14 Landslide disaster on high-resolution shading relief and SBAS surface displacement (LoS) map in site 5, Menhong, Xindian area

fall of Typhoon Soudelor and the annual average rainfall shows that the rainfall of the typhoon event is 24.6% (R/MAP; R: cumulative rainfall [mm], MAP: maximum annual precipitation [mm]) of the annual average rainfall in the Fushan area. Therefore, the short delay of (3, 6, 12 h) heavy rainfall and high cumulative rainfall are important factors to induce regional soil and sand disasters. Figure 5.15c is the I_{ave}-D chart which considers the average intensity-rainfall delay relationship of typhoon rainfall events that caused landslides in northern Taiwan in the past. Except for typhoon Xiangshen in 2000, which deviated from the data of disaster-causing rainfall field due to the low total cumulative rainfall (about 200 mm), the other disastrous typhoon events all fell above the rainfall critical line (averaged rainfall intensity: $I_{ave} = 269D$-0.75; D: duration [h]), which can suggest directions for future research on rainfall induction. The preliminary warning for rainfall conditions of landslides in the study area can be used as a reference. In addition, from the landslide rainfall analysis of Typhoon Morakot (2009), it takes about 1200–1500 mm of long-delay rainfall to create the new deep-seated landslide. On the other hand, if it is in the existing deep-seated landslide and the cumulative rainfall is more than 750 mm, it is very likely to induce the possibility of sliding again. Therefore, a method to further evaluate the recent activities of landslides is one of the important issues to prevent large-scale landslides.

Table 5.3 Field validations of sediment-related hazard based on the result of time-series SBAS 552 analysis

No.	Location	Topography and hazard characteristic	Type of natural hazard[a]	Description (additional remark)
1	Yun-Hsien Park	Slopeland (shallow landslide and debris flow, debris flood)	SL, DF, SD	A large SLS/DF occurred in 2015
2	Laka village	Slopeland	SL, DSL	It could be a developing DSLS site
3	Xiluoan	Slopeland (deep-seated landslide)	SL, DSL	Official report shows that sliding trend for DSLS is evident
4	Yorai	Confluence of rivers (sediment deposition of debris flow)	DF, SD	DF hazard occurred twice in 2015
5	Menhong	Slopeland (shallow landslide)	SL, DF	DF event occurred in the past 20 year
6	Xiaokeng creek	Debris flow torrent	SL, DF, SD	DF event occurred in 2015
7	Taipei Huacheng village	Slopeland (shallow landslide)	SL	Dip slope area, some artificial disturbances on the hillslope
8	Tutam	Piedmont (factory buildings)	SL, SD	Sediment transportation after typhoon events (2015)
9	Wantam	Plain	SD	Sediment transportation after typhoon events (2015)

[a]*SL* shallow landslide, *DF* debris flow, *DSL* deep-seated landslide, *SD* sediment deposition (transportation) on the creek/river bed

Regional Landslide Activity Assessment

Together with all the validation results mentioned above, we created a regional landslide activity map as shown in Fig. 5.16. In which an example of landslide activity is overlapped with time-series InSAR ground deformation at Wulai during 2015/9–2016/10. Based on this, and considering the feasibility of hazard management, four different landslide activities for each terrain unit have been identified. The red terrain slope (active) reflects frequent occurrence of geohazards and remarkable surface motion in the specific stage. Actually, the competent authority can forecast the hotspot of sediment-related hazard while combining the real-time landslide rainfall early warning system (LREWS) in the operating procedures for disaster warning (Lee et al. 2016). With embedding the LREWS on web-GIS platform, we believe that landslide hazard information can be announced more accurately to the public.

Degree of hazard for building or infrastructure upon the landslide-prone area is worth mentioning. As shown in Fig. 5.17, we assigned the relative intensity of ground movement to the manmade structure when located in the creeping region.

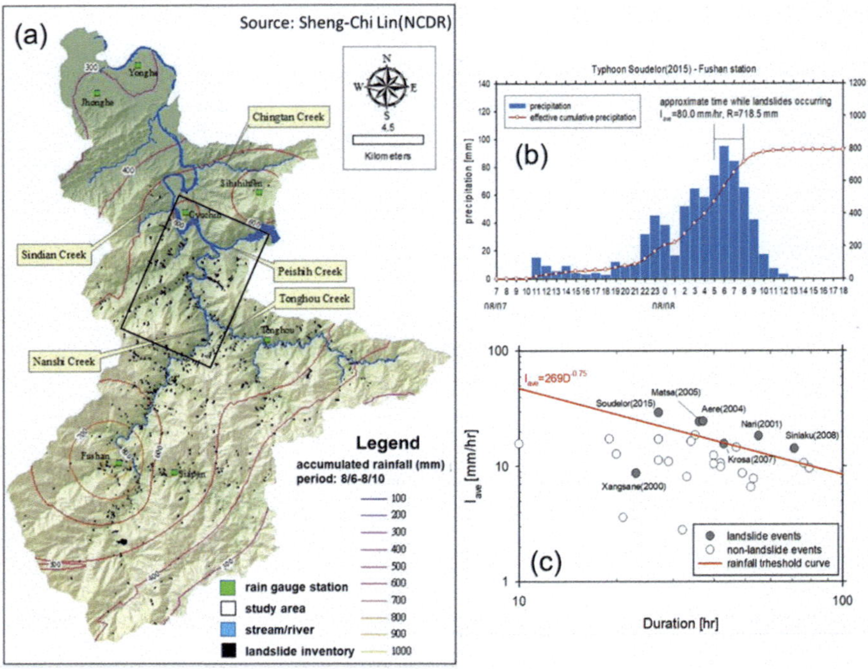

Fig. 5.15 (**a**) The cumulative rainfall contour map of Typhoon Soudelor in Nanshih River overlapped with landslide inventory; (**b**) Hourly intensity and effective cumulative precipitation for Typhoon Soudelor; (**c**) The critical rainfall line for landslide at Nanshih River watershed: the diagram of critical rainfall line with mean rainfall intensity versus duration

Table 5.4 The rainfall for different return periods at Fushan gauge and the accumulative rainfall 555 during Typhoon Soudelor (2015)

Fushan rain gauge station (N:24.7764941 E:121.5028080)						
Return period	5 year	10 year	25 year	50 year	100 year	200 year
3 h rainfall (mm)	156.3	178.3	202.8	219.2	234.3	248.4
6 h rainfall (mm)	248.1	279.6	312.2	332.8	350.9	367.3
12 h rainfall (mm)	389.2	450.0	515.7	558.6	597.4	633.2
24 h rainfall (mm)	583.2	678.7	782.0	849.4	910.6	967.0
Accumulated rainfall (mm)	3 h	6 h		12 h	24 h	
	253.0	442.5		670.0	768.5	

Once the corresponding population information, such as a list of residents is prepared, it will assist in making great progress on disaster emergency response and evacuation planning. Moreover, the decision to safeguard temporary evacuation shelters for high hazard susceptibility will be performed.

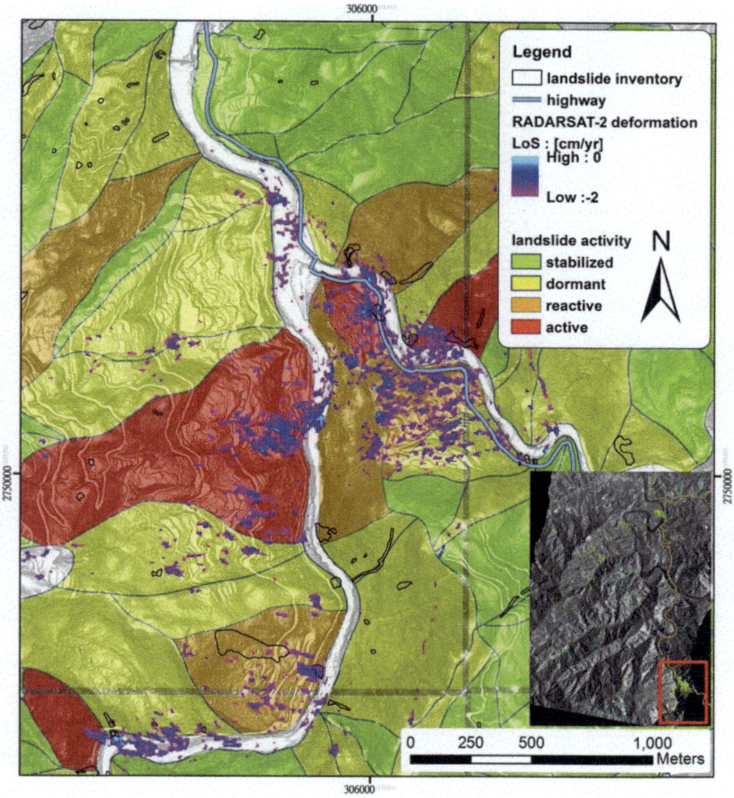

Fig. 5.16 Landslide activity map in Wulai area

5.5 Conclusions

This study proposes a time-series InSAR analysis method, which combines high-precision airborne LiDAR terrain interpretation, landslide analysis, and a UAV 3D aerial photography model to extract potential areas of mountain slope and sediment-related disasters. It also attempts to study the recent recurrent activities and topographic evolution of landslide hotspot areas.

It is important that one should begin with creeping motion if we aim to monitor in landslide-prone areas with high susceptibility. According to the model validation, it is certain that high-resolution InSAR analysis can provide definitive information of time-series ground deformation in dense vegetation areas and mountainous regions in Taiwan. In the study, some past deep-seated landslide areas such as Laka and Xilouan show a significant long-term sliding tendency without typhoon or torrential rainfall triggers. Hotspots of debris flow, pre-existing shallow landslides, and river sedimentation can be also mapped near deep-seated landslides. The proposed

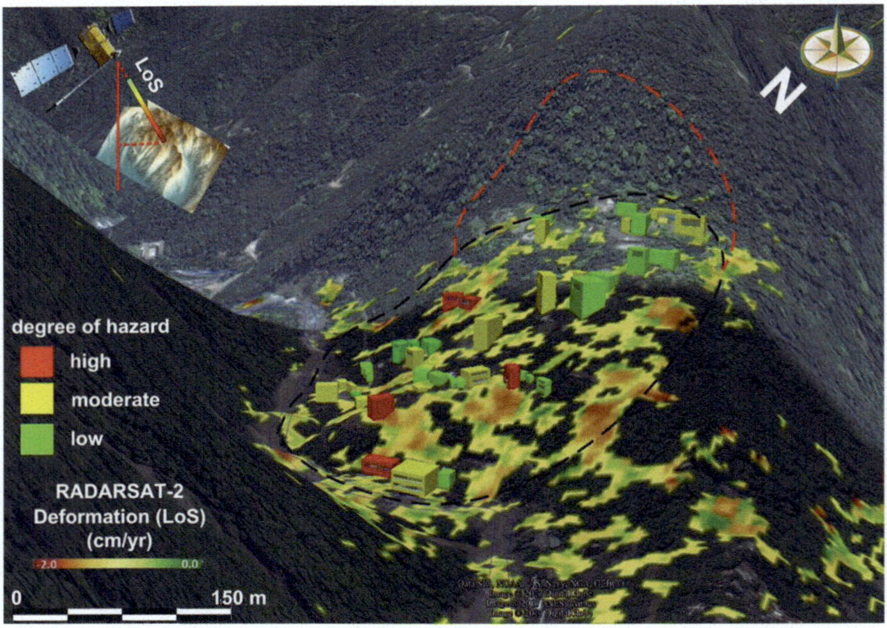

Fig. 5.17 Hazardous degree for building integrated time-series SBAS analysis at Laka area

assessment approach which integrates landslide frequency and terrain units can contribute to the hazard prevention to determine geohazard monitoring sites.

In the application we use InSAR analysis and evaluation to determine the potential sliding type, sliding rate, and failure mechanism of slope hazards before a drilling operation or monitoring system has been introduced. After the InSAR analysis, we recommended field investigation should be performed to confirm the ground surface movement. In addition, using the corner reflectors on low coherence vegetated areas would improve our monitoring of known large-scale landslide sites.

Acknowledgments The authors sincerely thank Central Geological Survey, MOEA (Taiwan) for providing us access to the high-resolution shaded reliefs. The authors thank Canada Centre for Remote Sensing for conducting the InSAR analysis and Sinotech Engineering Consultants, Inc. for financial support on this study. We also acknowledge support from the Ministry of Science and Technology, Taiwan (Grant No. MOST 106-2420-H-004-015- 406MY3).

References

Bonì, R., M. Bordoni, C. Meisina, A. Colombo, and L. Lanteri. 2017. Intergration of multi-sensor a DinSAR data for landslide inventory update. In *Advancing Culture of Living with Landslides. WLF 2017*, ed. M. Mikos, B. Tiwari, Y. Yin, and K. Sassa, 133–139. Cham: Springer.

Bovenga, F., G. Pasquariello, R. Pellicani, A. Refice, and G. Spilotro. 2017. Landslide monitoring for risk mitigation by using corner reflector and satellite SAR interferometry: The large landslide of Carlantino (Italy). *Catena* 151: 49–62.

Bozzano, F., P. Caporossi, C. Esposito, S. Martino, P. Mazzanti, S. Moretto, G.S. Mugnozza, and A.M. Rizzo. 2017. Mechanism of the Montescaglioso Landslide (Southern Italy) inferred by geological survey and remote sensing. In *Advancing Culture of Living with Landslides. WLF 2017*, ed. M. Mikos, B. Tiwari, Y. Yin, and K. Sassa, 97–106. Cham: Springer.

Calvello, M., D. Peduto, and L. Arena. 2017. Combined use of statistical and DInSAR data analyses to define the state of activity of slow-moving landslides. *Landslides* 14 (2): 473–489.

Center for Research on the Epidemiology of Disasters. 2015. The human cost of weather-related disasters 1995–2015. In *The United Nations Office for Disaster Risk Reduction*, 24.

Ciampalini, A., F. Raspini, S. Bianchini, W. Frodella, F. Bardi, D. Lagomarsino, A. Traglia, et al. 2015. Remote sensing as tool for development of landslide databases: The case of the Messina Province (Italy) Geodatabase. *Geomorphology* 249: 103–118.

Cruden, D.M., and D.J. Varnes. 1996. Landslide Types and Processes, Transportation Research Board, U.S. National Academy of Sciences, Special Report, 247: 36–75.

García-Davalillo, J., G. Herrera, D. Notti, T. Strozzi, and I. Á lvarez-Fernández. 2014. DInSAR analysis of ALOS PALSAR images for the assessment of very slow landslides: The Tena Valley Case Study. *Landslides* 14 (2): 225–246.

Gariano, S.L., and F. Guzzetti. 2016. Landslides in a changing climate. *Earth-Science Reviews* 162: 227–252.

Herrera, G., F. Gutiérrez, J.C. García-Davalillo, J. Guerrero, D. Notti, J.P. Galve, J.A. Fernández-Merodo, and G. Cooksley. 2013. Multi-sensor advanced DInSAR monitoring of very slow landslides: The Tena Valley Case Study (Central Spanish Pyrenees). *Remote Sensing of Environment* 128: 31–43.

Kokalj, Ž., and R. Hesse. 2017. *Airborne Laser Scanning Raster Data Visualization: A Guide to Good Practice*, 88. Ljubljana: Založba ZRC.

Lee, C.F., C.M. Huang, T.C. Tsao, L.W. Wei, W.K. Huang, C.T. Cheng, and C.C. Chi. 2016. Combining rainfall parameter and landslide susceptibility to forecast shallow landslide in Taiwan. *Geotechnical Engineering Journal of the SEAGS & AGSSEA* 47 (2): 72–82.

Lee, C.F., W.K. Huang, C.M. Huang, and C.C. Chi. 2017. Deep-seated landslide mapping and geomorphic characteristic using high resolution DTM in Northern Taiwan. In *Advancing Culture of Living with Landslides*, ed. M. Mikos, B. Tiwari, Y. Yin, and K. Sassa, vol. 2, 767–777. Cham: Springer.

Lo, C.M., C.F. Lee, and J. Keck. 2017. Application of sky view factor technique to the interpretation and reactivation assessment of landslide activity. *Environmental Earth Sciences* 76: 375.

Martire, D.D., S. Tessitore, D. Brancato, M.G. Ciminelli, S. Costabile, M. Costantini, G.V. Graziano, F. Minati, M. Ramondini, and D. Calcaterra. 2016. Landslide detection integrated system (LaDIS) based on In-situ and satellite SAR interferometry measurements. *Catena* 137: 406–421.

Mateos, R.M., J.M. Azañón, F.J. Roldán, D. Notti, V. Pérez-Peña, J.P. Galve, J.L. Pérez-García, et al. 2017. The combined use of PSInSAR and UAV photogrammetry techniques for the analysis of the kinematics of a coastal landslide affecting an urban area (SE Spain). *Landslides* 14 (2): 743–754.

Oliveira, S.C., J.L. Zêzere, J. Catalão, and G. Nico. 2014. The contribution of PSInSAR interferometry to landslide hazard in weak rock-dominated areas. *Landslides* 12 (4): 703–719.

Raspini, F., A. Ciampalini, S.D. Conte, L. Lombardi, Massimiliano Nocentini, Giovanni Gigli, Alessandro Ferretti, and Nicola Casagli. 2017. Mapping rapid-moving landslide with satellite SAR images: The case of Montescaglioso (South Italy). In *Advancing Culture of Living with Landslides. WLF 2017*, ed. M. Mikos, B. Tiwari, Y. Yin, and K. Sassa, 171–177. Cham: Springer.

Righini, G., V. Pancioli, and N. Casagli. 2012. Updating landslide inventory maps using persistent scatterer interferometry (PSI). *International Journal of Remote Sensing* 33 (7): 2068–2096.

Rosi, A., V. Tofani, L. Tanteri, C.T. Stefanelli, A. Agostini, F. Catani, and N. Casagli. 2017. The New Landslide Inventory of Tuscany (Italy) updated with PS-InSAR: Geomorphological features and landslide distribution. *Landslides* 15: 5–19.

Schlögel, R., B. Thiebes, M. Mulas, G. Cuozzo, C. Notarnicola, S. Schneiderbauer, M. Crespi, A. Mazzoni, V. Mair, and A. Corsini. 2017. Multi-temporal X-Band radar interferometry using corner reflectors: application and validation at the Corvara Landslide (Dolomites, Italy). *Remote Sensors* 9: 739.

Singhroy, V., F. Charbonneau, C. Froese, and R. Couture. 2012. Guidelines for InSAR monitoring of landslides in Canada. In *Landslides and Engineered Slopes*, ed. Eberhardt et al., 1281–1287. Boca Raton, FL: CRC Press.

Singhroy, V., and F.J. Charbonneau. 2014. RADARSAT: science and applications. *Physics in Canada* 70 (4): 212–217.

Singhroy, V., and K. Molch. 2004. Characterizing and monitoring rockslides from SAR techniques. *Advances in Space Research* 33 (3): 290–295.

The International Geotechnical Society. 1993. *Multilingual Landslide Glossary*, 17–18. Richmond, BC, Canada: Bitech Publishers Ltd.

The World Bank. 2005. *Natural Disaster Hotspots a Global Risk Analysis*, 3–14. Washington, DC: World Bank.

Vecchiotti, F., D. Peduto, and T. Strozzi. 2017. Multi-sensor a priori PSI visibility map for nationwide landslide detection in Austria. In *Advancing Culture of Living with Landslides. WLF 2017*, ed. M. Mikos, B. Tiwari, Y. Yin, and K. Sassa, 45–52. Cham: Springer.

Ventisette, C.D., G. Righini, S. Moretti, and N. Casagli. 2014. Multitemporal landslides inventory map updating using spaceborneSAR analysis. *International Journal of Applied Earth Observation and Geoinformation* 30: 238–246.

Zakšek, K., K. Oštir, and Ž. Kokalj. 2011. Sky-view factor as a relief visualization technique. *Remote Sensors* 3: 398–415.

Chapter 6
Remote Sensing of Deformation and Disturbance to Monitor and Assess Infrastructure in Urban Environments

Andrea Donnellan

6.1 Introduction

Remote sensing can provide rapid and synoptic observations of land surface processes for monitoring and assessing infrastructure and urban environments. Remote sensing can facilitate hazard assessment, forecasting of disasters, and possible detection of initial or imminent failure of the land surface or infrastructure. Understanding hazards can be used to predict how disasters may cascade into other disasters or damaging events. Various remote sensing techniques can be used to observe different processes that affect infrastructure and built environments. Integrating multiple data types can help improve understanding, assessment, and forecasts by sampling different aspects of the processes that lead to disasters. Data products from various remote sensing techniques can be non-intuitive to interpret, however. Simplifying data products and presenting them in map-based tools that allow overlays of multiple products eases interpretation for non-specialists in the various techniques. This chapter focuses on land surface and solid Earth processes that pose hazards or lead to disasters that can impact urban infrastructure and environments and remote sensing techniques that are applicable to observe and understand them.

6.2 Cascading Disasters

Earthquake, fire, landslides, and debris flows all pose a threat to urban environments and infrastructure. The individual hazards can interact leading to disaster cascades that can increase loss of life and property (Fig. 6.1). For example, earthquakes can

A. Donnellan (✉)
Jet Propulsion Laboratory, California Institute of Technology, Pasadena, CA, USA
e-mail: andrea.donnellan@jpl.nasa.gov

V. Singhroy (ed.), *Advances in Remote Sensing for Infrastructure Monitoring*, Springer Remote Sensing/Photogrammetry,
https://doi.org/10.1007/978-3-030-59109-0_6

cause fires and landslides; wildfires followed by heavy rains can cause debris flows. Aging infrastructure such as old and fragile water or gas lines can be affected by land surface disruption from natural disasters. The number of water or gas main breaks from an earthquake, for example, can be exacerbated by weak or eroding infrastructure. This chapter focuses on earthquakes as the main natural disaster, but volcanoes pose similar threats and can cascade into similar disasters.

Emergency responders must make rapid decisions based on clear and concise information. Analysis and accurate interpretation of remotely sensed observations often requires domain experts who have deep experience. A challenge is to bring the strengths of remote sensing observations to practitioners in other fields in an intuitive and often simplified form. For example, products displaying fewer colors presented in red, yellow, and green can be more useful for emergency response than products showing subtle gradations of colors. The latter is useful for science interpretation, but less useful when resources must be prioritized rapidly following a disaster. Lower accuracy quick look products can be more valuable to disaster response than precise science quality products. Disasters that pose a threat of cascading into other disasters require rapid analysis of the event and other factors that may contribute to the follow-on disaster.

6.3 Remote Sensing Methods

Remote sensing has the advantage of providing a synoptic view of an affected area from which a general overview can be derived. Orbiting remote sensing platforms can provide global access with a broad field of view. The latency of products depends on the revisit time of the platform in general, its proximity to the area to be

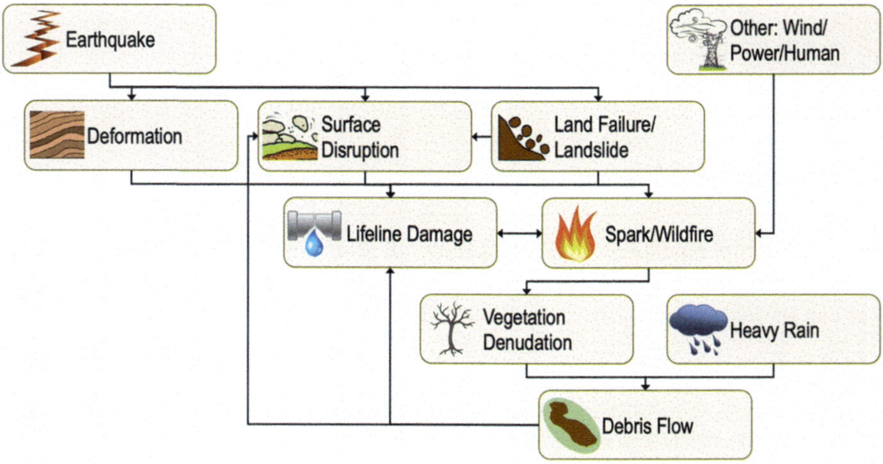

Fig. 6.1 Disasters are often interrelated and can cascade from one disaster to another

targeted, the downlink, and data processing capability for the resource or instrument. Airborne platforms may provide higher resolution products and the ability to provide targeted and configurable observations, but the availability of the platform may be an issue. Small Uninhabited Aerial Systems (sUAS) or drones can provide extremely high-resolution targeted products. However, airspace may not be available, or the area affected by the disaster might be too large to be adequately covered by near-surface observations. Both the spatial and temporal scale as well as product latency should be considered when evaluating a remote sensing technique for monitoring or assessing a land surface process or the impact on the environment.

Observations

The remote sensing techniques discussed here cover two broad categories of observations: surface deformation and disturbance. Deformation results from forces acting within the Earth, which change the shape of or distort the land surface. Deformation near the land surface can affect infrastructure. It can cause natural disasters such as earthquakes or be indicative of an impending natural disaster such as inflation before a volcanic eruption. Disturbance is the disruption of the land surface. Wildfires, debris flows, floods, and earthquakes can all cause surface disturbance.

Technologies

Global Navigation Satellite System (GNSS), Interferometric Synthetic Aperture Radar (InSAR) from NASA's airborne UAVSAR or various spaceborne platforms, and optical imagery through short-wavelength infrared (SWIR) have been used to observe infrastructure and various cascading disasters in California. Each technique provides differing resolution, coverage, and measurements, which address different aspects of hazards, disasters, and urban infrastructure (Table 6.1). This section describes technologies that are applicable to monitoring urban infrastructure. The next section describes various hazards to infrastructure and urban environments that the remote sensing methods can be applied to monitoring deformation and/or disturbance related to the hazard.

GNSS

Global Navigation Satellite System (GNSS) refers to any satellite constellation that provides positioning, timing, and navigation. The Global Positioning System (GPS) is the US navigation and positioning system. The satellites broadcast time and spacecraft position on a known frequency. From this position and time delay

Table 6.1 Typical measurements, accuracy, resolution, coverage, and application. Y = Yes, M = Maybe, N = No. Resolution is dependent on the sensor quality. Particularly for airborne measurements, resolution and field of view or coverage is dependent on the altitude of the platform

Method	Displacement, change resolution	Ground resolution	Coverage	Deformation	Disruption	Offsets	Fire
Fieldwork	cm–m	Local	Local	N	Y	Y	M (post event)
GNSS	mm, mm/year	≥10 km	Global	Y	N	M	N
UAVSAR	cm, cm/year	1–7 m	15 × 90 km	Y	Y	Y	M (polarimetry?)
InSAR	cm, cm/year	10–100 m	≥100 × 100 km	Y	Y	Y	Y
Imaging	≥10 cm (spaceborne)	≥30 cm–1 m (spaceborne)	Global	N	M	Y	M (SWIR)
Lidar	10 cm, 10 cm	1 m	1 km × 10 km	N	N	M	N
sUAS imaging	≤2 cm, 2 cm	2 cm	1 × 1 km	M	Y	Y	N

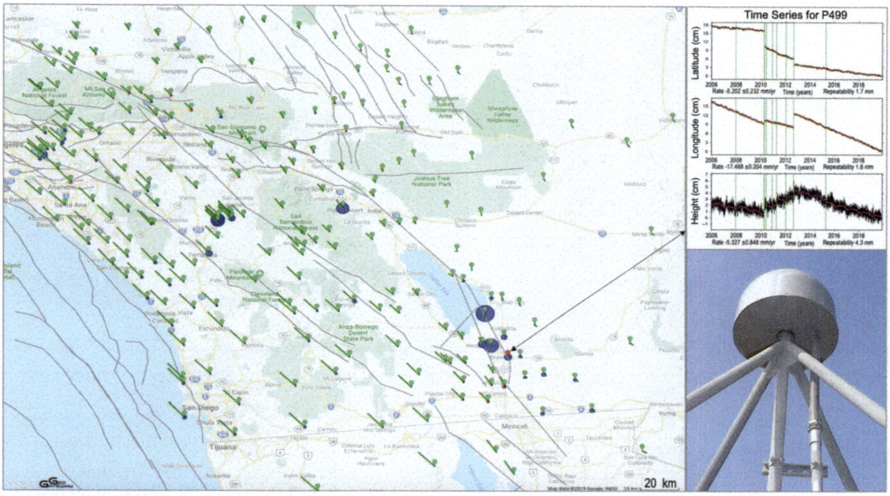

Fig. 6.2 Continuous GPS velocities for stations in southern California relative to P499 shown as the red icon. The Pacific Ocean is to the west. The US/Mexican border is marked by a light near horizontal line in the southern part of the image. The Salton Sea is the inland water body. The tail of the arrow is at the icons. The head has an error ellipse, but the velocities are so precise that the ellipses are too small to be seen. Time series for station P499 is shown in the upper right corner and a typical drilled braced GNSS station is shown in the lower left. Velocities are generated in GeoGateway (http://geo-gateway.org using the algorithm of Heflin et al. 2020)

between the spacecraft and a point on the ground the distance between the two can be computed. Distances to multiple satellites can be combined to compute a precise position and time of the receiver on the ground. GPS measurements first started being used for scientific measurement in the mid-1980s. A few continuous GPS stations were first deployed in the early 1990s as the California Permanent GPS Geodetic Array (PGGA; Lindqwister et al. 1991). Displacements from the Landers earthquake sequence were detected using the array (Bock et al. 1993). Today, precise GPS measurements can also be used to measure crustal deformation to a precision of about 1 mm/year for north, east, and up components of motion (Fig. 6.2).

InSAR

Interferometric Synthetic Aperture Radar (InSAR) bounces radar waves off the surface of the Earth from orbiting or airborne platforms (Fig. 6.3). An advantage of radar is that it is all-weather, can see through clouds, and can image at night. InSAR measurements require pairs of images from which an interferogram can be constructed. Creating interferograms that accurately reflect motion of the surface of the Earth requires precise repeat of the first and second passes of the instrument. Knowledge of the time of transmission as well as position of the instrument allows the phase of the radar wave from each pass to be compared. If the instrument is in exactly the same position on the first and second passes and the ground does not

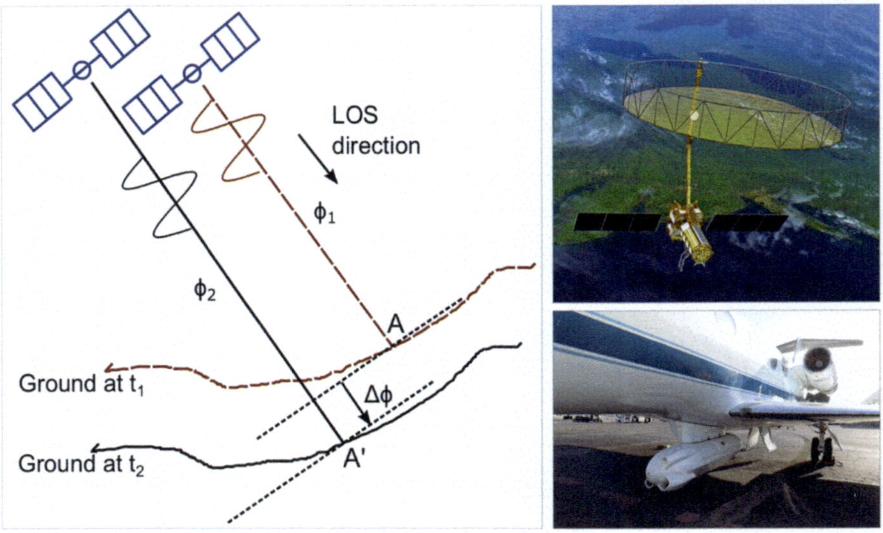

Fig. 6.3 Interferometric Synthetic Aperture Radar (InSAR) measures the phase difference between image pairs. If the positions between the first and repeat pass of the instrument are the same and the ground moves, the phase difference reflects ground surface motion (NISAR 2018). The phase difference and the wavelength of the radar signal can be used to determine the line-of-sight ground surface change between the two passes. Upper right: NISAR design. Bottom right: UAVSAR pod mounted to the bottom of a Gulfstream-III aircraft

move, then the waves for each pass would overlay directly. However, if the ground does move, then the radar waves would be slightly offset. This phase offset and the wavelength of the radar instrument can be used to compute line-of-sight displacements between a point on the ground and the instrument. InSAR antenna points sideways, perpendicular to the path of the instrument platform, at an angle to the ground of typically about 30–60°. The line-of-sight range change measurement, then, is neither horizontal or vertical, but for an oblique angle relative to the ground. Multiple looks make it possible to decompose the measurement into vertical and horizontal motions.

Spaceborne InSAR missions have been flown since the early 1990s when the first image of an earthquake was produced for the 1992 M7.3 Landers earthquake in southern California (Massonnet et al. 1993). Data collected before and after the earthquake were used from the European Radar Satellite ERS-1. ERS-1 and 2 were C-band radars. Radars used to measure surface deformation from interferometry are typically C-band with a wavelength of 3.75–7.5 cm or L-band with a wavelength of 15–30 cm. The upcoming NASA-ISRO Synthetic Aperture Radar (NISAR) mission, scheduled for launch in early 2022 will have both L-band and S-band capability. S-band has a wavelength of 7.5–15 cm. Range direction line-of-sight displacement accuracy is typically about 1 cm for InSAR measurements. Unmodeled water vapor path delay can degrade the measurement by up to 5 cm. InSAR measurements also depend on a high correlation of the radar waves for the first and second passes to form an interferogram. Where the surface is too disrupted the interferogram decorrelates, leaving gaps in the interferogram. As shown in examples below, decorrelation can create data gaps or be used as an indicator of damage or burn area.

If the ground does not move and there is an offset of the instrument between passes, then topography can be determined from the measurements. In this case, the range differences from a point on the ground to two different locations of the aircraft allow for stereo imaging of topography using radar. The Shuttle Radar Topography Mission (SRTM) had an antenna in the Space Shuttle payload bay and a second antenna on a mast deployed 60 m out. The mission flew in 2000 and produced 30 m topographic postings over most of the Earth surface from 54°S to 60°N latitude. Data are available from the United State Geological Survey (USGS) EarthExplorer (https://earthexplorer.usgs.gov/). The German mission, Tandem-X, is a set of satellites that began producing 12 m posting digital elevation models with 2–4 m height accuracy globally. WorldDEM is a commercial product produced from Tandem-X.

UAVSAR is NASA's Airborne Interferometric Synthetic Aperture Radar Platform. UAVSAR is currently flown on a piloted Gulfstream-III aircraft with a precision autopilot. The UAVSAR data are acquired from 12.5 km altitude. Repeat pass airborne interferometry requires the aircraft to fly within 5 m between the two flights. UAVSAR has been operational since 2009. It has extensively covered California, in particular the San Andreas Fault system and southern California. It has also covered parts of the Gulf coast near Houston, Texas and New Orleans, Louisiana, part of the Mississippi River, Florida, New England, Haiti, the Aleutians, Greenland, Iceland, and glaciers in the Andes of South America.

Synthetic Aperture Radar (SAR) can produce an image from a single pass. SAR is particularly useful when the radar wave is polarized. Waves can be vertically, horizontally, or circularly polarized when transmitted. The ground scatters the radar waves, and different surface characteristics polarize the waves differently. For example, leaves on trees that are moving in the wind scatter the waves in a random way. Strongly aligned surface features may change the returned polarization to reflect the waves in a consistent manner. For this reason, SAR polarimetry can be used to determine characteristics of the Earth surface.

Imaging

Optical imaging provides a direct image of the ground surface in any number of spectral bands. Visible imaging provides an intuitively interpretable image of the ground surface, though the resolution and field of view can dictate its usefulness for understanding infrastructure or damage. Visible wavelengths are 0.45–0.48 μm for blue, 0.5–0.565 μm for green, and 0.625–0.74 μm for red colors. Longer wavelength spectral bands, such as short-wavelength infrared imaging (SWIR), at 0.9–1.7 μm, provide high contrast imaging. Water vapor, smoke, and fog appear transparent at SWIR wavelengths, making it possible to image wildfire fronts or reduce haze in

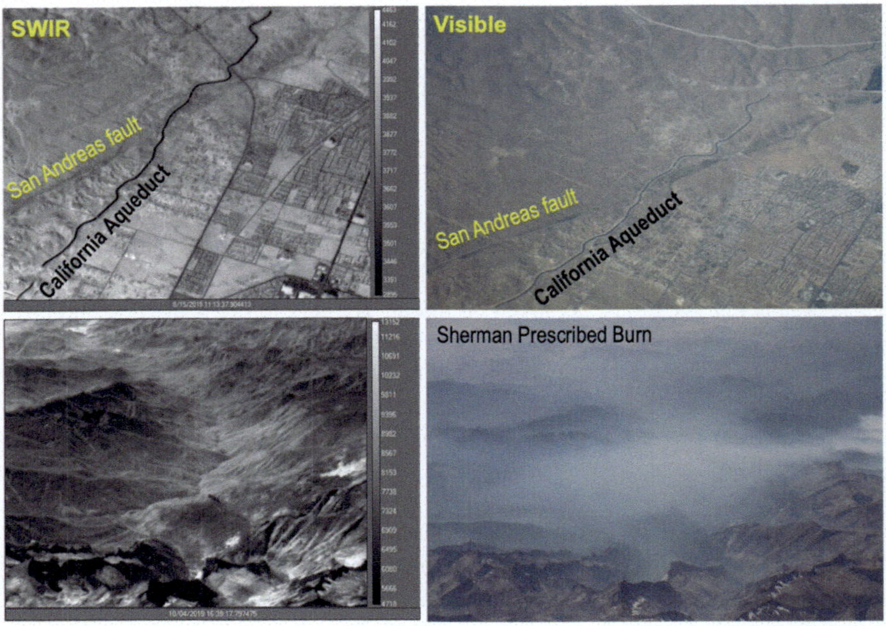

Fig. 6.4 SWIR and visible images of the San Andreas Fault and California Aqueduct near Palmdale California (top) and for the Sherman prescribed burn near Giant Forest, Sequoia National Park (bottom). The contrast is higher in the SWIR image, which also sees through the smoke from the prescribed burn. The visible image was hazier when the data were collected on August 15, 2019 (top) and smoky near Sequoia National Park on October 4, 2019 (bottom). Ground sample distance of the SWIR images is about 9 m and visible images collected out the side window about 2 m from an altitude of 12.5 km. Stereo PhotogrammetryStructure about 2 m from an altitude of 12.5 km

images (Fig. 6.4). There are numerous missions that provide data at a variety of spectral bands, which can be processed in various ways to highlight surface properties.

Stereo Photogrammetry

Topography can be calculated from two or more images of a subject from different vantage points. Topographic reconstruction from many images collected at different vantage points is accomplished with a stereo photogrammetric technique called Structure from Motion (SfM; Fig. 6.5). SfM has the advantage of reducing the vertical uncertainty so that vertical and horizontal resolutions are comparable. SfM measurements can also be suitable to imaging steep slopes. The method takes advantage of geometric constraints provided by image frames, as opposed to images produced by a line array, which is common for spaceborne pushbroom imagers with line array sensors. Data products include point clouds where 3D positions and color information are assigned to each point, digital surface models (Zhou 2016), and orthomosaic images. The flight path for collecting multiple images can be unstructured,

Fig. 6.5 Structure from motion (SfM) measurements are carried out by observing a target from multiple vantage points (left). In this example a small UAS (sUAS) or drone flies around two rock pinnacles collecting images from different positions and elevations (left). The resulting 3D model depicts the target of interest as a point cloud, digital surface model, and orthomosaic image. In this example a boulder fell from the top of the pinnacle during the 2019 Ridgecrest earthquake sequence

making rapid response and data collection fairly easy. Cameras are often flown on sUAS or can be flown on airplanes. sUAS flights in disaster areas must be coordinated with response and government agencies, however, and are often difficult to collect during an emergency.

Lidar

Light Detection and Ranging (lidar) uses laser scanning of a surface to produce highly accurate topography of the ground surface. Lidars are commonly flown on aircraft and terrestrial laser scanners are also used to develop 3D representations. 3D positions of points in the cloud are also assigned intensity values of the reflected light. Multiple returns occur from scatterers throughout the lidar shot, such as from vegetation and the ground, so it is possible to produce a digital surface model from the point cloud as well as a bare earth model (Fig. 6.6). This improves interpretation of faults of other geomorphic features.

Data Product Formats and Access Interfaces

In the following section we describe data products and their use in various examples. Several interfaces allow access to data products that can be useful for analysis. The interfaces described here provide a start for obtaining data products, but the list is not comprehensive and new interfaces are expected to come online in the future. Interfaces exist for archives of primary data sets or products. Higher level products enable easier interpretation for practitioners outside the specialist discipline of the data and several interfaces provide these products. Sites mentioned here provide examples of types of products that can be accessed, but the list is not exhaustive and new products and interfaces regularly become available.

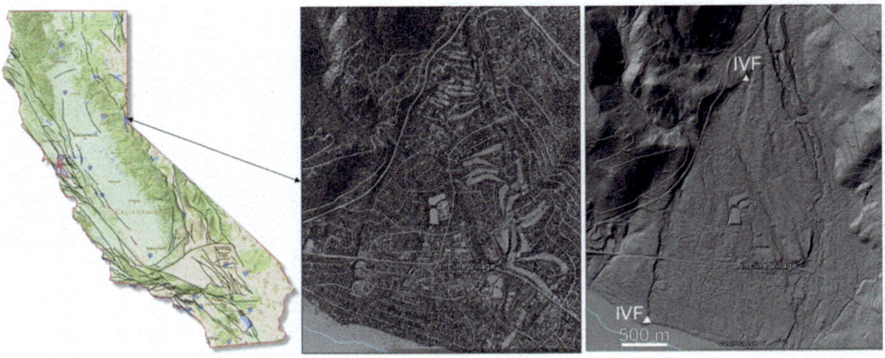

Fig. 6.6 Lidar image of Incline Village in Nevada at Lake Tahoe on the north side. Center image shows shaded relief full feature image that includes the built environment. Right image shows bare Earth model. White triangles mark the Incline Village fault (IVF) scarp. Data provided by the Tahoe Regional Planning Agency (https://doi.org/10.5069/G9PN93H2) and images created by Ramon Arrowsmith for Donnellan et al. (2016)

The authoritative source for NASA Earth Science data products are the NASA Distributed Active Archive Centers (DAACs, https://earthdata.nasa.gov/eosdis/daacs). NASA mission and field data are distributed across DAACs that host different types of data. The goal of the DAACs is to provide reliable and robust data services to both expert and non-expert users. There are currently twelve NASA Earth Science DAACs hosting a variety of data products.

The Advanced Rapid Imaging and Analysis (ARIA, https://aria.jpl.nasa.gov/) project distributes GNSS products and derived data products from Sentinel 1 A/B. In the future it will distribute products from the NASA—ISRO Synthetic Aperture Radar (NISAR) mission, slated for launch in the 2022 or 2023 timeframe. The goal of ARIA is to bring geodetic imaging capabilities to an operational level in support of local, national, and international hazard science and response communities. Data products can be downloaded from https://aria-products.jpl.nasa.gov/.

The United States Geological Survey (USGS) partners with NASA on some of the DAACS. The USGS also hosts The National Map (https://www.usgs.gov/core-science-systems/national-geospatial-program/national-map) as part of the USGS National Geospatial Program. The USGS partners with other Federal as well as State and local partners to deliver topographic data. Users can use the National Map Viewer to view and download landcover, topographic geologic, and other layers. For the topographic data users can display various hillshaded data, slope maps, contour maps, etc.

OpenTopography (http://opentopography.org) facilitates access to high-resolution topographic data. Users can both upload their own data products and download posted products. OpenTopography provides KML formatted products, raster digital elevation model (DEM) layers, common Geographic Information System (GIS) formatted products, and lidar and structure from motion point clouds. OpenTopography also provides on-demand processing for users to select their region of interest and generate custom derived products such as DEMs or point clouds to a specified resolution.

Science Gateways allow science and engineering communities access to shared data, software, computing services, and other discipline specific resources. GeoGateway (http://geo-gateway.org) is a web map-based science Gateway for modeling, analysis, and response of crustal deformation and Earth surface observations. The focus is on earthquake fault systems, though wildfire and debris flow products from common data have also been posted to GeoGateway. Users can overlay multiple data products and models. Data products include UAVSAR airborne Interferometric Synthetic Aperture Radar (InSAR) observations, Global Navigation Satellite System (GNSS) time series and vectors, earthquake faults, seismicity, and custom KML products. The goal is to simplify routine workflow, making observations and derived products available for further custom analysis.

6.4 Hazards and Urban Environments

Several examples demonstrate how remote sensing can address components of the hazard cycle and hazard cascades. The scenarios outlined here provide case studies representative of how hazards and urban infrastructure can be addressed with remote sensing in general. While the focus is on California scenarios the techniques can be applied to other disasters such as volcanic eruptions or flooding.

Earthquakes

Remote sensing, and in particular geodetic imaging, can be used to observe the complete earthquake cycle. Unlike seismology, which measures waveforms or the shaking part of earthquakes, geodetic imaging can measure the seismic and aseismic components of earthquake processes. These processes include tectonic strain accumulation and release, coseismic deformation, and postseismic deformation. Understanding of each part of the earthquake cycle informs earthquake hazard assessment.

Crustal deformation measurements were made using a variety of ground- and space-based geodetic survey methods in the latter part of the 1900s. In the mid-1980s the Global Positioning System (GPS) was first applied to measurement of crustal deformation in California. GPS measurements were collected for several years across the Ventura basin northwest of Los Angeles to test estimates of shortening across the basin from geologic methods and to model the geometry and characteristics of buried thrust faults bounding the basin. GPS was used to measure a shortening rate of 7 mm/year across the Ventura basin. The models that best fit the data showed steeply dipping faults that rooted in the lower crust rather than faults that flattened into decollements (Fig. 6.7). Hazard assessment that estimated seismic

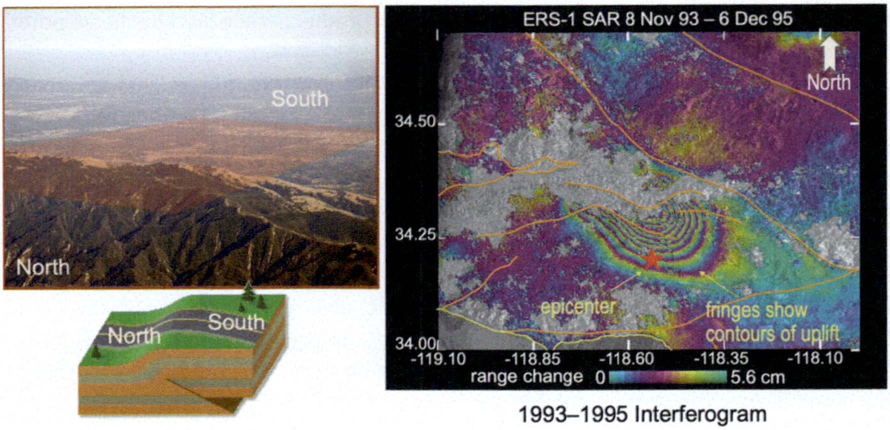

Fig. 6.7 Aerial view of the San Fernando Valley with fault rupture projection from the 1994 M6.7 Northridge earthquake (top left). Lower left shows schematic of the preferred modeled fault in the pre-Northridge study and of the Northridge rupture. Right image shows ERS-1 interferogram of the cumulative surface deformation from the fault rupture and the postseismic rupture in a two-year time frame spanning the earthquake

potential based on the geometry and slip rates of the bounding faults suggested that they could produce earthquakes on the order of M6.4 (Donnellan et al. 1993). No timeframe was placed on the potential earthquake, but on November 17, 1994 the M6.7 Northridge earthquake occurred (Jones et al. 1994) on a fault similar to that modeled using the GPS results demonstrating the value of using observation of crustal deformation for estimating fault activity and earthquake potential, critical to assessing vulnerabilities of urban infrastructure.

Today, rupture forecast models for policy and setting insurance rates incorporate geodetic crustal deformation observations. The Uniform California Earthquake Rupture Forecast version 3 (UCERF-3) is the state-of-the art estimation of fault activity in California (Field et al. 2014, 2017). UCERF-3 incorporates geologic and GNSS crustal deformation data to estimate fault activity on earthquake faults segmented into smaller sections. UCERF-3 focuses on faults presumed to be active during the Holocene (~10,000 years). GNSS data are inverted for fault slip rates (Zeng and Shen 2016) or combined with geologic information to infer fault slip rates and off fault deformation (Parsons et al. 2013). Continuous GNSS networks have now provided up to decades of daily or sub-daily position time series allowing for measurement of time-dependent crustal deformation processes. Stations within the networks are widely spaced, however, with the average 10 km spacing in southern California or Japan providing better spacing than other places in the world.

Interferometric Synthetic Aperture Radar (InSAR) provides measurements that are less frequent in time, but spatially continuous. The 1992 Landers earthquake was first earthquake to be measured using InSAR (Massonnet et al. 1993). There have been several spaceborne InSAR missions. The platforms have near polar orbits with spatial resolutions of about 100 m and crustal deformation measurements accurate to about 1 cm. Radar interferograms show fringes that relate surface deformation to line-of-sight range changes. Because the radar is side looking, orbiting InSAR platforms are most sensitive to east-west crustal deformation. Airborne InSAR, such as UAVSAR, has an accuracy of 7 m for unwrapped products and has a configurable flight path.

InSAR has been used to measure earthquakes down to ~M5. The 2014 M5.1 La Habra earthquake (Fig. 6.8) was observed with Radarsat-2 (Fielding et al. 2014) and UAVSAR (Donnellan et al. 2015). The Radarsat-2 image is least sensitive to north-south motions, but shows the primarily vertical motion from the main rupture in the interferogram. UAVSAR imagery was collected along east-west flight paths and is therefore most sensitive to vertical and north-south motion. A substantial amount of horizontal displacement of up to 10 cm was observed and attributed to slip on an unconformity at about 500 km depth. The UAVSAR measurements guided field observations of damage and InSAR measurements are often now used to find damage in the field (e.g. Rymer et al. 2011). Interferograms for the 2014 M6.0 earthquake showed clearly that the main rupture occurred 2 km west of the mapped West Napa fault (Fig. 6.9; Yun et al. 2014; DeLong et al. 2016). Latency of the observa-

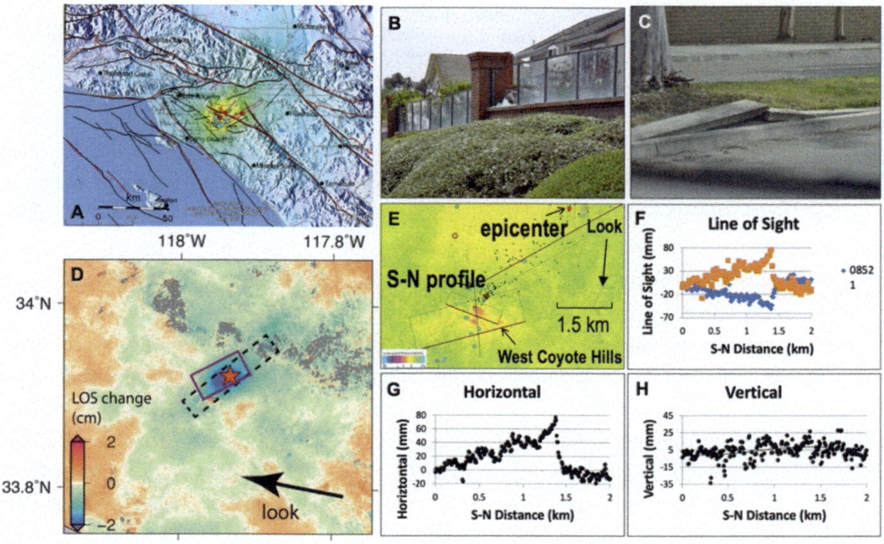

Fig. 6.8 Observations of the 2014 M5.1 La Habra earthquake. (**a**) Shake map of the earthquake from the USGS with yellow experiencing the most shaking. Los Angeles is just west of the maximum shaking. Black lines indicate active faults. Water main breaks are shown as blue dots. (**b**) Damage to a glass panel near the maximum displacement observed with UAVSAR. (**c**) Tented curb in the same location of maximum displacement. (**d**) Radarsat-2 interferogram looking west showing maximum displacement over the epicenter (after Fielding et al. 2014). (**e**) South looking UAVSAR interferogram showing displacement in the West Coyote Hills 2 km south of the main rupture. (**f**) Line-of-sight profiles for north and south looking products showing gradient that decomposes into (**g**) horizontal motion of up to 10 cm and (**h**) a small amount of vertical motion. (All images except D after Donnellan et al. 2015)

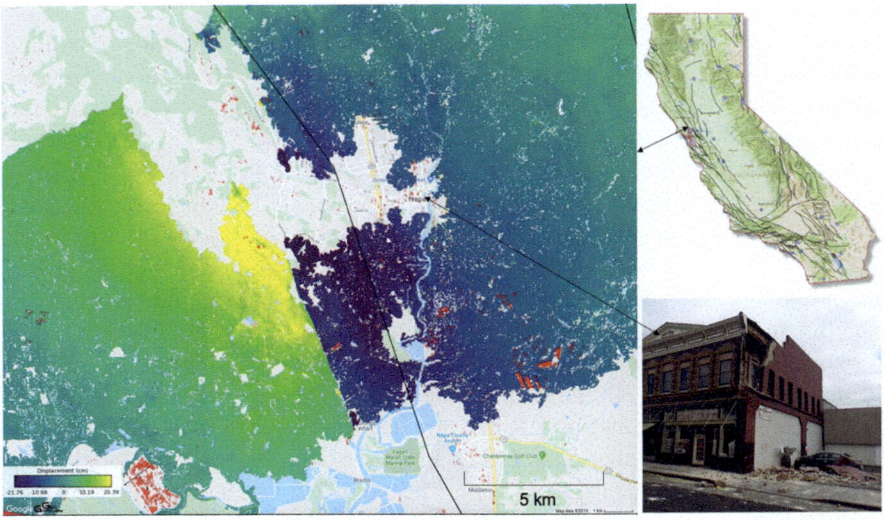

Fig. 6.9 UAVSAR interferogram for the 2014 M6.0 South Napa earthquake shown as purple to yellow color gradient (DeLong et al. 2016). The east side of the image moved south along the right-lateral fault marked by the discontinuity in the image. The previously mapped West Napa fault is marked by the black line about 2 km east of the rupture. Damage Proxy Map (DPM) identified in X-band COSMO-SkyMed (Yun et al. 2014) is the red overlay on the image. Damaged building (photo USGS) can be noted as a red dot in the DPM

tions and production InSAR products need to be low to be useful for field damage assessment. Simplifying products to display a color range that corresponds to displacement (Fig. 6.9) makes the images easier to interpret than an interferometric image displayed as fringes that correspond to cycles of the radar waves that are dependent on wavelength.

Interferograms for large earthquakes can decorrelate where damage is greatest; radar waves between the first and second pass cannot be matched where the ground is too disrupted. This information can be exploited to produce a Damage Proxy Map (DPM) by comparing coherence from the before and after images (Yun et al. 2015a). This method was successfully applied in 2011 to the M6.3 Christchurch and M9.0 Tohoku-oki earthquakes (Yun et al. 2011). The radar signal decorrelated for the South Napa earthquake, but the DPM could be used to find damage in this decorrelated region shown as red on the left image. The DPM is a powerful method for quickly mapping possible damage and can be used to guide ground assessment of damage. It was used for the 2015 M7.8 Gorkha earthquake in Nepal (Yun et al. 2015b) and has been used to map damage from typhoons and other natural disasters. Phase gradient maps (Price and Sandwell 1998) in synthetic aperture radar imagery can also direct practitioners to areas of greatest damage and can be combined with radar interferograms, DPMs, and seismic information (Ross et al. 2019a, b).

Structure from motion (SfM) measurements from small uninhabited aerial systems (sUAS) are also useful for mapping rupture and damage as long as sUAS are

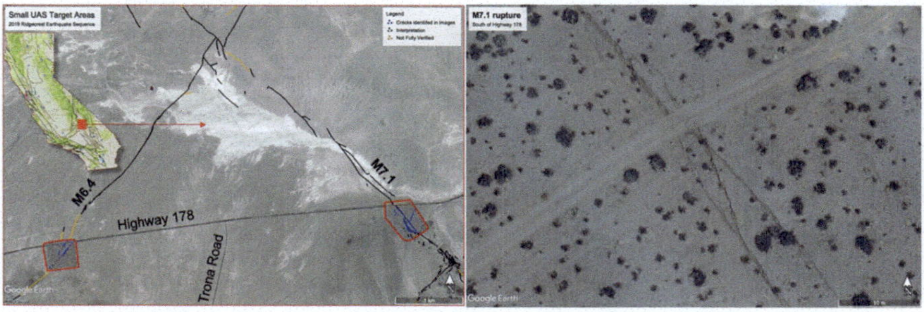

Fig. 6.10 2019 Ridgecrest earthquake sequence ruptured in a M6.4 and M7.1 earthquake. High-resolution topographic products were produced using sUAS for two targeted areas (left). Ground cracking can be observed in the final products (right) and identified in the images for field verification

permitted to operate in the necessary airspace. The ruptures of the M6.4 and M7.1 earthquakes in the 2019 Ridgecrest earthquake sequence were mapped along the fault and in targeted sections of the two rupture (Brandenberg et al. 2019; Donnellan et al. 2020). Ground cracking is easily observed in the 1.5 cm resolution 3D products (Fig. 6.10) and the technique can be used to assess damage to infrastructure following events.

Wildfires, Landslides, and Debris Flows

Short-wavelength infrared (SWIR; Fig. 6.4) is used for seeing through smoke to fight wildfires. NASA is combining SWIR with UAVSAR to explore whether radar polarimetry and SWIR can be used to image active fire fronts. Combining techniques is often the best way to identify and separate processes and changes. UAVSAR and optical observations of the 2018–2019 Thomas Fire and ensuing Montecito debris flows highlight how derived products can be used to separate the fire scar from rocky and muddy debris flows (Donnellan et al. 2018). When fire burns away vegetation the radar return of the denuded fire scar is polarized differently than the unburned vegetated area. Only a single image is required to create a polarimetric radar map. With the 15 km wide swath of UAVSAR fire scars could be mapped quickly after a fire with meter resolution. Orbiting platforms should be able to map fire scars to 10–50 m resolution (Fig. 6.11).

Pairs of radar images where available show surface disruption in interferograms and correlation between the two images. For the Montecito, California area this surface disruption included the 2018–2019 Thomas wildfire scar and the 2019 ensuing debris flows that occurred when heavy rain followed the fire. The Thomas fire destroyed 280 structures and the ensuing debris flows destroyed 73 homes and killed 20 people. The radar products were simplified to highlight the areas disrupted by the fire and debris flows to facilitate their use by emergency responders. The products

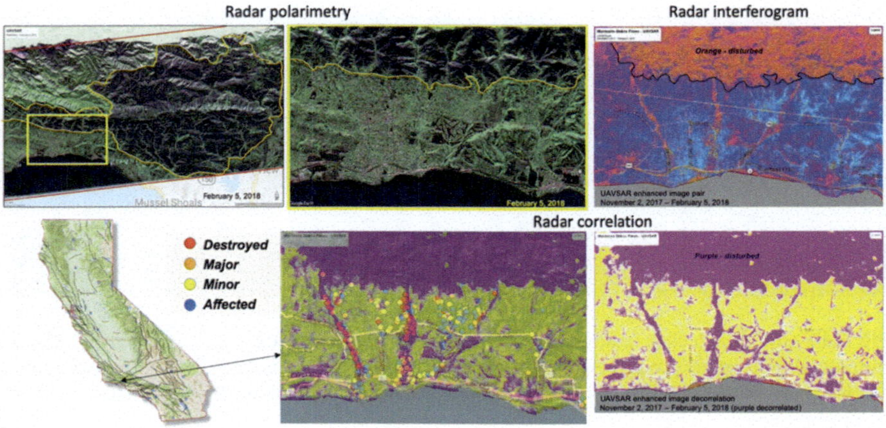

Fig. 6.11 UAVSAR was flown November 2, 2017 and February 5, 2018 over the area of the Thomas fire and Montecito debris flows (Donnellan et al. 2018). Polarimetric images show that the fire scar appears as darker purple areas. The radar interferogram and radar correlation show the fire scar and debris flows. The debris flows identified in the radar image pairs correlate well with structures that were destroyed or had major damage

were downloaded by the State of California from GeoGateway. The identified disrupted areas in the radar products correlate well with the areas where structures were destroyed or severely damaged (Donnellan et al. 2018). The advance of radar is that it is an all-weather, day and night platform that can image through clouds or when it is dark. A combination of polarimetric products and image pairs can be used to separate wildfire scar from the debris flows. Different bands of optical imagery can be used to further discriminate between rocky debris flows and scoured slopes from muddy deposits at the toe of the debris flow (Donnellan et al. 2018).

Underground Disturbances

Processes occurring underground can affect the ground surface, often in the form of subsidence or other surface disruption. These disruptions can result from fragile infrastructure, be naturally occurring, or be caused by humans. Remote sensing has been useful for studying water main breaks in Los Angeles, studying subsidence, and for studying the impacts of a moving mud pot in southern California.

Water pipeline breaks can be a result of aging infrastructure, corrosive environment, and external disturbances. In the summer of 2009 Los Angeles experienced a spate of pipeline breaks in its water system. The main cause was determined to be due to restricting sprinkling to a Monday/Thursday watering schedule that caused hammering in the pipes and particular damage to the City's older pipelines (Bardet et al. 2010). A pilot remote sensing study indicated that continuous GNSS stations in close proximity to water pipes may show transient motion prior to surface failure that could be due to the pipe leaking prior to failure. Measurement of surface displacements due to earthquakes or landslides can aid in locating and assessing pipe breaks (e.g. Brandenberg et al. 2019).

Deformation and subsidence can affect pipeline and infrastructure as well as longer term availability of the resource such as water. Removal of annual signatures in InSAR data shows that long-term subsidence of the Los Angeles basin is occurring as water is withdrawn (Riel et al. 2018). Water depletion in aquifers affects urban infrastructure over the long-term. Subsidence and the built environment can interact and affect subway tunnels (Khorrami et al. 2019). When water is being depleted the elevation of the ground surface remains relatively stable. InSAR can be used to study long-term trends of the land surface using a method called stacking. Interferograms are made for pair combinations of a series of geographically aligned images, which averages out atmospheric error and allows users to create time series of pixels in the images. Though the San Fernando and San Gabriel Valleys show little subsidence associated with groundwater withdrawal there are three areas that show fairly localized subsidence that continues over time (Fig. 6.12). There is subsidence at the northwest edge of the San Fernando Valley, near the Aliso Canyon gas production field. Localized subsidence occurs at the end of the metro red line in North Hollywood, where it had been terminated in 1995 due to extensive subsidence (Buenker and Robinson 2014). The subsidence continues to the present. Subsidence has been observed previously at the San Jose fault near Pomona, California in the northeast Los Angeles basin (Land Subsidence Committee 2014) and also continues to present. Combining InSAR to measure subsidence with electromagnetic data to better represent the subsurface improves predictive models of future subsidence (Smith and Knight 2019).

Tectonic processes including earthquakes, volcanoes, and fault creep can impact urban infrastructure. In southern California, east of the Salton Sea an ambient temperature mud spring or mud pot has increased in size and migrated to the southwest, undermining the railroad, causing relocation of a gas line, and threatening a fiber optic line and State Highway 111. The mud spring, which appeared in 2005 and began moving in 2016, is CO_2 driven and as of fall 2019 was producing 150 m^3 of water per day (Lynch and Deane 2019). SfM measurements track the change of the mud pot over time, showing the migration of the leading edge and increase in size over time (Fig. 6.13). Thermal measurements show that the spring is ambient temperature at about 27 °C. UAVSAR decorrelates over this agricultural area, but regional analysis of GNSS data suggest that a transtensional northeast striking left-lateral creep event may have occurred during 2016. An earthquake swarm occurred in the Salton Sea during 2016 (Ross et al. 2019a, b), and again in 2020 consistent with the GNSS observations.

Fig. 6.12 Three areas in the northern Los Angeles basin show continued subsidence over time in stacked UAVSAR data. Time series of pixels in the data show long-term decadal scale trend of subsidence near North Hollywood

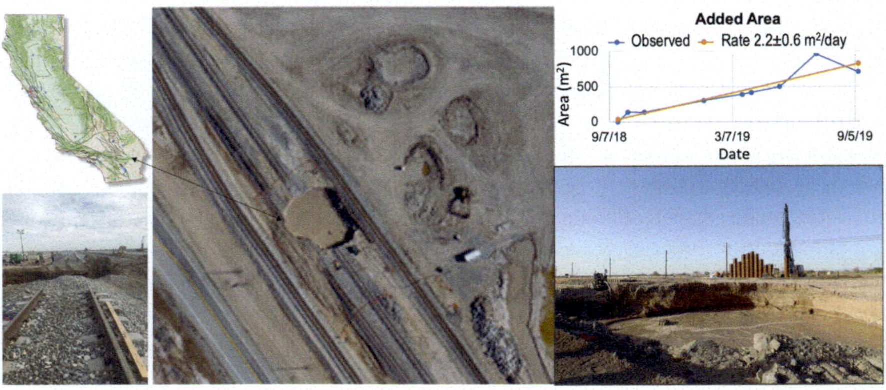

Fig. 6.13 Moving Mundo mud pot or mud spring just east of the Salton Sea in southern California. Center image shows the mud pot on 17 May 2019 from an orthomosaic image composite from sUAS flight over the area. The Union Pacific Railroad tracks can be seen as well as highway 111 in the lower left part of the center image. Lower left shows tracks on 29 October 2018 that were undermined by the mud pot. Lower right image shows mud spring on 10 September 2019. The area of the mud pot is increasing over time (upper right)

Levees

Levees protect urban and agricultural infrastructure from flooding. Levees are susceptible to subsidence of the surrounding area, from compaction of underlying peat layers (Dixon et al. 2006; Brooks et al. 2012; Bennett et al. 2017). Sea level rise exacerbates the vulnerabilities of levee systems. The atmosphere around levee systems is often humid, adding tropospheric delay to radar measurements and a resulting large source of error. The ground surface can also decorrelate making it difficult to measure surface deformation. Permanent scatterers in the images provide a means of determining deformation (Dixon et al. 2006; Zebker et al. 2007).

Certain points within the image such as road intersections are present in each image and can form a network of fiducial points to constrain the images and remove atmospheric noise over long-time spans. Persistent or permanent scatters can be used to apply radar interferometry to other built infrastructure such as dams and within cities.

6.5 Conclusions

Remotely sensed data products provide powerful tools for monitoring and assessing infrastructure in urban environments. Multiple data types provide insight into different processes and temporal and spatial scales. Measurement of broad scale deformation over time provides a predictive capability for understanding how infrastructure will behave in the future. Optical imagery is often easy to interpret, but data collection can be hampered by clouds or time of day. Radar has the advantage of being all-weather day and night capable, but can be complicated to interpret. Data products of all types need to be simplified for disaster responders or non-discipline experts. Simplified products facilitate ease of understanding and also enable rapid decision making and prioritization from highlighted results. Numerous resources are becoming available to search for and analyze remote sensing data products. Remotely sensed data can be used to observe cascading disasters and interactions between natural processes and urban infrastructure.

Acknowledgements The work was carried out at the Jet Propulsion Laboratory, California Institute of Technology, under a contract with the National Aeronautics and Space Administration (80NM0018D0004). The work has been supported by the following NASA programs: Earth Surface and Interior (ESI), Advance Information Systems Technology (AIST), Geodetic Imaging, and Advancing Collaborative Connections for Earth System Science (ACCESS). This work could not have been accomplished without the contributions of numerous other investigators. These include Co-Investigators Gregory Lyzenga, Jay Parker, John Rundle, Lisa Grant Ludwig, Jun Wang, Marlon Pierce, Michael Heflin, and Robert Granat. Work on the Mundo Mud Spring is led by David Lynch of Shannon and Wilson. Numerous interns have contributed to work described here including Lindsay LaBrecque and Nathan Pulver for the landslide work, and Juan Carlos Beltran for the subsidence study. Cathleen Jones assisted with processing and analysis of UAVSAR for constructing time series of subsidence. Postdoctoral fellow Chris Milliner led development of optical techniques for separating types of materials in debris flows.

References

Bardet, J.P., D. Ballantyne, G.E.C. Bell, A. Donnellan, S. Foster, T.S. Fu, J. List, R.G. Little, T.D. O'Rourke, and M.C. Palmer. 2010. Expert review of water system pipeline breaks in the City of Los Angeles—Summer 2009. *Report to Los Angeles City Council* 13: 2010.

Bennett, V., C. Jones, D. Bekaert, J. Bond, A. Helal, J. Dudas, M. Gabr, and T. Abdoun. 2017. Deformation monitoring for the assessment of Sacramento Delta Levee performance. In *Geo-Risk 2017*, pp. 22–31.

Bock, Y., D.C. Agnew, P. Fang, J.F. Genrich, B.H. Hager, T.A. Herring, K.W. Hudnut, R.W. King, S. Larsen, J.B. Minster, and K. Stark. 1993. Detection of crustal deformation from the Landers earthquake sequence using continuous geodetic measurements. *Nature* 361 (6410): 337.

Brandenberg, S.J., P. Wang, C.C. Nweke, K. Hudson, S. Mazzoni, Y. Bozorgnia, K.W. Hudnut, C.A. Davis, S.K. Ahdi, F. Zareian, and J. Fayaz. 2019. Preliminary report on engineering and geological effects of the July 2019 Ridgecrest earthquake sequence. Geotechnical Extreme Event Reconnaissance Association, pp. 1–69.

Brooks, B.A., G. Bawden, D. Manjunath, C. Werner, N. Knowles, J. Foster, J. Dudas, and D. Cayan. 2012. Contemporaneous subsidence and levee overtopping potential, Sacramento-San Joaquin Delta, California. *San Francisco Estuary and Watershed Science* 10 (1): 1–18.

Buenker, J.M., and R.A. Robinson. 2014. A history of tunneling in Los Angeles, North American tunneling: 2014 *Proceedings*, Los Angeles, p. 1119.

DeLong, S.B., A. Donnellan, D.J. Ponti, R.S. Rubin, J.J. Lienkaemper, C.S. Prentice, T.E. Dawson, G. Seitz, D.P. Schwartz, K.W. Hudnut, and C. Rosa. 2016. Tearing the terroir: Details and implications of surface rupture and deformation from the 24 August 2014 M6. 0 South Napa earthquake. *California. Earth and Space Science* 3 (10): 416–430.

Dixon, T.H., F. Amelung, A. Ferretti, F. Novali, F. Rocca, R. Dokka, G. Sella, S.W. Kim, S. Wdowinski, and D. Whitman. 2006. Space geodesy: Subsidence and flooding in New Orleans. *Nature* 441 (7093): 587.

Donnellan, A., B.H. Hager, and R.W. King. 1993. Discrepancy between geologic and geodetic deformation rates in the Ventura basin. *Nature* 366: 333–336.

Donnellan, A., L. Grant Ludwig, J.W. Parker, J.B. Rundle, J. Wang, M. Pierce, G. Blewitt, and S. Hensley. 2015. Potential for a large earthquake near Los Angeles inferred from the 2014 La Habra earthquake. *Earth and Space Science* 2 (9): 378–385.

Donnellan, A., R. Arrowsmith, and V. Langenheim. 2016. Select airborne techniques for mapping and problem solving. In *Applied geology in California*, ed. R. Anderson and H. Ferriz, 541–566. Belmont: Star Publishing.

Donnellan, A., J. Parker, C. Milliner, T.G. Farr, M. Glasscoe, Y. Lou, and B. Hawkins. 2018. UAVSAR and optical analysis of the Thomas fire scar and Montecito debris flows: Case study of methods for disaster response using remote sensing products. *Earth and Space Science* 5: 339–347.

Donnellan, A., G. Lyzenga, A. Ansar, C. Goulet, J. Wang, and M. Pierce. 2020. Targeted high-resolution structure from motion observations over the M6.4 and M7.1 ruptures of the Ridgecrest earthquake sequence. *Seismological Research Letters* 91 (4): 2087–2095.

Field, E.H., J.R. Arrowsmith, G.P. Biasi, P. Bird, T.E. Dawson, K.R. Felzer, D.D. Jackson, K.M. Johnson, T.H. Jordan, C. Madden, A.J. Michael, K.R. Milner, and M.T. Page. 2014. Uniform California earthquake rupture forecast, Version 3 (UCERF3)—The time-independent model. *Bulletin of the Seismological Society of America* 104 (3): 1122–1180. https://doi.org/10.1785/0120130164.

Field, E.H., T.H. Jordan, M.T. Page, K.R. Milner, B.E. Shaw, T.E. Dawson, G.P. Biasi, T. Parsons, J.L. Hardebeck, A.J. Michael, and R.J. Weldon. 2017. A synoptic view of the third uniform California earthquake rupture forecast (UCERF3). *Seismological Research Letters* 88 (5): 1259–1267.

Fielding, E.J., M. Simons, S. Owen, P. Lundgren, H. Hua, P. Agram, Z. Liu, A. Moore, P. Milillo, J. Polet, and S. Samsonov. 2014. Rapid imaging of earthquake ruptures with combined geodetic and seismic analysis. *Procedia Technology* 16: 876–885.

Heflin, M., A. Donnellan, J. Parker, G. Lyzenga, A. Moore, L. Grant Ludwig, J. Rundle, J. Wang, and M. Pierce. 2020. Automated estimation and tools to extract positions, velocities, breaks, and seasonal terms from daily GNSS measurements: Illuminating nonlinear salton trough deformation. *Earth and Space Science*. 7 (7): e2019EA000644.

Jones, L., K. Aki, M. Celebi, A. Donnellan, J. Hall, R. Harris, E. Hauksson, T. Heaton, S. Hough, K. Hudnut, K. Hutton, M. Johnston, W. Joyner, H. Kanamori, G. Marshall, A. Michael, J. Mori, M. Murray, D. Ponti, P. Reasenberg, D. Schwartz, L. Seeber, A. Shakal, R. Simpson, H. Thio, M. Todorovska, M. Trifunic, D. Wald, and M.L. Zobak. 1994. The magnitude 6.7 Northridge California, earthquake of January 17, 1994. *Science* 266: 389–397.

Khorrami, M., B. Alizadeh, E. Ghasemi Tousi, M. Shakerian, Y. Maghsoudi, and P. Rahgozar. 2019. How groundwater level fluctuations and geotechnical properties lead to asymmetric subsidence: A PSInSAR analysis of land deformation over a transit corridor in the Los Angeles metropolitan area. *Remote Sensing* 11 (4): 377.

Land Subsidence Committee. 2014. 2013 Annual Report of the Land Subsidence Committee. Wildermuth Environmental, Inc.

Lindqwister, U.J., J.F. Zumberge, F.H. Webb, and G. Blewitt. 1991. Few millimeter precision for baselines in the California permanent GPS geodetic array. *Geophysical Research Letters* 18 (6): 1135–1138.

Lynch, D.K., and T. Deane. 2019. A moving mystery. *Civil Engineering* 89 (7): 54–63.

Massonnet, D., M. Rossi, C. Carmona, F. Adragna, G. Peltzer, K. Feigl, and T. Rabaute. 1993. The displacement field of the Landers earthquake mapped by radar interferometry. *Nature* 364 (6433): 138.

NISAR, 2018. *NASA-ISRO SAR (NISAR) mission science Users' handbook*. NASA Jet Propulsion Laboratory. 261 pp.

Parsons, T., Johnson, K.M., Bird, P., Bormann, J., Dawson, T.E., Field, E.H., Hammond, W.C., Herring, T.A., McCaffrey, R., Shen, Z.K. and Thatcher, W.R., 2013. Appendix C—Deformation models for UCERF3.

Price, E.J., and D.T. Sandwell. 1998. Small-scale deformations associated with the 1992 Landers, California, earthquake mapped by synthetic aperture radar interferometry phase gradients. *Journal of Geophysical Research: Solid Earth* 103 (B11): 27001–27016.

Riel, B., M. Simons, D. Ponti, P. Agram, and R. Jolivet. 2018. Quantifying ground deformation in the Los Angeles and Santa Ana Coastal Basins due to groundwater withdrawal. *Water Resources Research* 54 (5): 3557–3582.

Ross, Z.E., B. Idini, Z. Jia, O.L. Stephenson, M. Zhong, X. Wang, Z. Zhan, M. Simons, E.J. Fielding, S.H. Yun, and E. Hauksson. 2019a. Hierarchical interlocked orthogonal faulting in the 2019 Ridgecrest earthquake sequence. *Science* 366 (6463): 346–351.

Ross, Z.E., D.T. Trugman, E. Hauksson, and P.M. Shearer. 2019b. Searching for hidden earthquakes in Southern California. *Science* 364 (6442): 767–771.

Rymer, M.J., J.A. Treiman, K.J. Kendrick, J.J. Lienkaemper, R.J. Weldon, R. Bilham, M. Wei, E.J. Fielding, J.L. Hernandez, B.P. Olson, and P.J. Irvine. 2011. Triggered surface slips in southern California associated with the 2010 El Mayor-Cucapah, Baja California, Mexico, earthquake (No. 2010–1333). US Geological Survey.

Smith, R., and R. Knight. 2019. Modeling land subsidence using InSAR and airborne electromagnetic data. *Water Resources Research* 55 (4): 2801–2819.

Yun, S., P.S. Agram, E.J. Fielding, M. Simons, F. Webb, A. Tanaka, P. Lundgren, S.E. Owen, P.A. Rosen, and S. Hensley. 2011. Damage proxy map from InSAR coherence applied to February 2011 M6. 3 Christchurch Earthquake, 2011 M9. 0 Tohoku-oki Earthquake, and 2011 Kirishima Volcano Eruption. In *AGU Fall Meeting Abstracts*.

Yun, S.H., S.E. Owen, H. Hua, P. Milillo, E.J. Fielding, K.W. Hudnut, T.E. Dawson, T.P. Mccrink, M.J. Jo, W.D. Barnhart and G.J.M. Manipon. 2014. Rapid response products of the ARIA project for the M6.0 August 24, 2014 South Napa Earthquake. In *AGU Fall Meeting Abstracts*.

Yun, S.H., E.J. Fielding, F.H. Webb, and M. Simons, and California Institute of Technology. 2015a. Damage proxy map from interferometric synthetic aperture radar coherence. U.S. Patent 9,207,318.

Yun, S.H., K. Hudnut, S. Owen, F. Webb, M. Simons, P. Sacco, E. Gurrola, G. Manipon, C. Liang, E. Fielding, and P. Milillo. 2015b. Rapid damage mapping for the 2015 M w 7.8 Gorkha earthquake using synthetic aperture radar data from COSMO–SkyMed and ALOS-2 satellites. *Seismological Research Letters* 86 (6): 1549–1556.

Zebker, H., P. Shankar, and A. Hooper. 2007. InSAR remote sensing over decorrelating terrains: Persistent scattering methods. In *2007 IEEE Radar Conference*. IEEE, pp. 717–722.

Zeng, Y., and Z.K. Shen. 2016. A fault-based model for crustal deformation, fault slip rates, and off-fault strain rate in California. *Bulletin of the Seismological Society of America* 106 (2): 766–784.

Zhou, Q. 2016. Digital elevation model and digital surface model. In *International encyclopedia of geography: people, the earth, environment and technology: People, the earth, environment and technology*, 1–17. Hoboken: Wiley.

Chapter 7
Remote Monitoring of Highway Bridges with RADARSAT-2 Satellite: Validation Case Study on Jacques Cartier and Victoria Bridges in Montreal, Canada

Daniel Cusson, Istemi Ozkan, Fernando Greene Gondi, and Jayson Eppler

7.1 Introduction

Highway bridges are critical to the national road transportation networks of many countries. A good percentage of them are either structurally or functionally deficient and in urgent need of rehabilitation owing to aging, increased demands, limited capital investment, and insufficient inspection and maintenance (Transport Canada 2012). This deficiency is further exacerbated by the effects of climate change and extreme weather events (NCHRP 2007, 2017). Remote monitoring technologies can provide extremely useful data on the performance and reliability of bridges and the effectiveness of climate change adaptation measures, as bridge owners have been struggling for decades with how to cost effectively rehabilitate their portfolios of bridges (U.S. D.O.T. et al. 2007). By conducting interferometric synthetic aperture radar (InSAR) analysis on acquired radar satellite imagery, the movement at hundreds or even thousands of point targets on a single bridge structure and its nearby foundation can be monitored with sub-centimeter accuracy (Marinkovic et al. 2007). Such satellite-based monitoring can also be conducted at the network level to cost effectively screen several bridges at once for stability issues and flag them for inspection (Acton 2013).

Using satellite imagery from Canada's RADARSAT-2 satellite, a previous validation study on a new bridge made of continuous spans of a concrete deck on steel box girders found a good match between satellite-measured thermal displacements and numerical predictions calibrated on sensor measurements (Cusson et al. 2018).

D. Cusson (✉) · I. Ozkan
National Research Council Canada, Ottawa, ON, Canada
e-mail: daniel.cusson@nrc-cnrc.gc.ca

F. Greene Gondi · J. Eppler
MDA Geospatial Services Inc., Vancouver, BC, Canada

V. Singhroy (ed.), *Advances in Remote Sensing for Infrastructure Monitoring*, Springer Remote Sensing/Photogrammetry,
https://doi.org/10.1007/978-3-030-59109-0_7

With the aim of validating the approach on different structural systems, this paper presents the findings of two additional validation case studies on historical bridges made of a deck supported by a steel truss superstructure. In future phases of the project, a decision-support tool based on radar satellite data will be developed to assist engineers in performance assessment and maintenance planning of highway bridges.

7.2 Bridge Sites and Acquired Radar Imagery

Jacques Cartier Bridge

The 90-year old Jacques Cartier Bridge—owned by Jacques Cartier and Champlain Bridges Incorporated (JCCBI)—crosses the St. Lawrence River and links the cities of Montreal and Longueuil in the province of Quebec, Canada, which also provides access to St Helen's Island. The bridge has five traffic lanes over 40 spans of a combined length of approximately 3 km. Its main section, with a length of 590 m, is made of a steel truss superstructure carrying a high-performance concrete (HPC) deck on concrete piers (Fig. 7.1a). The bridge deck was designed such that its slab behaves independently from the steel superstructure so that thermal loads are not transferred from one another.

Victoria Bridge

The 160-year old Victoria Bridge—owned by Canadian National Railway (CN)—crosses the St. Lawrence River, and links the cities of Montreal and Saint-Lambert in Quebec, Canada. The 3 km long bridge consists of through

a b

Fig. 7.1 Typical spans of the monitored bridges in Montreal (QC), Canada. (**a**) Jacques Cartier Bridge (main section). (**b**) Victoria Bridge (typical spans)

trusses, deck trusses, girder spans, and two lift spans that carry two railway tracks and four narrow roadway lanes. The railway tracks are supported by a deck, which is mostly made of timber bridge ties, while the roadway riding surface is mostly steel grating. The roadway is carried on cantilevered floor beam extensions on the through trusses, which are carried under the tracks on deck trusses (Fig. 7.1b).

Acquired Satellite Imagery

For typical InSAR analysis, scenes can be obtained from RADARSAT-2 at a 24-day frequency for two satellite directions (ascending, when satellite navigates from south to north poles; and descending, from north to south poles), for different beam modes (with various resolutions and footprints), and for different viewing angles (20° to 50° from vertical).

For monitoring the Jacques Cartier and Victoria bridges, the Spotlight (SLA) beam mode was selected, which is best-suited for bridge monitoring applications since it provides the best horizontal ground resolution (1 m × 3 m). More details on available RADARSAT-2 beam modes can be found in MDA (2018). Two image stacks with opposite viewing geometries were selected. The first one was the SLA26-asc stack in ascending direction; looking sideways to the East with a 48° incidence angle (Fig. 7.2). The second one was the SLA12-des stack with a descending direction; looking sideways to the West with a 39° incidence angle (Fig. 7.2). A total of 19 scenes were available for InSAR analysis from the SLA-26-asc stack and 28 scenes from the SLA12-des stack, covering a two-year monitoring period from May 2016 to August 2018.

It should be noted that only one single stack of radar images is normally required to monitor a bridge on an operational basis; however, two independent stacks were analyzed in this study since one of the goals was to investigate the feasibility of performing 2D decomposition on the line-of-sight (LOS) displacement data from two independent stacks of satellite imagery.

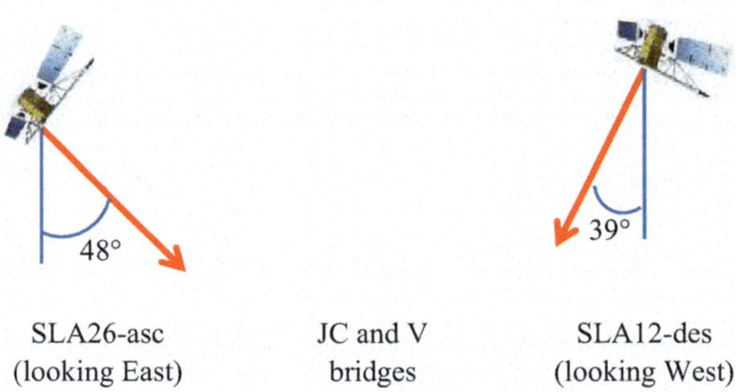

Fig. 7.2 Viewing geometries selected for monitoring the Jacques Cartier and Victoria Bridges

7.3 Overview of InSAR Analysis of Satellite Imagery

Synthetic aperture radar (SAR) is a remote sensing technique where a radar image is formed by collecting and processing a series of transmitted coded microwave pulses which have reflected or scattered from elements on the earth's surface. Interferometric synthetic aperture radar (InSAR) is a technique where the amplitude and phase of two complex radar images acquired at two different times are combined to measure deformation. Each pixel in an interferogram comprises a phase difference (from 0° to 360°) between two distinct SAR snapshots of a given resolution cell. The interferogram phase is cumulatively sensitive to all geometric and physical variables that affect the return path of the microwaves between the satellite and Earth's surface. Hence, the phase is proportional to surface topography, surface displacement along satellite LOS, atmospheric pressure, and water vapor. Knowing the phase difference and the radar wave length (5.66 cm for RADARSAT-2), the relative displacement of a given target over time can be estimated. The persistent scatterers interferometric synthetic aperture radar (PS-InSAR) method allows the detection of point-like coherent targets in the radar image that are dominated by a stable reflector called a persistent scatterer. This method was selected because it preserves the full resolution of selected targets, and provides discrimination between high and low quality targets (e.g., separating bridge elements from water elements.

The analysis of radar backscatter returns from bridges can be complicated due to the combination of layover, shadow, and multi-bounce effects. Layover effects can occur when two or more spatially separated scattering elements are at the same radar range and may thus appear to overlap in the radar image. For example, at the Victoria Bridge, the near-range portion of the steel truss superstructure is in layover with a portion of the near-range roadway, and a portion of the near-range roadway is in layover with unstable water surface, which may act to reduce coherence. Shadow effects can occur when incident radar signals are blocked by other parts of the structure. For example, at the Victoria Bridge, a portion of the far-range roadway is in the shadow of the central steel truss superstructure over the railway. Multi-bounce effects can occur when radar signals bounce off more than one physical object before returning to the satellite. At the Victoria Bridge, strong double bounce returns were observed from the deck railings and the sides of the steel truss superstructure. Similar effects were also found in the data analysis of the Jacques Cartier Bridge. More details on these effects can be found in Ferretti et al. (2007).

A linear phase model was fit to the PS-InSAR derived unwrapped phase using a least square regression analysis to estimate the following parameters:

- Residual height (m)—fitted against perpendicular baseline (stereo sensitivity) data;
- Thermal sensitivity of displacement (mm/°C)—fitted against ambient temperature data;
- Linear time rate of displacement (mm/year)—fitted against time from the first image acquisition; and

- Coherence (ranging 0–1)—confirming quality of fit and reliability of the processed data.

In this case study, targets with a coherence of at least 0.8 were kept for the final analysis.

In a final step of the analysis, 2D decomposition of the LOS data acquired from the two opposite stacks was performed to obtain the vertical and longitudinal components of the measured displacements at each bridge.

InSAR Analysis Results for the Jacques Cartier Bridge

Figure 7.3 shows the residual height, linear time rate, and thermal sensitivity components estimated over the Jacques Cartier Bridge for the descending stack SLA12-des. Similar results from the SLA26-asc stack were also obtained but are not shown here due to space limitation. The residual height component shows two height peaks located at the towers of the steel superstructure of the western portion of the bridge. The access ramp located in St Helen's Island shows a gradual increase in height consistent with the true height.

The linear time rate component does not reveal any significant displacement over the superstructure, along any portions of the bridge, or in the bridge surroundings. The linear time rate data are consistent between the ascending and descending stacks, supporting the absence of near-vertical displacement.

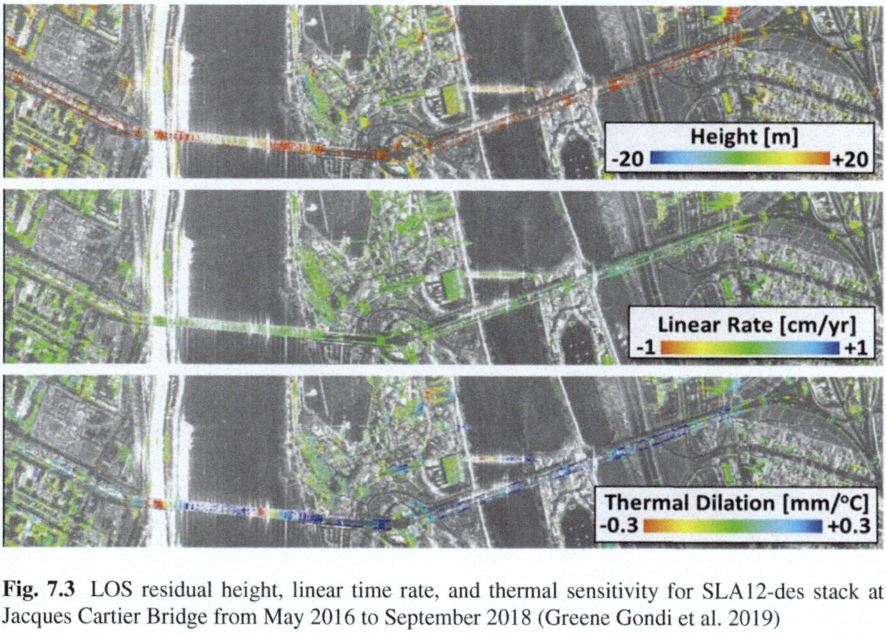

Fig. 7.3 LOS residual height, linear time rate, and thermal sensitivity for SLA12-des stack at Jacques Cartier Bridge from May 2016 to September 2018 (Greene Gondi et al. 2019)

The thermal sensitivity component shows thermal variations ranging from approximately −0.3 to +0.3 mm/°C along the bridge longitudinal axis in the western portion of the bridge, with the highest thermal values found at the two towers. The eastern portion of the Jacques Cartier Bridge displays repeating patterns with a smaller range, from approximately 0.0 to +0.3 mm/°C, which seem to correspond to the different spans of the bridge. This last set of data (i.e., thermal sensitivity) is the set of measurements that will be compared to the thermal displacement predictions in the next section.

InSAR Analysis Results for the Victoria Bridge

Figure 7.4 shows the residual height, linear time rate, and thermal sensitivity components estimated over the Victoria Bridge for the SLA26-asc stack. Similar results from the SLA12-des stack were also obtained but are not shown here due to space limitation. The residual height reveals large height estimates along the entire bridge. An increase in height is observed from each access ramp onto the bridge. Variations of height can be observed along the bridge from span to span, which may be attributed to multiple bridge elements being radar illuminated (e.g., steel truss superstructure with height of 12 m vs. railings at deck level). The linear time rate does not reveal any significant movement over the superstructure, deck, or any other areas

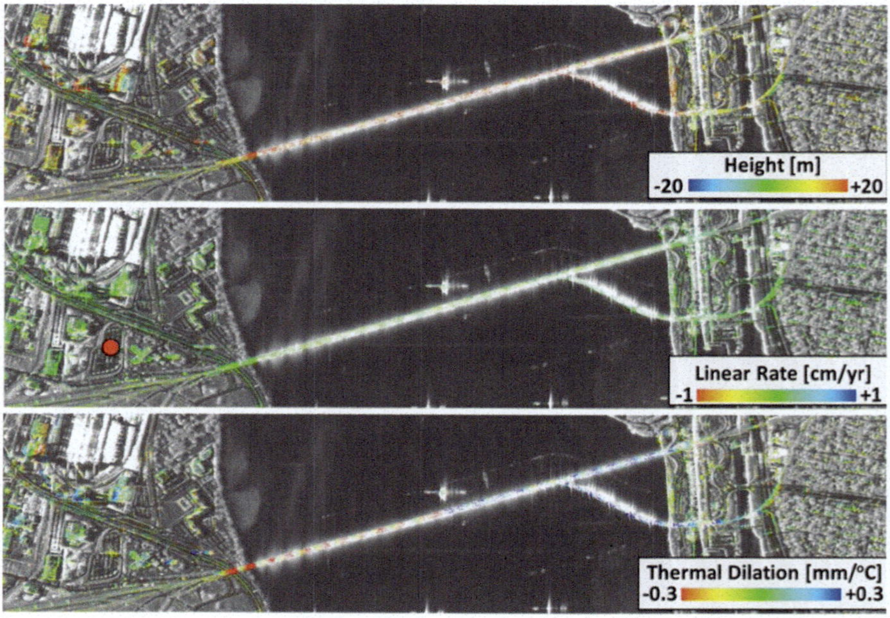

Fig. 7.4 LOS residual height, linear time rate, and thermal sensitivity for SLA26-asc stack at Victoria Bridge from May 2016 to August 2018 (Greene Gondi et al. 2019)

near the bridge, including the approach slabs and the nearby roads. The thermal sensitivity shows regular cycles of thermal movement along the bridge, with a change in polarity near the center of the bridge. This last set of data (i.e., thermal sensitivity) is the set of measurements that will be compared to the thermal displacement predictions in the next section.

7.4 Bridge Thermal Displacement Predictions

In a previous case study (Cusson et al. 2018), it was demonstrated that the use of finite element models (FEM) for predicting the bridge thermal behavior is a reliable means for validating InSAR-derived thermal displacement results. A similar approach is adopted here for the Jacques Cartier Bridge. For the Victoria Bridge case study, simple analytical calculations are made assuming linear static structural behavior and isotropic linear materials.

Jacques Cartier Bridge Numerical Model

The numerical model for the Jacques Cartier Bridge was originally developed for JCCBI by their bridge consultant, RWDI, using the CSIBridge software, and was graciously shared with NRC for this project. This 3D model (Fig. 7.5) was therefore used to simulate the thermal behavior of the bridge. The main section superstructure is made of steel elements, for which an elastic modulus of 200 GPa and a thermal expansion coefficient (CTE) of 0.000012/°C are assumed. The deck width is 24.1 m and the three continuous spans have lengths of 128 m, 334 m, and 128 m (including a 115-m drop-in section). As discussed above, the concrete slab is not included in the model since it does not impose deformations or stresses on the supporting steel

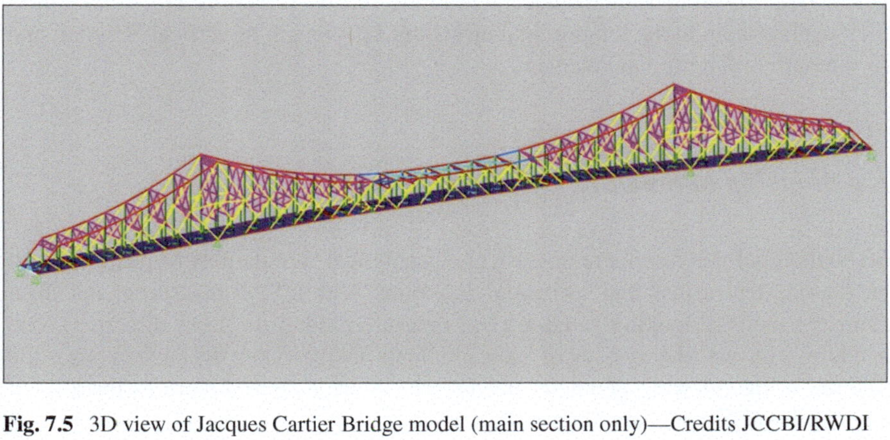

Fig. 7.5 3D view of Jacques Cartier Bridge model (main section only)—Credits JCCBI/RWDI

superstructure but is rather accounted for as additional dead load. The supports at the extremities are free in the longitudinal direction and fixed in the other directions, while the supports at the two towers are fixed in all directions. A structural linear static analysis was conducted, which is valid under service loads.

Victoria Bridge Theoretical Model

The main structural material for the Victoria Bridge is steel, for which an elastic modulus of 200 GPa and a thermal expansion coefficient (CTE) of 0.000012/°C are assumed. The deck width is 20.4 m and the steel truss height above the roadway is 12 m, while 21 of the 26 spans have a length of 77.4 m. The supports at each simply-supported span are a roller at one end and a pin at the other end.

Considering displacements in the vertical direction, a bridge where the same structural material is used in the deck and superstructure will exhibit negligible displacement due to bending at the deck level from changes in temperature, assuming negligible thermal gradients across the section depth. However, vertical elements of the bridge are assumed to expand or contract in the axial direction due to temperature variations (Eq. 7.1). Considering displacements in the longitudinal direction, the bridge will display thermal movement proportional to the length of the bridge span under consideration and the change in ambient temperature (ΔT), assuming that the bridge temperature is in equilibrium with the ambient temperature (Eq. 7.2). Based on the above assumptions, the vertical and longitudinal components of the axial thermal displacement at the Victoria Bridge can be obtained as follows:

$$D_{\mathrm{Vth}} = L_V \times CTE \times \Delta T \tag{7.1}$$

$$D_{L\mathrm{th}} = L_L \times CTE \times \Delta T \tag{7.2}$$

where D_{Vth} = axial thermal displacement in the vertical direction; $D_{L\mathrm{th}}$ = axial thermal displacement in the longitudinal direction; L_V = length of vertical element; and L_L = length of longitudinal element.

Relationship to Satellite Viewing Geometry

Since the goal is to compare numerical or analytical calculations (typically calculated along the vertical and horizontal directions) with InSAR measurements taken from the inclined-looking satellite (i.e., measured along the LOS direction), one needs to convert one system of measurements to the other. In such a case, the satellite-bridge viewing geometry needs to be considered and requires the knowledge of the following parameters:

- α: Incidence angle of the satellite line-of-sight (48° for SLA26 and 39° for SLA12)
- β: Angle between satellite track heading and bridge azimuth
- (Jacques Cartier Bridge: 113° for SLA26 and 274° for SLA12)
- (Victoria Bridge: 73° for SLA26 and 234° for SLA12)

With the above two parameters, Eq. (7.3) can be used to determine the LOS displacement from the calculated vertical and horizontal components of displacement, as follows:

$$D_{\text{LOS}} = -D_V \cos\alpha - D_L \sin\alpha \sin\beta \tag{7.3}$$

where D_{LOS} is the line-of-sight displacement (or range); D_V is the vertical component of displacement; D_L is the longitudinal component of displacement; and the α and β angles are as defined above. As far as the horizontal displacement is concerned, Eq. (7.3) indicates that InSAR measurements from RADARSAT-2 will be most sensitive to the longitudinal component for East-West oriented bridges (β ~90° or 270°), such as the Jacques Cartier and Victoria bridges in Montreal, and to the transverse component for North-South oriented bridges (β ~0° or 180°).

7.5 Assessment of Satellite-Based Monitoring Approach

As it was found in previous study by Cusson et al. (2018), one particular set of InSAR data obtained from the RADARSAT-2 satellite can be directly compared to thermal displacement predictions, which is the thermal sensitivity expressed in units of displacement per temperature, and is typically plotted as a function of the position along the longitudinal axis of the bridge. The thermal sensitivity represents the bridge displacement resulting from a global temperature increase of 1 °C. This comparative analysis of InSAR measurements and analytical predictions of thermal sensitivity is used here to explain and validate the InSAR data sets measured by RADARSAT-2 over the monitored bridges.

Validation of Results for the Jacques Cartier Bridge

A first comparison is to look at the longitudinal component of thermal sensitivity along the bridge layout line from the 0 m position (West) to the 590 m position (East) over the main section of the bridge. The InSAR data set was created from corresponding point targets of the SLA26-asc and SLA12-des image stacks. As a result, a total of 1856 valid points were selected for the comparison with the FEM predictions (blue curve), as illustrated in Fig. 7.6.

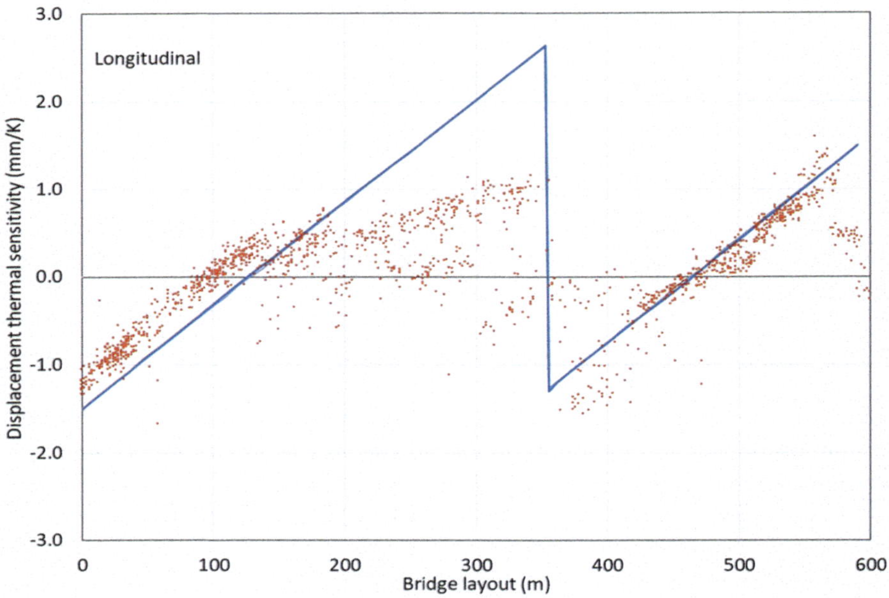

Fig. 7.6 Comparison between InSAR and FEM thermal sensitivity data for the longitudinal direction decomposed from the SLA26-asc and SLA12-des stacks (Jacques Cartier Bridge; May 2016–August 2018)

A good match is found in terms of the extent and slope for two portions of the bridge included between the 0 m position (west end) and the 128 m position (west tower), and between the 462 m position (east tower) and the 590 m position (east end). However, the InSAR measurements seem to deviate significantly from the expected trend outlined by the FEM predictions for the portion of the bridge between the two towers (128 m and 462 m positions) with smaller than expected thermal displacement in the longitudinal direction. The reason for this observation is not fully understood at the moment but is likely due to violation of one or more of the following assumptions:

(1) The 2D decomposition of the LOS results from the 2 stacks into the bridge longitudinal/vertical coordinate system is not fully constrained and requires an assumption regarding the transverse component—in this case assumed to be zero.
(2) The 2D decomposition requires that the results of each of the LOS stacks be spatially mapped into a common geometry such that LOS results are combined from the same bridge elements. This assumption can be violated by errors in geolocation of LOS results—which depend on accurate estimation of target heights from the InSAR linear regression analysis.
(3) The thermal expansion model relies on thermal equilibrium between the bridge elements and ambient air. This may be violated to some extent by direct solar illumination. The sun-synchronous orbit of RADARSAT-2 results in ascending

images being acquired at ~6 pm local time and descending images at ~6 am local time (at Montreal's latitude).

Violation of Assumption 1 can be expected to have a similar effect along the full section length, whereas violation of Assumptions 2 and 3 would be expected to cause larger differences near the towers. For these reasons, field measurements may be necessary to confirm the satellite measurements on the central span between the towers.

A second comparison is to look at the vertical component of thermal sensitivity along the bridge layout line. The InSAR data set was created from corresponding point targets of the SLA26-asc and SLA12-des image stacks. As a result, a total of 1856 valid points were selected for the comparison with FEM predictions, as illustrated in Fig. 7.7. In the figure, two FEM predictions are illustrated for the vertical thermal sensitivity at deck level (blue curve) and along the top chord of the bridge superstructure (green curve) to account for the axial thermal displacement of the vertical elements of the superstructure. Since it is difficult to know exactly from which part of the structure the satellite readings are coming (e.g., deck, railings, or anywhere on the steel truss superstructure), one would expect most satellite measurements to fall between the deck level (blue curve) and the top chord of the steel superstructure (green curve). Similar to the above observation for the horizontal data, a relatively good match is found over the two portions of the bridge included between the west end and the west tower of the bridge, and between the east tower

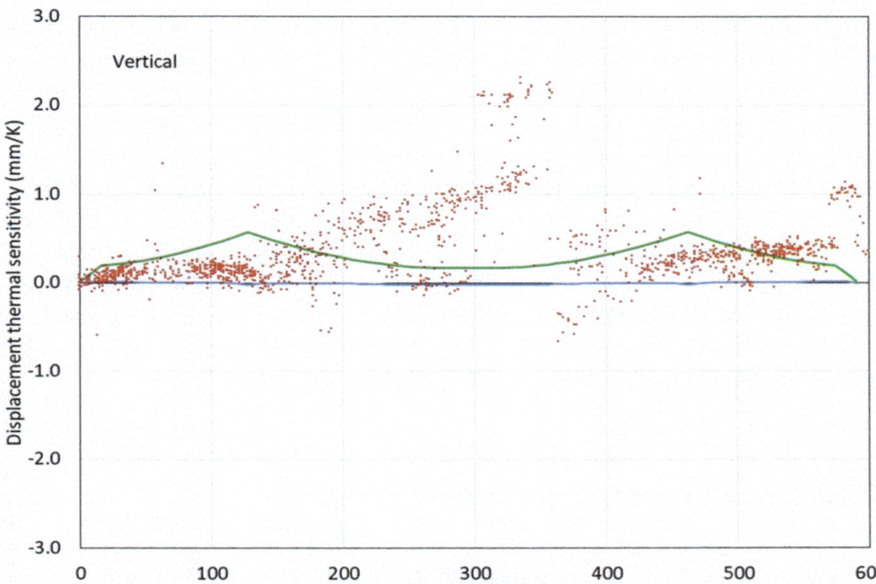

Fig. 7.7 Comparison between InSAR and FEM thermal sensitivity data for the vertical direction decomposed from the SLA26-asc and SLA12-des stacks (Jacques Cartier Bridge; May 2016–August 2018)

and the east end of the bridge. However, the InSAR measurements seem to lie outside the expected range outlined by the FEM predictions for the bridge portion between the two towers with larger than expected thermal displacement in the vertical direction (which could have been affected by geo-referencing issues).

This is consistent with the assertion that geolocation errors are likely the cause of this difference. Field measurements might be necessary to confirm or disprove the satellite measurements on the central span between the towers. However, the points above 1.3 mm/K may be considered as outliers and perhaps should be rejected immediately.

A third observation is to look at the same measurement points directly from each stack of satellite imagery before any 2D decomposition calculations were done. Viewing the data from each satellite LOS may give additional perspective on the interpretation of displacement detected by each satellite stack, since the north side of the bridge is mainly seen by the SLA26-asc stack, whereas the SLA12-des stack sees primarily the south side of the bridge. InSAR data were filtered to select only data points with high coherence values (from 0.8 to 1.0, as defined in Sect. 7.3) and located within close proximity of the bridge footprint. This process resulted in 6169 points from the SLA26-asc stack and 6264 points from the SLA12-des stack. The FEM predictions were recalculated using Eq. (7.3) for the satellite viewing geometry for each image stack.

The predominant importance of the longitudinal component of thermal sensitivity compared to the vertical component is clearly seen in both stacks shown in Figs. 7.8 and 7.9. As it was observed above in Fig. 7.6 for the longitudinal component, the InSAR data do not follow the FEM prediction for the section of the bridge between the two towers. Data from the SLA12-des stack, however, allow to make a new interesting observation, where a very good match is observed for the entire length of the bridge, except for some minor groups of outliers. This new information may provide two additional preliminary observations: (i) the SLA12-des stack may have provided a better, unobstructed view of the bridge compared to the other stack while, for some reason, the view from the SLA26-asc stack might have been partially obstructed or affected by layover and shadow effects or (ii) the thermal displacement of the bridge between the two towers might have been different between each side of the bridge. The first alternative is more plausible than the second one since a field survey that measured the elevation profile on each side of the bridge deck confirmed that both sides have nearly identical elevations.

Validation of Results for the Victoria Bridge

A first comparison is made by looking at the longitudinal component of thermal sensitivity along the bridge layout line. The InSAR data set was calculated from corresponding point targets of the SLA26-asc and SLA12-des image stacks. A total of 4978 valid points were available for the comparison with the analytical prediction as shown in Fig. 7.10. An excellent match in terms of extent and slope is found over

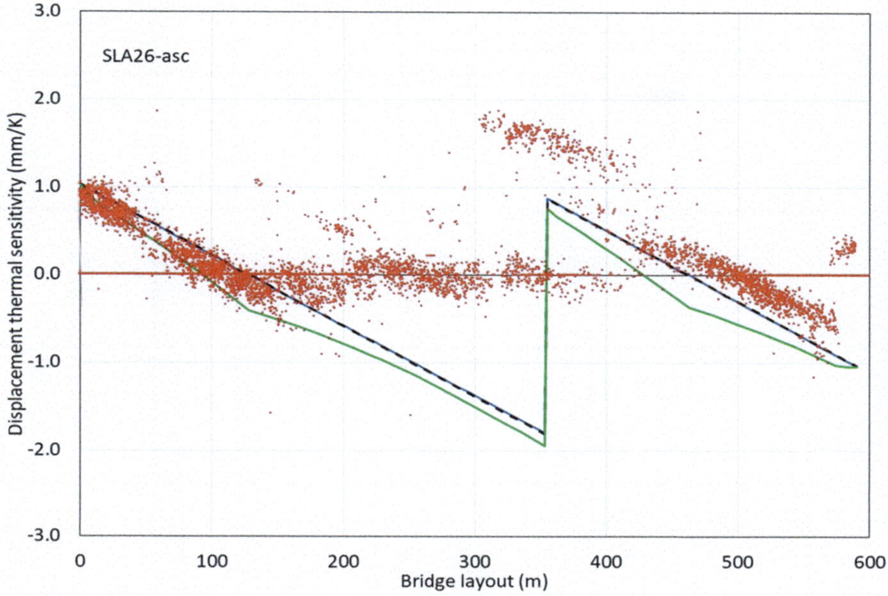

Fig. 7.8 Comparison between InSAR and FEM thermal sensitivity data for the LOS direction obtained from the SLA26-asc stack (Jacques Cartier Bridge; May 2016–August 2018)

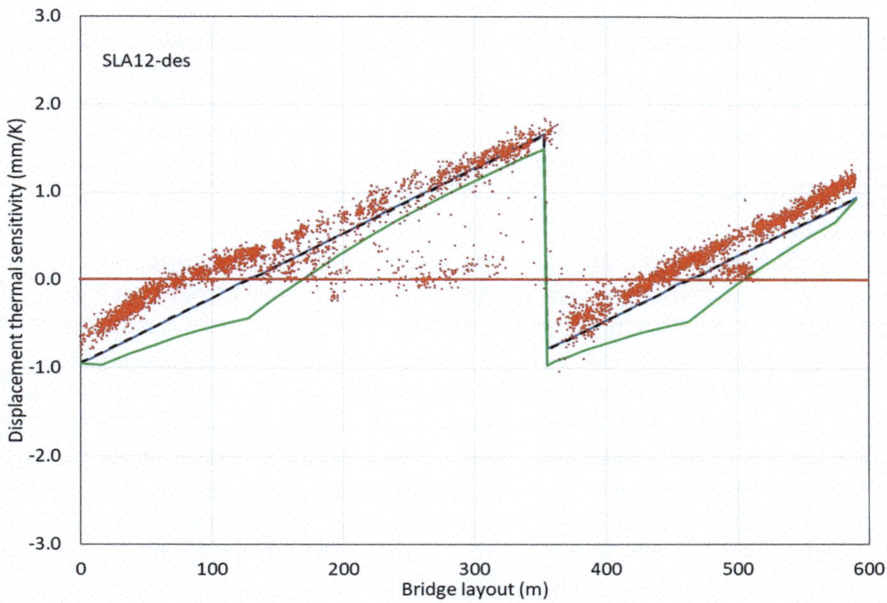

Fig. 7.9 Comparison between InSAR and FEM thermal sensitivity data for the LOS direction obtained from the SLA12-des stack (Jacques Cartier Bridge; May 2016–September 2018)

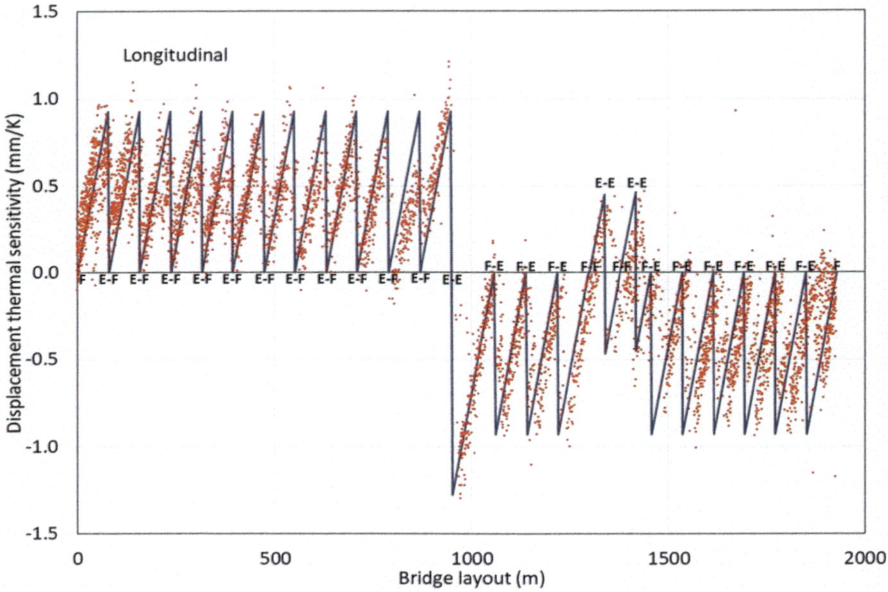

Fig. 7.10 Comparison between InSAR and predicted thermal sensitivity data for the longitudinal direction decomposed from the SLA26-asc and SLA12-des stacks (Victoria Bridge; May 2016–August 2018)

the entire length of the bridge. It is worth noting the well-captured effects on longitudinal thermal sensitivity from: (i) span length (three lengths: 38.1 m, 77.4 m, and 106.1 m) and (ii) movement direction according to the support conditions (see Fig. 7.10 where F = Fixed and E = Expansion). Data over Span 11 (between 790 m and 868 m), however, seem to indicate lower values of longitudinal thermal sensitivity compared to either the prediction or the InSAR measurements over identical adjacent spans.

A second comparison is to look at the vertical component of thermal sensitivity along the bridge layout line. The InSAR data set was created from corresponding point targets of the SLA26-asc and SLA12-des image stacks. As a result, a total of 4978 valid points were selected for the comparison with the theoretical predictions as illustrated in Fig. 7.11. In the figure, two predictions are illustrated for the vertical thermal sensitivity at deck level (blue curve) and along the top chord of the bridge superstructure (green curve) to account for the axial displacement of the vertical elements of the superstructure (which is proportional to the vertical shape of the superstructure). Somewhat similar to the above observation for the horizontal data, a relatively good match is found over the entire length of the bridge for the vertical component of thermal sensitivity including the 24 simply-supported spans, except for several groups of points lying above or below the expected range of values. It is not clear from which parts of the bridge structure these outlier points may be coming. However, the cause of the departure from the predicted envelope may be due to geolocation errors or violation of Assumption 3 as detailed in Sect. 7.5.1.

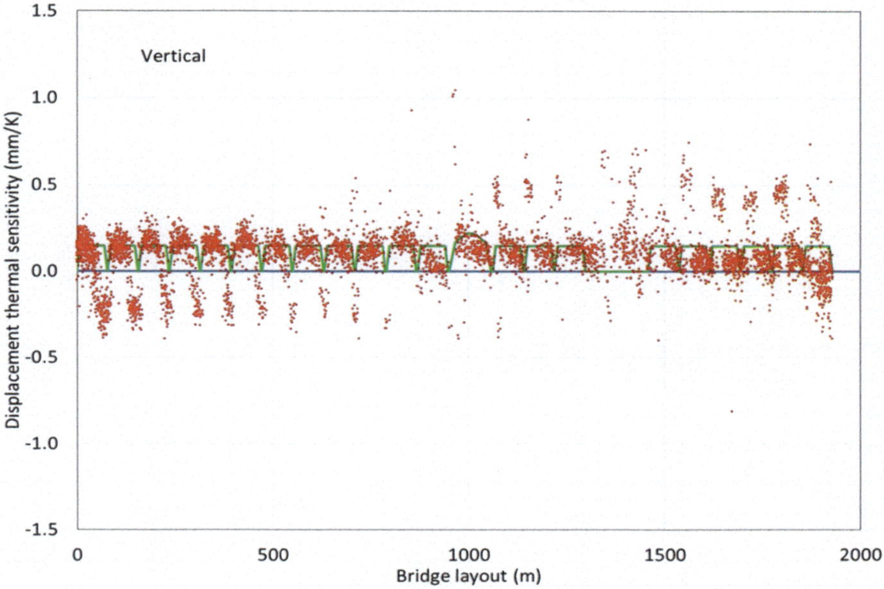

Fig. 7.11 Comparison between InSAR and predicted thermal sensitivity data for the vertical direction decomposed from the SLA26-asc and SLA12-des stacks (Victoria Bridge; May 2016–August 2018)

A third comparison is made by looking at the LOS measurement points taken directly from each stack of satellite imagery. InSAR data with high coherence values above 0.8 were selected, resulting in 16,703 points for SLA26-asc stack (Fig. 7.12) and 32,493 points for SLA12-des stack (Fig. 7.13). The LOS predictions were calculated with Eq. (7.3) for the respective satellite viewing geometry, where the blue and green lines represent the predictions at the deck level and top chord of the steel truss superstructure, respectively. Apart from the very good match observed, the predominant importance of the longitudinal component of thermal sensitivity is clearly seen.

As it was observed in Fig. 7.10 above, the trend at Span 11 with the slightly lower values of thermal sensitivity is observed again in Fig. 7.12; however, it is not seen in Fig. 7.13 from the SLA12-des stack. Another observation concerns the second half of the bridge (i.e., spans 13 and up), which appears to have slightly higher values than expected for each specific span. These two observations seem to indicate that it is more of an issue related to the stack viewing geometry, rather than an issue with the thermal behavior of the bridge. Viewing measured data from each satellite stack may give additional perspective on the interpretation of movement detected by each satellite stack since the north side of the bridge was predominantly seen by the SLA26-asc stack and the south side of the bridge was predominantly seen by the SLA12-des stack.

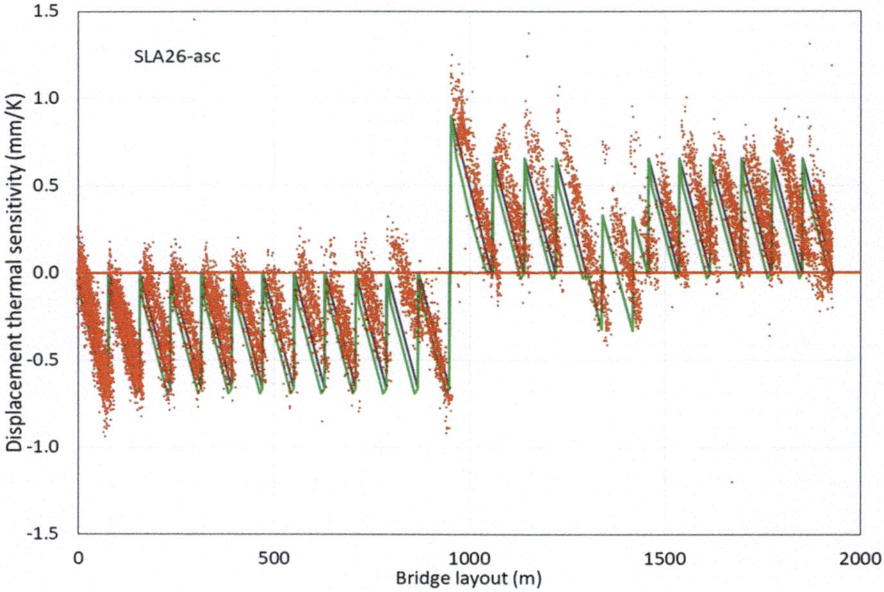

Fig. 7.12 Comparison between InSAR and predicted thermal sensitivity data for the LOS direction obtained from the SLA26-asc stack (Victoria Bridge; May 2016–August 2018)

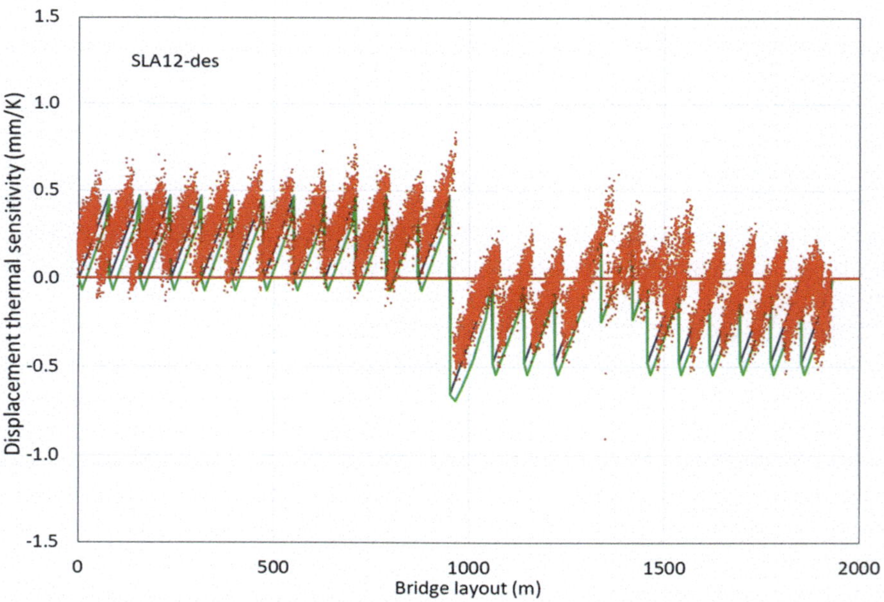

Fig. 7.13 Comparison between InSAR and predicted thermal sensitivity data for the LOS direction obtained from the SLA12-des stack (Victoria Bridge; May 2016–September 2018)

7.6 Lessons Learned from Satellite Monitoring of Bridges

The goal of this section is to discuss what has been learned so far in this ongoing research that has taken place at NRC in recent years on remote InSAR monitoring of bridges. In this research work, a total of four Canadian bridges have been monitored and satellite results assessed and validated, including:

- Jacques Cartier Bridge in Montreal, QC (as described above) https://goo.gl/maps/Ca3MQxk227gVraWk9
- Victoria Bridge in Montreal, QC (as described above) https://goo.gl/maps/YAGSe1ZG5s2yY2JT9
- North Channel Bridge in Cornwall, ON (Cusson et al. 2018) https://goo.gl/maps/TbCLYZAidHQazL436
- Confederation Bridge in PEI (Cusson and Ozkan 2019) https://goo.gl/maps/xnkvw9DStHghvmT86

Lessons Learned from FEM Studies

As part of the ongoing research work, FEM parametric studies have been conducted in order to demonstrate the thermal behavior of different bridge types and the effects of various parameters on bridge thermal sensitivity. In this section, some of the outcomes are presented briefly. More detailed descriptions of the FEM can be found in Cusson and Ozkan (2019).

Single Material Bridges vs. Composite Bridges

The effect of having structural materials with different thermal expansion rates in bridges on the thermal sensitivity was investigated. For the evaluation, the finite element model of the main section of the Jacques Cartier Bridge (outlined in Sect. 7.4.1) was used. A comparison was conducted between the original model of the bridge superstructure made of a single primary material (structural steel), which is the accurate structural representation of the steel superstructure, and an NRC-modified model with a concrete slab (shell elements) attached to the bottom of the steel superstructure, thus increasing its overall stiffness (however, in reality, the concrete slab of the Jacques Cartier Bridge is not rigidly connected to the steel superstructure).

From the finite element analysis, it was observed that the vertical component of thermal sensitivity of the all-steel bridge model at the deck level was practically negligible. For the modified bridge model, even with the increased stiffness due to the attached concrete deck, the vertical thermal sensitivity at the deck level was significantly larger in comparison, which is due to the difference between the CTEs of the steel and concrete members of the modified model. In the longitudinal direction, both models demonstrated similar linear thermal expansions. As expected, the modified model predicted slightly smaller longitudinal expansion due to the slightly lower coefficient of thermal expansion of concrete compared to that of steel. Results show that the thermal behavior of composite bridges, as seen by the satellite in the LOS direction, may be significantly affected by their vertical thermal sensitivity; however, that of single material bridges may be mostly affected by the longitudinal thermal sensitivity.

Effect of Span Length

In order to demonstrate the effect of span length on the vertical component of thermal sensitivity of bridges made of concrete deck on steel girders, the North Channel Bridge finite element model outlined in Cusson et al. (2018) was used. The North Channel Bridge has 4 continuous spans with lengths of 78 m, 90 m, 90 m, and 78 m. The parametric study was conducted by varying these span lengths by ±10% with the assumption that the bridge design remains valid. The modified models demonstrated the expected increase in vertical thermal sensitivity with increasing span lengths. Changes in vertical displacement were found to be between 21% and 37% for a 10% change in span length (due to bending effects caused by a difference between the CTE of steel girders and that of concrete deck). It was concluded that span length is a very significant factor affecting the vertical thermal sensitivity of bridges made with a concrete deck on steel girders.

Effect of Span Continuity

A comparison of thermal responses of multiple-span bridges with either continuous vs simply-supported spans was conducted using the four-span model of the North Channel Bridge (Cusson et al. 2018). The original model was used to represent a continuous span bridge (which is the true representation of the bridge) and a modified model to simulate the bridge with four simply-supported spans. Significantly different thermal displacement patterns were observed. As expected, the deflections of the simply supported span model were much larger than those of the continuous span counterpart demonstrating that vertical thermal sensitivity (due to bending effects caused by a difference between the CTE of steel girders and that of concrete deck) is significantly affected by the connection type between the spans of a bridge.

Effect of Temperature Gradient

The effect of temperature changes with and without a temperature gradient across the depth of the bridge structure on the thermal sensitivity of concrete-slab-on-steel-box-girder bridges was evaluated by using a simplified version of Cornwall's North Channel Bridge model outlined in Cusson et al. (2018). In a first step, three runs were conducted for uniform temperature changes from 0 °C to T_1 (being 5 °C, 10 °C, or 15 °C). In a second step, three more runs were conducted for the same temperature changes but with a 30 °C gradient across the depth of the concrete slab ($T_2 - T_1 = 30$ °C), as determined for Type B structures according to the Canadian Highway Bridge Design Code (CSA-S6 2018) for summer conditions, while the temperature across the depth of the girders would remain uniform (see Fig. 7.14).

As expected, larger changes in uniform temperature resulted in larger vertical deflections due to bending effects. The presence of a temperature gradient was found to significantly modify the deflection patterns. Longitudinal displacements for the same set of temperature changes with and without the above specified temperature gradient were also investigated. As expected, higher increases in longitudinal displacement along the bridge length were observed with increasing uniform temperature changes. The presence of a temperature gradient in the concrete slab also resulted in higher increases in the longitudinal displacement along the bridge length.

Using Eq. (7.3) above, the LOS thermal sensitivity for the SLA74-asc satellite viewing geometry was calculated from the vertical and longitudinal components of displacement obtained from the numerical model for the two cases of temperature changes (i.e., with and without a temperature gradient in the slab). The process led to exactly the same LOS thermal sensitivity regardless of the gradient effect on thermal displacement. From this observation, it may be concluded that the thermal sensitivity obtained from InSAR analysis of satellite imagery is not influenced by temperature gradients across the depth of the bridge structure. This greatly simplifies the assumptions and calculations when comparing FEM predictions to InSAR measurements of thermal sensitivity. In fact, for InSAR measurements obtained from descending satellite passes (occurring at around 6 am local time in Cornwall), the bridge temperature profile may be assumed to be similar to the uniform temperature case; while for InSAR measurements obtained from ascending satellite passes (occurring at around 6 pm local time in Cornwall), the bridge temperature profile may be assumed to be similar to the gradient temperature case (with the steel girders still at ambient temperature). Both cases should result in similar temperature sensitivities detected by the satellite, according to this parametric study.

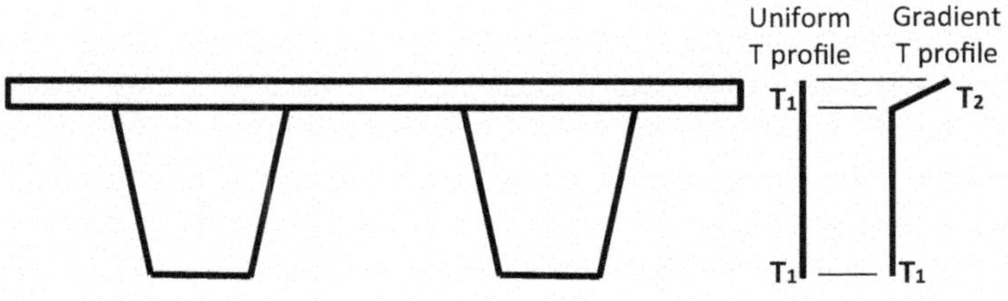

Fig. 7.14 Temperature profiles applied across the depth of bridge structure

Lessons Learned from Satellite Data Assessment

Bridge Geometry and Materials

The backscatter analysis of the Confederation Bridge reported in Cusson and Ozkan (2019) revealed two important characteristics of the bridge response to C-band radar satellites. The first one is that all the bridge signals were in layover with the water for both ascending and descending stacks of satellite imagery analyzed. This means that the phase stability of all observed features on the bridge was reduced because of a combined response with the constantly changing water surface. The second observation was that only very few elements of the bridge could be expected to provide an intense response because of the type of construction (e.g., smooth concrete surfaces with only few elements providing strong backscatters) and the orientation of the illumined features. The double bounce responses from the far-range lamp posts were the most intense responses observed on the bridge. The result of these two characteristics was that there were very few coherent features available for InSAR analysis, with lamp posts being the primary coherent elements of the bridge, which are not ideal for bridge monitoring purposes.

In contrast to the Confederation Bridge, there were significantly more coherent features present on the Jacques Cartier Bridge and the Victoria Bridge. The steel truss superstructures of these two bridges provided many additional elements from which a phase-stable response can be expected. In particular, the double bounce response from the vertical elements of the steel superstructures appeared to provide highly intense and coherent responses.

Number and Type of Stacks of Satellite Imagery

In this project, each bridge was monitored and analyzed using SAR data from two independent stacks of satellite imagery with opposite viewing geometry (ascending vs. descending). Typically, only one stack is required to conduct such analysis; however, having access to two stacks proved to be useful on several occasions. It helped clarifying some specific observations by allowing to conduct 2D decomposition in discretizing displacement into vertical and longitudinal components. It also helped confirming or disproving trends observed from one stack; however, at twice the computational cost.

For the Confederation Bridge, Ultra-Fine stacks (with 3 m × 3 m ground spatial resolution) were used instead of Spotlight stacks (with 1 m × 3 m ground spatial resolution) for InSAR analysis since the bridge—being 13 km long—could not fit entirely within the smaller footprint of the higher resolution Spotlight stacks. An open question is whether the InSAR analysis accuracy could be improved by using two adjacent Spotlight stacks from the same viewing geometry to cover the entire bridge with the highest resolution imagery available from RADARSAT-2. Another advantage maybe that a greater portion of the land at both ends of the bridge could be covered, helping to obtain stable ground references for the InSAR analysis.

Summary of Main Bridge Features Related to Satellite InSAR Monitoring

Based on what was learned in the past two years from the case studies on the four above-mentioned bridges, Table 7.1 summarizes the advantages and disadvantages that each bridge has when it comes to satellite-based monitoring. It can be seen from Table 7.1 that all four bridges have both advantages and disadvantages to various degrees with regard to their suitability of being easily monitored by radar satellites. In some cases, a certain bridge feature can be both an advantage and a disadvantage depending on given factors; for example, a bridge with a steel truss superstructure over the deck can offer many scattering elements for InSAR monitoring, which however can also be an issue due to the complex truss geometry making the bridge structure prone to several layover, shadow, and multi-bounce effects.

Another general observation is regarding the specific features of the Confederation Bridge. Although the bridge has some interesting features similar to those found on the other bridges, their benefits had been overshadowed by the specific disadvantages found at the Confederation Bridge. For example, one may think that the Confederation Bridge, with long spans, no complex structures over the deck, and being away from any adjacent structures would have all the required features for being easily monitored by satellites. Unfortunately, it was not found to be the case for the reasons indicated in Table 7.1. One question remains open however about whether the satellite monitoring could be improved by using two adjacent ascending (or two adjacent descending) *Spotlight* stacks instead of the lower resolution *Ultra-Fine* stack. A mitigation strategy may be the installation of trihedral corner reflectors at strategic locations of interest on the bridge deck to provide a good number of clear targets for InSAR monitoring. These reflectors should be sized, placed, and oriented to provide sufficient backscatter amplitude to ensure good coherence in the presence of any layover effect from other scattering elements. This may be the subject of future research.

7.7 Conclusions

Two independent stacks of satellite imagery from RADARSAT-2 were collected over the Jacques Cartier and Victoria bridges for a period of two years, then processed and analyzed with the InSAR method to extract the bridge displacement measurements. The validation consisted of comparing the InSAR measurements of displacement thermal sensitivity to numerical and analytical predictions of thermal displacement for each stack of satellite imagery. The comparisons showed good

Table 7.1 Main advantages and disadvantages of studied bridges related to satellite InSAR monitoring

Bridge monitored	Advantage/Disadvantage	Impact
North Channel Bridge Advantages	• Moderately long spans (78–90 m) • Steel railings on sidewalks • Slow traffic (50 km/h max)	• Large longitudinal displacement detectable by satellite • Many returns to satellite for displacement detection • Low vibration amplitudes, which reduce noise level
Disadvantages	• Adjacent to old bridge structure • North-south orientation	• Shadow effect reducing number of returns • Detection less sensitive to longitudinal displacement • Layover and multi-bounce effects with water surface
Jacques Cartier Bridge Advantages	• Long spans (128–334 m) • Steel truss superstructure • East-west orientation	• Large displacements easily detectable by satellite • Many returns to satellite for displacement detection • Detection sensitive to longitudinal displacement
Disadvantages	• Steel truss superstructure	• Complex geometry prone to layover & shadow effects
Victoria Bridge Advantages	• Steel truss superstructure • East-west orientation	• Many returns to satellite for displacement detection • Detection sensitive to longitudinal displacement
Disadvantages	• Steel truss superstructure	• Complex geometry prone to layover & shadow effects
Confederation Bridge Advantages	• Long spans (155–250 m) • SW-NE orientation	• Large displacements easily detectable by satellite • Detection moderately sensitive to longitudinal displacement
Disadvantages	• Lack of elements providing strong backscatters • Changing water level • Very long bridge over water	• Few returns to satellite for displacement detection • Multi-bounce effects with incoherent water surface • Lower resolution imagery covering larger area • Small land area in imagery offering no stable reference • Large body of water increasing risk of layover effects

agreement, in general, between the thermal displacement measurements and their analytical predictions both based on available ambient temperature data for the given bridge locations. It was therefore possible to explain most of the trends observed in the satellite measurements, which, for example, depended on factors such as span length, support type, superstructure height, and thermal expansion of the main structural materials.

A few outlier groups of InSAR data points, however, could not be explained readily, which may be due to InSAR processing difficulties related to 3D geo-location, atmospheric phase residuals, layover effects, and other factors, especially when it is difficult to know from exactly which element of the bridge a given signal is originating. It was found useful to have two stacks of satellite imagery to confirm a given observation made from one stack with a second independent image stack looking at the bridge from a different angle. Another benefit from having InSAR data from two opposite image stacks is the possibility to perform a 2D decomposition on the combined data sets in order to obtain the true vertical and true horizontal (longitudinal or transverse) components of displacement, which can help identify the source of movement. However, these benefits come at more than twice the computational cost. A second type of displacement data available from the satellite imagery was the displacement time rate (units of mm/year), which does not depend on temperature, and thus can be the result of mechanical effects, soil settlement, etc. For the Jacques Cartier and Victoria bridges, the displacement time rate (or velocity) data coming from the two stacks of satellite imagery indicated very little movement. In summary, the results so far show great promise and value in applying satellite-based technology for the remote monitoring of highway and railway bridges in order to optimize preventive maintenance management, extend structural service life, minimize traffic disruptions due to late repairs, and ensure structural integrity.

Acknowledgments The authors would like to acknowledge the financial contributions from Transport Canada (with special thanks to Daniel Hébert and Howard Posluns) and Infrastructure Canada through NRC's Initiative entitled Climate-Resilient Buildings and Core Public Infrastructure. The technical assistance from the owners/operators of the monitored bridges is greatly appreciated, including Emanuel Chênevert from Jacques Cartier and Champlain Bridges Inc., and Hoat Le from Canadian National Railway. This contribution was critically reviewed by David Huntley (Geological Survey of Canada).

References

Acton, S. 2013. Sinkhole Detection, Landslide and Bridge Monitoring for Transportation Infrastructure by Automated Analysis of Interferometric Synthetic Aperture Radar Imagery, Final Report No. RITARS11-H-UVA, University of Virginia.

CSA-S6. 2018. *Canadian Highway Bridge Design Code*, 894. Toronto, Ontario: Canadian Standards Association.

Cusson, D., and I. Ozkan. 2019. Satellite-Based Remote Structural Health Monitoring of Bridges for Performance Management and Climate Change – Year 2, NRC Report A1-014011.1, March 2019, 92.

Cusson, D., K. Trischuk, D. Hébert, G. Hewus, M. Gara, and P. Ghuman. 2018. Satellite-based InSAR Monitoring Validated for Highway Bridges – Validation Case Study on the North Channel Bridge in Ontario, Canada. *Transportation Research Record*: 11. https://doi.org/10.1177/0361198118795013.

Ferretti, A., A. Monti-Guarnieri, C. Prati, F. Rocca, and D. Massonnet. 2007. *InSAR Principles: Guidelines for SAR Interferometry Processing & Interpretation, European Space Agency Publication TM-19*, 48. Noordwijk, The Netherland: European Space Agency.

Greene Gondi, F., J. Eppler, P. Oliver, and M.A. McParland. January 2019. *Satellite Monitoring of Highway Bridges for Disaster Management Under Extreme Weather – Year 2 Technical Report to NRC*, 112. Richmond, BC, Canada: MDA Geospatial Services.

Marinkovic, P., G. Ketelaar, F. Van Leijen, and R. Hanssen. 2007. InSAR Quality Control - Analysis of Five Years of Corner Reflector Time Series, Fifth International Workshop on ERS/ Envisat SAR Interferometry, 'FRINGE07', Frascati, Italy, Nov 26–30.

MDA. 2018. RADARSAT-2 Product Description, Report Number RN-SP-52-1238, Available at https://mdacorporation.com/docs/default-source/technical-documents/geospatial-services/521238_rs2_product_description.pdf?sfvrsn=10, Issue 1/4, September 10, 91.

NCHRP. 2007. *Bridge Inspection Practices, Synthesis 375, National Cooperative Highway Research Program*. Washington DC: Transportation Research Board.

———. 2017. *Post-Extreme Event Damage Assessment and Response for Highway Bridges, Synthesis Report 497, National Cooperative Highway Research Program*, 91. The National Academies Press.

Transport Canada. 2012. Road Transportation, https://www.tc.gc.ca/eng/policy/anre-menu-3021.htm.

Chapter 8
Interferometric Synthetic Aperture Radar (InSAR) in the Context of Bridge Monitoring

S. Selvakumaran, C. Rossi, E. Barton, and C. R. Middleton

8.1 Introduction

Our cities are a complex mix of diverse ecosystems, institutions, assets and infrastructure. Within this ecosystem, bridges form key links within transport networks. A study by the RAC Foundation (2019) found that within the UK alone, 71,652 bridges were managed by local authorities in 2018, of which 3177 (4.4% of the total) are categorised as 'substandard' (meaning that they are unable to carry the heaviest vehicles now seen on roads, including lorries of up to the legal load limit of 44 tonnes) and will require weight restrictions, increased monitoring or even managed decline. The different regional Highway authorities were found to have 1 to 8% of their bridge stock as 'substandard'. Within the USA, the American Society of Civil Engineers found 9.1% of bridges to be structurally deficient in 2016, and in that year on average there were 188 million trips across a structurally deficient bridge each day (American Society of Civil Engineers (ASCE) 2017). The implications of bridge asset failure can be profound in terms of safety, economic prosperity and social well-being of communities.

Current standard practice for the monitoring of bridges in most countries is to periodically schedule visual inspections, relying on inspectors to be able to spot signs of problems or unusual behaviours before they reach a catastrophic stage. The subjective nature of human judgement is useful to identify non-standard behaviours and to apply

S. Selvakumaran (✉) · C. R. Middleton
Engineering Department, University of Cambridge, Cambridge, UK
e-mail: ss683@cam.ac.uk; crm11@cam.ac.uk

C. Rossi
Satellite Applications Catapult, Didcot, UK
e-mail: Cristian.Rossi@sa.catapult.org.uk

E. Barton
National Physical Laboratory, Teddington, UK

V. Singhroy (ed.), *Advances in Remote Sensing for Infrastructure Monitoring*,
Springer Remote Sensing/Photogrammetry,
https://doi.org/10.1007/978-3-030-59109-0_8

a case-specific approach, but previous studies (Moore et al. 2001; Lea and Middleton 2002) highlight that this does not always provide reliable results. For example, a study by Moore et al. (2001) published by the Federal Highways Authority in the USA points out some of the weaknesses in the current approach in the USA where there was a significant amount of variation in assigned Condition Rating (scaled numerically from 0 to 9) during visual inspection. This variation was found both in terms of thoroughness of inspection records, as well as in terms of which sections of the bridge were inspected. The study found that the data collected depended on factors such as the level of training obtained by the inspector, whether the inspector was colour-blind, whether the inspector had a fear of traffic or heights, or what the weather was like on the day of inspection. This would suggest a need to supplement visual inspections with less subjective insights. A balance also needs to be struck between having frequent, regular asset monitoring and the cost of doing so, The cost of doing so refers to financial cost, as well as costs with regard to the disruption caused to the network (e.g. by bridge closure, Fig. 8.1). Consequently, inspections are typically carried out every few years.

Earth observation data from satellites is becoming more accessible, and with improving spatial and time resolutions. This data can provide a means of remote monitoring which covers large geographical regions and provides insights into infrastructure assets that are difficult to access or regularly monitor with conventional approaches. Visual inspection is invaluable in spotting potential issues in a localised means that is inaccessible to most technologies, but remote sensing can provide a different context by measuring ground movements on and around bridge assets that cannot be monitored by eye. For example, in Fig. 8.2, the deformation velocity of various points across several square kilometres shows that the wider area around single assets may have a large influence on the behaviour of the single asset, in this case settlement. Within the asset management context outlined, satellite monitoring provides the opportunity to remotely collect information related to an asset in an interval period of days. Thus, there are opportunities in being able to monitor for signs of unusual behaviour that develop in the periods between inspections or are not picked up visually. Remote satellite monitoring offers a further advantage over traditional in-situ sensor monitoring in that it does not require an electrical connection or power source at site, or the closure of and physical access to the bridge.

Interferometric Synthetic Aperture Radar (InSAR) techniques, in particular, have the capability to provide wide-area, high density, remote measurements of displacement at millimetre-scale (Bamler and Hartl 1998; Ferretti et al. 2001). A stack of Synthetic

Fig. 8.1 Closure of railway lines to inspect the top and side of the bridge deck

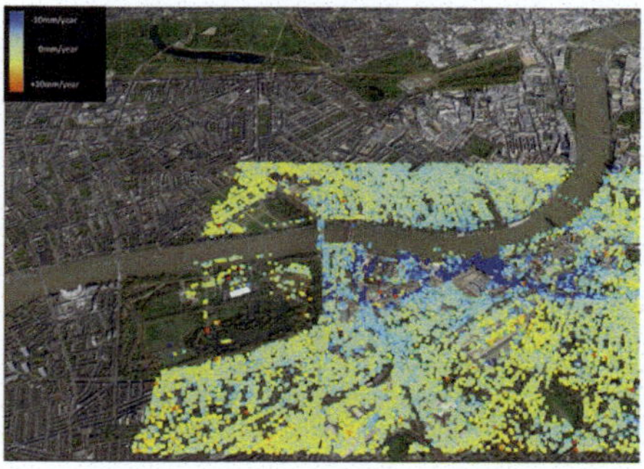

Fig. 8.2 InSAR results visualising velocity of numerous points over several kilometres where the ground in settling (blue points indicating settlement of ground)

Aperture Radar (SAR) images acquired at different time instances can be processed using InSAR techniques to monitor millimetre-scale changes over time of multiple discrete points over large spatial areas. These SAR images are obtained even in darkness and adverse weather conditions. Civil engineers stand to gain from research and development of InSAR, as such technologies may support them in being better equipped to predict and prevent structural failure. As part of a wider toolkit of monitoring resources, InSAR may support better inspection, operation and maintainence of bridge assets.

8.2 InSAR Techniques

SAR Imaging

The following section provides only a brief introduction to the theory of InSAR to provide an understanding of the key elements related to the application of InSAR for bridge monitoring. It avoids detailed descriptions and derivations, for which there are a number of suitable reference books. For further information on SAR image formation, books by Bamler and Schättler (1993) or Cumming and Wong (2005) provide useful descriptions. Bamler and Hartl (1998) and Hanssen (2001) are key literature references on Interferometric SAR (InSAR) processing. There are also various other tutorial papers and guides available on SAR principles and theory (Ferretti et al. 2007a, b, c; Moreira et al. 2013).

Satellites with the objective of SAR data collection are designed to follow a near-polar orbit (i.e. they travel in a north-south direction from pole to pole) around the Earth. The satellites travel northwards on one side of the Earth (the ascending pass) and then towards the south pole on the second half of their orbit (the descending pass). These SAR satellites have a side-looking imaging geometry and take images by means of radar pulses installed on a forward moving satellite. The impact of this imaging geometry is that the satellites are much more sensitive to capturing deformation changes in an east-west direction, and capture only a small part of any deformation that occurs in a north-south direction.

The pixels in a SAR image are associated with an area on the Earth's surface (the size of the resolution cell varies with each satellite) but taken in SAR imaging coordinates. A geocoding process is required to convert the SAR data into a map

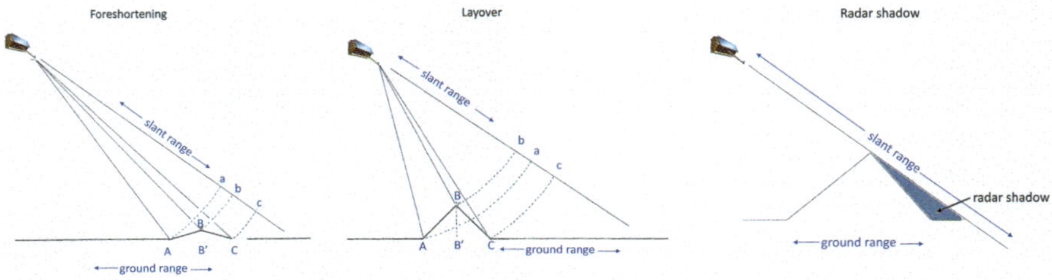

Fig. 8.3 SAR sideways-looking imaging illustrating foreshortening (left), layover (middle) and radar shadow (right) effects

projection so that each pixel in the SAR image is directly associated with the position on the ground. Before interpreting the InSAR results, it is important for bridge engineers to understand that there are SAR imaging effects and that the SAR images do not necessarily map directly to classical geographic references and optical images we are used to seeing. For example, SAR images typically appear geometrically distorted. In a radar image, the three-dimensional objects of the scene are mapped to a two-dimensional image in slant-range and azimuth. This results in effects such as foreshortening, layover and radar shadows (Fig. 8.3).

The values in each pixel comes from the received radar reflections and is complex in nature, with each pixel containing amplitude and phase values. Objects are monitored using the coherent processing of back-scattered signals from multiple distributed targets within a SAR resolution cell. Any objects causing a scattering of a SAR signal are called scatterers. The SAR satellites operate using specific wavelengths on the electromagnetic spectrum with L-band, C-band and X-band being the predominate wavelengths. X-band data (approximately 3 cm in wavelength) is more suited to bridge monitoring as it provides higher spatial resolution than L-band or C-band satellites, but is currently only commercially available (not available on open-access data such as the ESA Sentinel constallation). This brings an additional cost to overall monitoring that must be considered.

Multi-Temporal InSAR

The satellite generates the outgoing SAR signal (with known phase) which can then be compared to the phase of the return signal which is dependent on the distance to the ground. The complex phase vector information of one image is multiplied by the complex conjugate phase vector information of another image to form an 'interferogram'. This results in the common backscatter phase in each resolution element being cancelled whilst leaving a phase term proportional to the differential path delay. Thus, an 'interferogram' is a complex image with each pixel comprising a phase difference (from 0 to 2π radians) between two distinct SAR snapshots of a given horizontal resolution. Differential InSAR (Gabriel et al. 1989) makes use of differential interferograms (two interferograms) relevant to the area of interest, to compute the deformation of the observed area (Fig. 8.4).

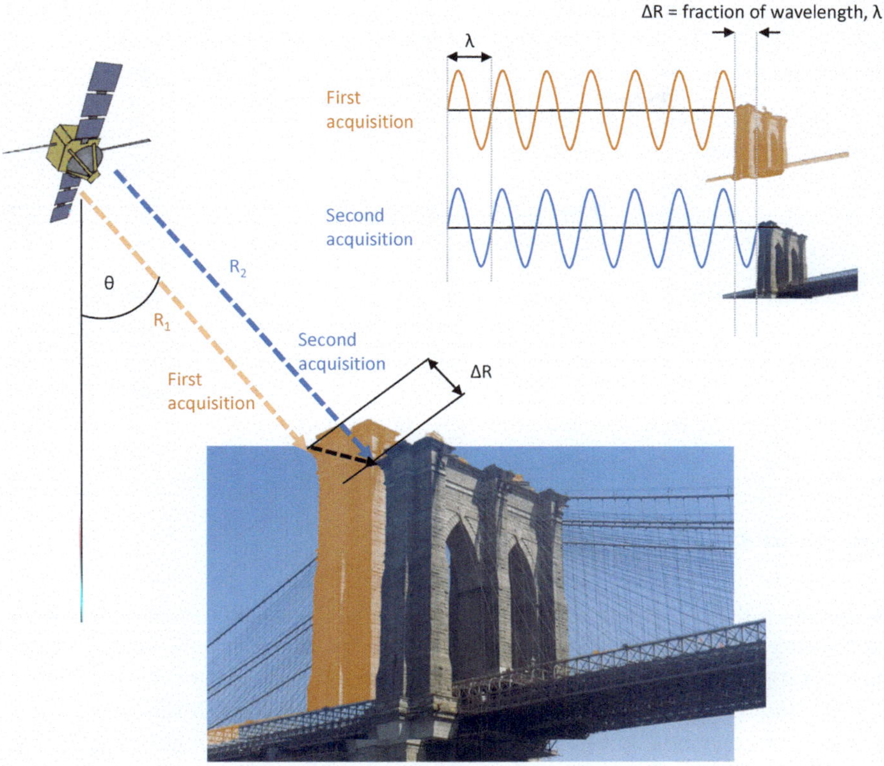

Fig. 8.4 Simplified explanation of satellite geometry for differential imaging

Multi-temporal InSAR techniques involve the processing of multiple SAR acquisitions over the same area to allow for the correction of phase noise terms and therefore reduce the errors associated with the deformation estimates. The end result is a time series measuring change in deformation in the line of sight (LOS) of the satellite over time. Several approaches have been described in the literature in order to extract the different contributions of interest. Two of the most established algorithms are used within this Chapter. The choice between these two methods is based on the scattering mechanism of the target being studied. The Persistent Scatterer Interferometry approach (Ferretti et al. 2000, 2001) is employed for single or dominant point scatterers, whilst the Small Baselines Subsets (SBAS) approach is used for distributed targets (Berardino et al. 2002a). There are several other algorithms which develop and vary these methods, and a review and comparison of different methods in the literature are presented by Crosetto et al. (2016).

Within the context of bridge monitoring, SAR imaging and InSAR processing techniques have limitations as to which types of deformation may be suitable for monitoring. For example, using an X-band SAR sensor with a wavelength of 31 mm limits the movements that can be unambiguously defined to measurements of

7.75 mm or less between acquisitions. To monitor deformation rates of ground surrounding bridge structures, the maximum detectable velocity depends on (Bovenga et al. 2018):

- the spatial gradient of the deformation;
- the spatial density of scatterers;
- the radar wavelength;
- the temporal sampling of the acquisitions defined by the revisit time of the SAR satellite.

Considering the wavelength and revisit time of the TerraSAR-X and Sentinel-1 satellites as an example, the maximum differential deformation rate measurable is 25.7 cm/year and 42.6 cm/year, respectively (Crosetto et al. 2016). These values are theoretical and in practice also depend on the noise level of the data and the specific phase unwrapping technique used to resolve phase ambiguities.

Applications

A review of the literature reveals that the potential for smaller scale, single asset monitoring using InSAR is growing, but bridges in particular have been developing as a topic of interest within the InSAR research field in recent years. These include studies on precursors to failure (Sousa and Bastos 2013; Milillo et al. 2019) as well as studies related to understanding bridge motion and behaviour (Cusson et al. 2017; Cusson and Ozkan 2019; Qin et al. 2017). The relevant studies for bridge applications also include those related to image interpretation (Qin et al. 2017; Soergel et al. 2006), Persistent Scatterer three-dimensional positioning (Dheenathayalan et al. 2018; Gisinger et al. 2015) and thermal expansion (Lazecky et al. 2015; Monserrat et al. 2011).

8.3 Monitoring Measurement Uncertainty for Bridge Applications: Waterloo Bridge

Introduction

For applied monitoring uses, it is important to consider the scale of measurement accuracy that InSAR can provide. There are several studies in the literature which deploy reflectors to create SAR reflection points (Ferretti et al. 2007d; Quin and Loreaux 2013). Whilst these studies assess resolution, sensitivity and uncertainty of InSAR measurement data, it is also important to consider these properties within an applied bridge engineering context—including the measurement of vertical and horizontal displacements of the bridge deck with time, specifically with respect to temperature expansion and other loads. A study was undertaken using Waterloo

Bridge over the River Thames in London, United Kingdom. Further details on this study can be found in (Selvakumaran et al. 2020).

It is important for end users to understand that not all bridges reflect SAR waves well. Depending on the context of the bridge's location (for example, bridges over water) the reflections may bounce off multiple sources before being returned to the SAR satellite (Qin et al. 2017). In the case of Waterloo Bridge, the concrete structure provided few natural persistent scatterers and the signal noise from other objects was very low. Taking advantage of such properties enabled an experiment to be set up, comparing satellite measurements at specific points (SAR reflections being artificially created by means of corner reflectors) with traditional surveying by means of an automated total station and survey prisms (Fig. 8.5). The corner reflectors were installed at each bridge pier and at the two expansion joints to monitor the key structural points of interest (6 in total on each side, Fig. 8.6). Reference surveying prisms were also installed on the river embankment (i.e. in locations not on the bridge) as reference points for the automated total station (ATS).

Corner Reflector Installation

A trihedral radar reflector, commonly known as a 'corner reflector', was selected for use in this study (Fig. 8.7). This type of reflector facilitates a triple-bounce of the incident radar energy from three mutually orthogonal plates such that the incoming

Fig. 8.5 Installed corner reflector and survey prism over one pier on the east side of the Waterloo Bridge. Image on bottom right presents a view of the installation of one of the corners as seen looking down from the bridge deck

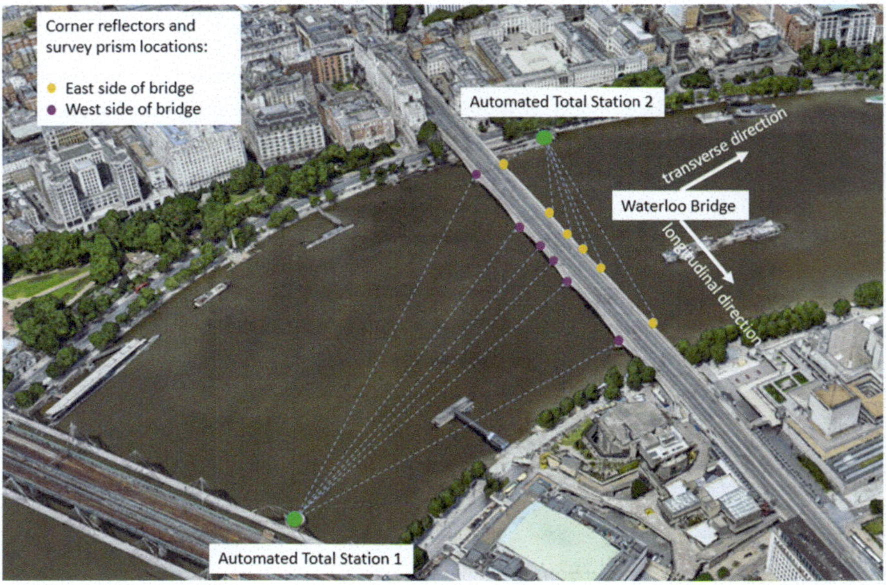

Fig. 8.6 Location of Automated Total Stations and corner reflectors with survey prisms on the bridge (base imagery and map data provided by Google Earth (2020). Reference prisms are not marked on this diagram

radar wave emitted from the satellite is reflected back to the satellite receiver, so that a lot of energy would be recorded by the satellite sensor, i.e., showing as a bright spot in the intensity map (Fig. 8.8). Corner reflectors are the most commonly used type of radar target deployed for Persistent Scatterer Interferometry (PSI) (Doerry 2008). The Radar Cross Section (RCS) is a measure of the reflectivity (i.e. brightness) of a point-like target in a SAR image. The RCS of the corner reflector was sized to be significantly larger than the background clutter. For a corner reflector, the theoretical peak RCS value was sized using the following (in m^2) (Ruck and Krichbaum 1968):

$$\mathrm{RCS} = \frac{4\pi l^4}{3\lambda^2}$$

where l is the length of the non-hypotenuse sides of the right-angled isosceles triangular plate (Fig. 8.7), and λ is the radar wavelength used by the target SAR sensor.

In this instance, a balance was struck between having a reflector large enough to serve this purpose but also small enough in size to be permitted for installation on the bridge by the bridge owners. Specifically, the reflectors were made to have a corner to edge length, $l = 35$ cm, and were made of a perforated aluminium in order to be lightweight, durable for a 1-year period, and provide drainage. They were installed and orientated to face the line of sight of the ascending TerraSAR-X satellite on the west side of the bridge. The corner reflectors on the east side of the bridge were pointed towards the line of sight of the descending TerraSAR-X satellite.

Fig. 8.7 Trihedral corner reflector and mounting bracket with dimension l marked (left) as installed for measurement experiment conducted on site at the National Physical Laboratory (middle and right)

Prior to installation on the bridge, a reference study was undertaken to test the reflectors and better understand measurement uncertainty (Fig. 8.7). The set-up of this study was similar to other classic studies in the literature undertaken to investigate measurement accuracy (Ferretti et al. 2007d; Quin and Loreaux 2013). This study enabled a testing of the corner signal response prior to installation on the bridge (which would require road closure and expense in having operatives suspended from the bridge to install the corner reflectors and brackets).

The experiment found that there was some variation in repeat readings, giving a range of up to 1.5 mm between readings and the mean line around which they were centred. An assessment of uncertainty in deriving the vertical component of these readings was undertaken, and found to have an uncertainty of vertical measurement ±0.7 mm or better. From this experiment, it was determined that the relative displacement measurements obtained from SAR acquisitions were of millimetre-scale in precision, up to 2 mm. This confirmed that the corner reflectors would be suitable for the bridge installation, and provide asset owners with some confidence that the InSAR measurements were of the right scale to be considered for use in bridge monitoring and other civil engineering applications.

Surveying Data Analysis

An automated total station works by taking a line of sight distance and level readings from the automated total station machine to each survey prism. It does so by taking readings from the reference prisms and then measuring the location of the bridge prisms relative to the reference. These readings were taken in three dimensions at Waterloo Bridge: the longitudinal, transverse and vertical directions of the bridge.

InSAR Data Analysis

TerraSAR-X Stripmap mode SAR images were tasked and collected for this study. The bridge was closed in December 2017 for the installation of corner reflectors and other instrumentation, and the satellite acquisitions were tasked in ascending and descending directions from December 2017 until November 2018. The acquisitions were taken every 11 days at the same time of day for ascending (17:44) and descending (06:17). All acquisitions were taken with an incidence angle range of 36.11–38.5 degrees.

The corner reflectors created point targets which were clearly visible above any other noise or clutter (persistent scatterers) within each pixel (Fig. 8.8). As such, the PSI method (Ferretti et al. 2000, 2001) was deemed to be the suitable InSAR processing method, rather than any method optimised for more distributed scatterers. A stack of SAR acquisitions in the ascending and the descending directions were each processed separately using the SARscape software package (Sarmap SA 2014).

Results

The stacks of SAR acquisitions were collected for a period of time after the corner reflectors were installed. Processing these stacks using the persistent scatterer interferometry method revealed new persistent scatterers on the bridge compared to the

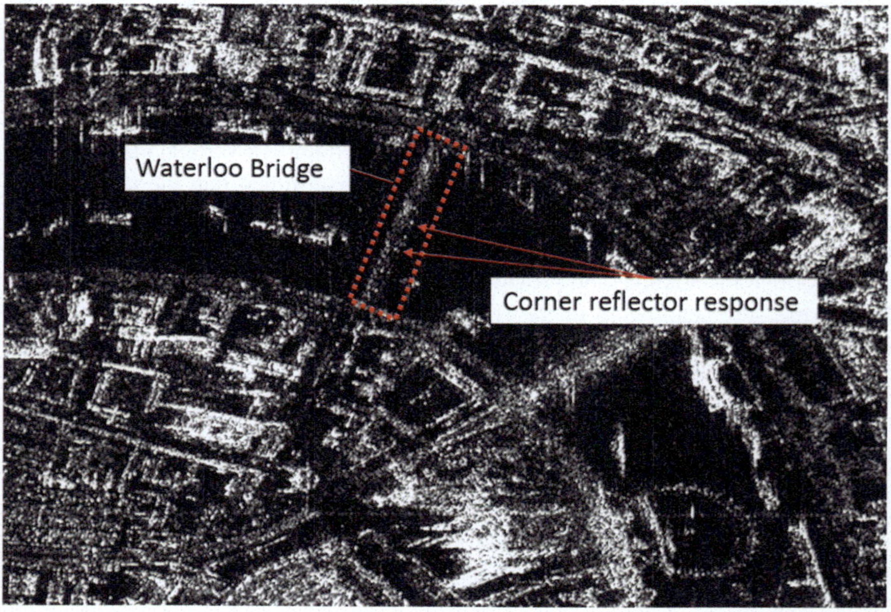

Fig. 8.8 SAR amplitude image after corner reflector installation with bridge marked and corner reflector response visible as bright points that appeared only in images following installation

processing of data stacks covering the same area during a period of time prior to the installation of corner reflectors. New persistent scatterers at each of the 12 corner reflector locations were expected, and after processing, 9 out of the 12 reflector locations were identified as persistent scatterers locations using InSAR processing. The reason that 3 corner reflectors were not picked up as persistent scatterers was not immediately apparent from inspection. There may be a number of different reasons for this. One possible reason is that these corners may have been obscured by debris or a bird nesting within it at the time of some of the acquisitions. Alternatively, may have become dirty (affecting coherence) or may have been installed with some misalignment by error during the installation. Each of these persistent scatterers had a line of sight measurement (measured at multiple points in time) associated with it. However, to a bridge engineer, this measurement in the line of sight of the satellite is arbitrary and not particularly useful. Useful measurements would be within the bridge's own reference system and in three dimensions in the longitudinal, transverse, vertical directions of the bridge.

Mathematically, the line of sight movement we have measured using SAR (d_{LOS}) can be considered as the equation below:

$$d_{\text{LOS}} = \underline{\text{A}\,\text{y}^{\text{T}}}$$

where

$$A = \left[\left(\cos\theta \quad \sin\theta\cos\alpha \quad \sin\theta\sin\alpha\right)\right]$$

$$y = \left[\left(d_{\text{V}}\, d_{\text{L}}\, d_{\text{T}}\right)\right]$$

α is the heading angle relative to the bridge (angle of the SAR satellite flight path, Fig. 8.9), θ is the incidence angle (angle between the SAR beam and the vertical, Fig. 8.9), and d_{V}, d_{L} and d_{T} refer to the vertical, longitudinal and transverse components of the displacement vector, respectively.

However, it is not possible to decompose a single line of sight measurement into three dimensions without some strong assumptions on the deformation behaviour. As a first step here, it was assumed that all bridge movement only occurred in the longitudinal direction (direction of road travel). Loading envelopes comprising various different loading scenarios were applied to a finite element structural model of Waterloo Bridge (using the software package LUSAS). This revealed that the predominant loading case was likely to be due to thermal expansion of the bridge in the longitudinal direction. This assumption was used to derive an estimation for the component of the line of sight displacement acting in the bridge longitudinal direction (Fig. 8.10). When plotting the ATS- and SAR-derived relative longitudinal movements as well as temperature over time (such as the example in Fig. 8.10), the seasonal variation seen in the SAR readings over the year matched the overall temperature profile and agreed with the broad season trend displayed in the ATS readings. However, on looking at the direct comparison of points (selecting the ATS readings at the time of SAR acquisition) there is variation between the two readings.

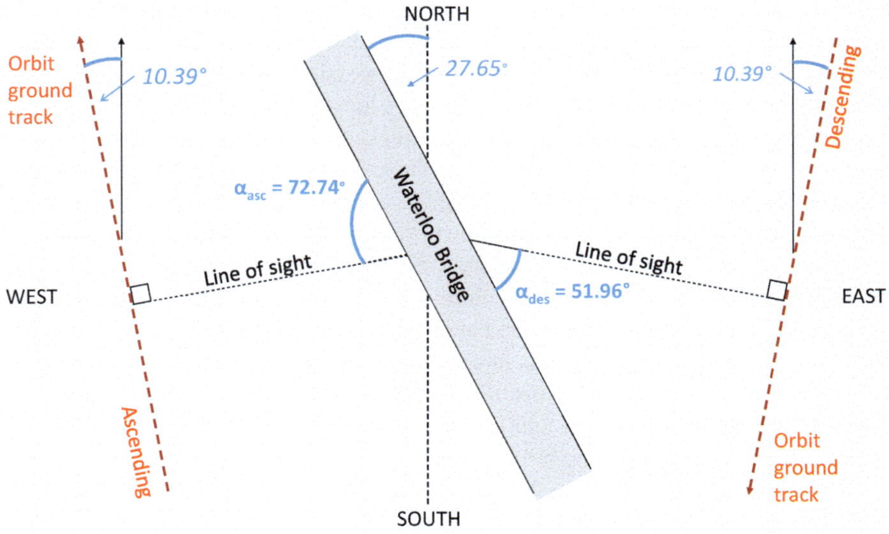

Fig. 8.9 Bridge orientation in relation to satellite imaging geometry. Image taken from Selvakumaran et al. (2020)

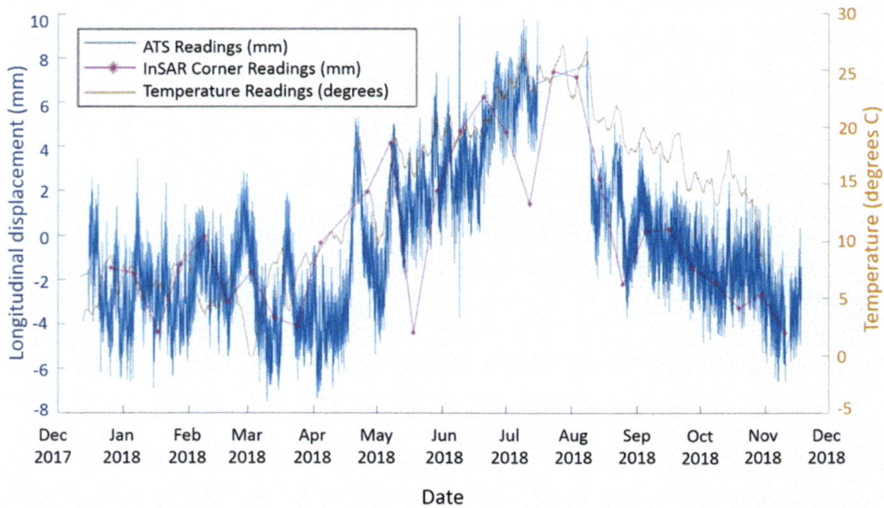

Fig. 8.10 Graph of ATS and SAR movements for a single corner reflector location on the east end side of the bridge relative to the end of the bridge

Other methods to determine three-dimensional surface displacement using InSAR include making use of InSAR line of slight measurements from different satellite viewing geometries (Hu et al. 2014). Studies in bridge monitoring have extended this three-dimensional deformation monitoring approach (Cusson and Ozkan 2019; Qin et al. 2018) but in this instance this was found to have limitations in using this approach for direct comparison with ATS readings. When considering the thermal response of the bridge, the observation period of the ascending and descending datasets (the two different look directions) was not the same. If the ascending direction acquisition is taken at a different time of day than the descending acquisition, the bridge will be subjected to different temperature at each of those times. This then means that the bridge will experience a different amount of movement according to each temperature. As such, the measurement readings could not be combined to form the total movement, and this approach was therefore not suitable for monitoring of bridges such as Waterloo Bridge.

As a first approximation of the Waterloo Bridge movement, it was assumed that all bridge movement was occurring in the longitudinal direction. Using a second line of sight direction would solve one further unknown. The bridge piers are concrete and of 35 m in length and 5 m in width, spanning the transverse length of the deck and a structural assumption of rigidity in the transverse direction can be employed. Using this assumption enabled the LOS measurements for the east and west sides of the bridge (from the ascending and descending acquisitions) to be combined by using acquisitions taken at the same temperature. If the structure is rigid in the transverse direction, zero transverse movement can be assumed and the LOS vector can be decomposed into longitudinal and vertical components. However, doing so here would have reduced the number of readings for which this approach could be taken (i.e. readings at the east and west side at same temperature instance) down to 10 instances. Then considering that the applied tidal load was not the same at each instance meant that there were fewer viable readings still.

In summary, the above study provided asset owners with an understanding of InSAR measurements in context with conventional bridge surveying methods and data. It showed that seasonal thermal expansion can be captured and monitored using InSAR measurements (but not daily temperature cycles which are limited by the current frequency of satellite acquisition being in the order of days rather than hours). The study also highlighted the possibility of using SAR measurements as a complementary measuring tool to aid traditional visual inspection regimes. Further relationships in comparing surveying measurements against InSAR measurements are investigated by Selvakumaran et al. (2020).

8.4 Monitoring Bridge Behaviour: Hammersmith Flyover

Introduction

In the context of many bridges, the presence of corner reflectors at key locations of structural interest is neither desirable nor practical. Here, the case study of Hammersmith Flyover provides an example of a bridge without corner reflectors or other retrofits, making use of 'natural SAR reflections' from the bridge for InSAR monitoring. Similar to the previous study, InSAR measurements of displacement were compared against traditional in-situ measurements, this time in the form of displacement sensors fitted onto the bridge rather than surveying methods.

Hammersmith Flyover is a 16-span segmental pre-cast concrete post-tensioned bridge structure carrying road traffic in and out of the west of London, United Kingdom. The bridge is able to expand and contract by means of roller bearings at the base of each of the 15 piers, with one expansion joint located near the centre of the bridge (and separating the bridge deck into two structurally independent sections). In recent years the bridge has experienced several problems. Some of these problems are not suited to being picked up by remote monitoring—for example, corrosion of the pre-stressing tendons did not result in visible movement of the bridge and do not cause any visible change on the surface level of the bridge. Other concerns can be monitored, such as problems in joints and articulation which prevent a bridge from moving in a normal manner.

Visual inspection of Hammersmith Flyover revealed concerns regarding bridge bearing performance at multiple locations. Bearings allow bridges to move under imposed loading, and the seizing or restricted movement of bearings can lead to significant problems due to restraint against expansion and contraction. This restraint could cause additional bending moments in the deck and induce stresses that the bridge was not designed to withstand. InSAR provides the opportunity to collect readings related to bridge movement in an interval period of days. Thus, it may enhance bridge monitoring by detecting signs of unusual behaviour that develop in the periods between visual inspections or are not picked up visually.

Deployed Traditional Monitoring Systems

From 2010 to 2018, the bridge was fitted with a number of different monitoring systems comprising a vast range of hundreds of sensors. For the purpose of this study, only those deployments providing information on thermal expansion of the bridge were considered:

- Temperature sensors embedded in the top, bottom and web of the deck at 4 mid-span locations of the bridge and
- Displacement transducers (Linear Variable Differential Transformers, LVDTs) used for measuring linear displacement at bearing locations at the base of each pier.

InSAR Data Analysis

Archive Cosmo SkyMed-X SAR data together with Cosmo SkyMed-X acquisitions tasked specifically for this research (both of the same pixel resolution of approximately 3 m by 3 m) were collected to provide a stack of remotely sensed records collected from 2011 to 2018. These images were taken in the descending direction with an incidence angle range of 25.1 to 27.9 degrees. Shuttle Radar Topography Mission (SRTM) data with resolution of 3 arc-second (90 m) was used as a Digital Elevation Model (DEM) during the interferometric processing (to determine the height information of PS). The SAR data (2011–2018 acquisitions) was processed using standard PSI algorithms as included in the SARscape software package.

Results

The InSAR processing results were then interpreted to extract the relevant PS of interest for this study. Figure 8.11 shows a typical output from PSI processing in the format of a shape file where PSs are marked as points, the location of which is determined during the geocoding stage of InSAR processing, where the SAR coordinates are mapped onto geographical coordinates. In this case the results were mapped onto the World Geodetic System 84 (WGS84) coordinate system. The shape file of PS locations can be opened in Geographic Information Systems (GIS) software packages to display the Cartesian coordinates of the PS points. Each of these data points contains a profile of line of sight movement over time, as well as characteristics describing the behaviour of the output products from the PSI processing (such as coherence value or linear velocity in mm/year).

The PS points can be overlaid onto an optical satellite image or map of the area (as in Fig. 8.11) to manually interpret the locations of the PS points as attributed to objects on the ground, such as buildings or bridges. It is worth reminding end users that each 'dot' represents an approximate location of the resultant reflection from a multi-looked pixel area.

It is clear from Fig. 8.11 that there are several regions of the bridge missing PS points. The bridge has a consistent form and structure, which would suggest that each section should be equally suitable for PSI application. However, the viewing geometry of a SAR image results in certain geometric distortions on the resultant imagery (Fig. 8.3). The side-view imaging in the slant-range direction causes imaging effects, such as foreshortening, layover and radar shadows previously mentioned. Another factor may be the reduced coherence of some pixels (thus rejection of potential PS candidates) due to seasonal effects (e.g. ice, snow or debris) and intermittent traffic or construction, as observed in Cusson and Ozkan 2019.

Combining Structural Models with InSAR

Overlay of PSI results onto three-dimensional representations of buildings and other assets to aid interpretation has become common practice in recent years. This is done within visualisation tools for the specific task of visualising PS data points. Within traditional civil engineering and asset management practice, the management and exchange of monitoring data are addressed by the information technology approaches such as building information modelling (BIM). The main purpose of BIM is to manage digital representations of all information related to a built asset during its entire life cycle, to improve productivity and quality while reducing costs (Davila et al. 2016). The concept of BIM is not simply to describe graphical repre-

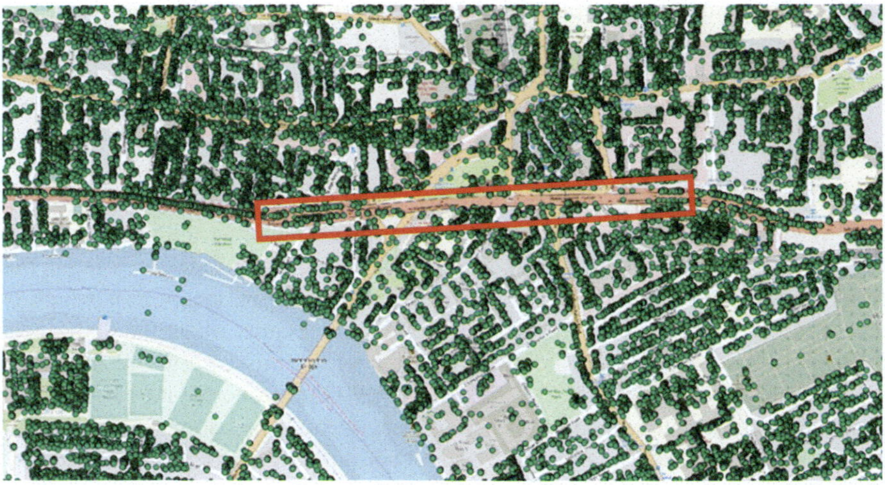

Fig. 8.11 Persistent scatterers on Hammersmith Flyover and surrounding area overlaid onto Open Street Map base imagery. Extent of Hammersmith Flyover is marked with a red box

sentations of assets, but also to link other associated datasets (for example, material specifications, costs, component identification, etc.). However, for this purpose, a simple three-dimensional model accurately depicting the outer surface geometry will suffice.

In the application of InSAR data into civil engineering environments, it is advantageous to make use of any existing data and model, as well as to integrate the satellite data within an established and existing system. The processed InSAR data (PS point output) can be geocoded to specific geographical locations (based on selected mapping and projection information) and output files can be opened and interpreted within GIS software. This allows for geometrical models of the bridge and surrounding buildings, as well as other environmental datasets (such as floodplains, planned construction activities such as excavation or dewatering, etc. in map layers) to be overlaid onto the asset in the GIS domain. Overlaying the different environmental datasets can provide further insights about the context in which the asset is sitting and enhance the interpretation of the PS point cloud. Here the approach was to integrate the standard 3-D geometry models into GIS domains (using software provided by Group BC). Openly available Ordnance Survey building height data was also imported into the same GIS domain (Fig. 8.12). This does not account for errors in geocoding (Montazeri et al. 2018) and some interpretation must be made.

Another tool to aid interpretation of SAR images that could be employed would be simulated reflectivity maps. The ability to predict backscattering effects in SAR images and to identify groups of strong PS points has been developed into software tools for simulating backscattering effects which occur in high resolution SAR images (Auer et al. 2010; Gernhardt and Hinz 2008). This simulation concept is based on ray tracing techniques which simulate emission of radar from a virtual SAR through model space and illuminate three-dimensional models of the real-world scenario to be studied.

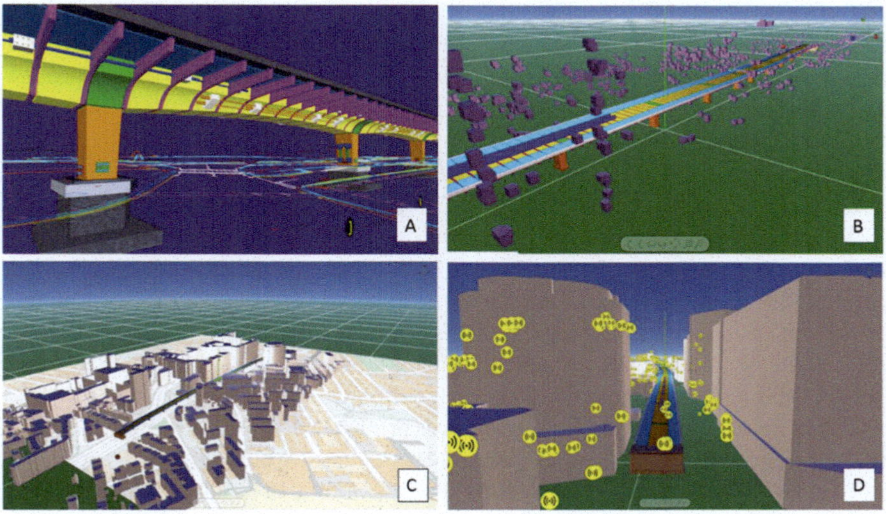

Fig. 8.12 Three-dimensional interpretation of PS points on Hammersmith Flyover: (**a**) BIM model stripped to pure geometry; (**b**) Bridge model imported into GIS space with PS points; (**c**) Ordnance Survey building elevation information also imported into GIS space; (**d**) Model and PS in GIS environment used to aid interpretation of PS points

A finite element model of Hammersmith Flyover (made using the software LUSAS) was used to model a series of different loading and temperature scenarios. Using structural modelling and understanding of the bridge behaviour, the primary cause of deformation of Hammersmith Flyover was determined to be thermal loading which caused longitudinal expansion and contraction. The largest movement occurred in the centre of the bridge, with decreasing magnitude in each pier moving from the centre to each abutment. This was confirmed by displacement gauge data collected from the base of each pier which showed horizontal movement of the bearing in the form of a sinusoidal oscillation over a yearly period, reprising the basic trend of the temperature measured.

Method for the Identification of Target Points

Using the above method enabled the user to isolate points that were most likely attributed to a given bridge pier location. Within these points, the finite element modelling of structural behaviour indicated that the pier motion facilitated by the bearings was an oscillation (directly proportional to the temperature). Therefore, the PSs of interest from the bridge (i.e. those that reflect the movement accommodated through the bearings) were those with a sinusoidal profile.

The basic method employed to identify the PS points of interest made use of the Fast Fourier Transform (FFT). Fourier Transforms are used to decompose a function of time into the frequencies that it is comprised of, i.e., it is the frequency domain representation of the original signal. To select the PS points attributed to a particular pier, each of the PS displacement over time profiles was passed through an FFT to identify which exhibited the sinusoidal profile. Those which exhibit a sinusoidal profile displayed clear spikes in the frequency domain and were plotted over time, whilst those that did not display sinusoidal behaviour were discarded from the considered set.

These selected points were then plotted over time against the movement measured by traditional in-situ displacement gauges attached to the bearings and bearing pit. Figure 8.13 shows one of the PS points attributed to a single bridge pier alongside the measured bearing displacement data and nearest temperature gauge (average deck temperature in the closest span with temperature sensors installed). The InSAR measurement plotted was the measured line of sight displacement.

In this instance, the bridge movement was primarily due to temperature changes, and the thermal-related displacement measurements would highlight potential problems with movement bearings and expansion joints. In terms of the application of SAR to bridge monitoring, SAR data can be used in the absence of other sensor measurements (in this instance to spot unusual or anomalous behaviours) but it can also be used to augment the information provided by a wider digital sensor network. However, these potential applications must be taken with an understanding that SAR measurements are not useful in all bridge monitoring cases, and will not provide measurements for all bridges. The suitability for use in InSAR is dependent on factors such as geometry, material, orientation and environment, which affect the reflection of SAR waves emitted and collected by the SAR sensor on the satellite.

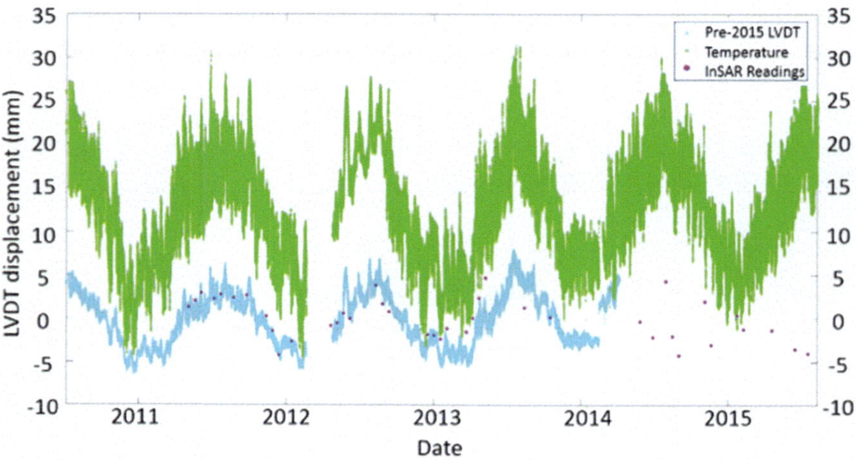

Fig. 8.13 Horizontal movement of a single bridge pier as measured by LVDT (pre-2015), selected InSAR PS (measurements in the satellite line of sight direction) attributed to the motion of the same pier and temperature from nearest deck sensors, over time

8.5 Pre-Cursors to Scour Failure: Tadcaster Bridge

Introduction

We have established that InSAR measurement technologies are able to provide displacement measurements at a scale that is relevant to the bridge-monitoring community (millimetres). These millimetre-scale movements may be due to a number of different loading cases, or deformation scenarios. One of the potential opportunities for the application of InSAR satellite measurement technology is to use it to identify signs or precursors of failure in bridge assets. As InSAR can only measure displacements at a measurement interval of multiple days, this limits the application of InSAR to identifying signs of problems expressed through slow moving displacements over long time periods, such as settlement or the seasonal thermal responses discussed earlier in this Chapter. Another failure mechanism which may show movements indicating unusual behaviour prior to failure is scour failure.

Scour is a natural phenomenon. It can be defined as the excavation and removal of material from the bed and banks of streams as a result of the erosive action of flowing water (Hamill 1998). This erosive action in the vicinity of bridge piers can lead to the removal of ground material on which bridges are founded, increasing the risk of undermining bridge piers and resulting in collapse (Fig. 8.14). Changes in water flow rates during flooding can make bridge piers particularly susceptible to scour. The collapse of bridges and other structures in or adjacent to water bodies highlights the essential importance in finding new methods to undertake inspection and structural health monitoring (SHM) of bridges to identify precursors indicating signs of impending failure.

When bridges have piers founded within water bodies, such that they are not able to be seen below the water surface, an additional inspection challenge is presented in inspecting foundations to detect scour or other damage. Standard practice is to conduct initial scour assessments at a minimum prescribed time interval (e.g. once every 3 years for UK rail structures), with those highlighted as being at risk having a special maintenance plan put in place, including inspections after major flood events. The traditional method of conducting such inspec-

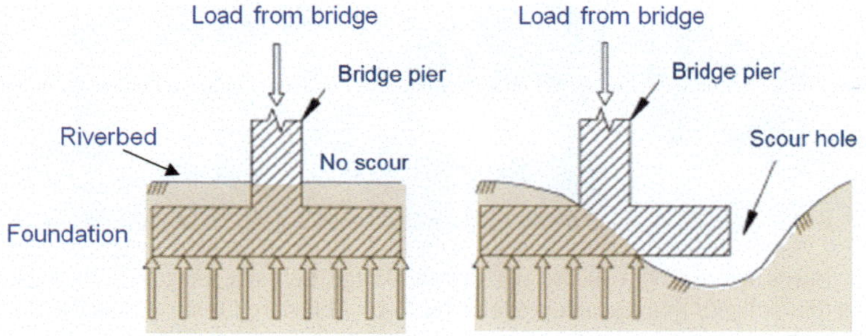

Fig. 8.14 Illustration of how scour might affect a bridge pier

tions is to send divers to visually assess damage. This procedure has several limitations. Diver inspections cannot be undertaken during flood events or during the recession of the flood (due to flow velocity, turbulence or debris accumulation at a structure) when bridges are especially vulnerable (Kirby et al. 2015). Diver safety is put at risk when working in hazardous water environments to look under foundations that could collapse on top of them, and even when a diver or other recording device is sent underwater, it may not be apparent that there is a problem (for example, when loose backfill material hides a scour hole) (Highways Agency 2012).

The case study of Tadcaster Bridge is presented as evidence of how InSAR techniques could potentially be used to monitor bridges at risk of scour. Tadcaster Bridge is a historic nine-arch masonry bridge over the River Wharfe in Tadcaster, United Kingdom. It is approximately 100 m long and 10 m wide, carrying a single lane of vehicular traffic in each direction and a pedestrian walkway on each side. On the evening of December 29th, 2015, following a period of severe rainfall and flooding, the upstream section of the fifth pier of Tadcaster Bridge collapsed into the River Wharfe, resulting in a partial collapse and closure of the bridge. Video footage capturing the collapse of Tadcaster Bridge in 2015 (Fig. 8.15) following a period of flooding highlights a pronounced dip in the bridge, presumably as the bridge pier falls into the scour hole. A study by Selvakumaran et al. (2018) was undertaken to investigate whether this vertical 'dip' in the bridge could be measured using InSAR prior to the day of the bridge collapse.

Data Analysis

To analyse the deformation behaviour in the period preceding collapse, 48 TerraSAR-X 'Stripmap' mode images (3 m by 3 m ground resolution) taken prior to the collapse in the 20-month period from ninth March 2014 to 26th November 2015 were analysed. The final acquisition in November was the last image available prior

Fig. 8.15 Stages before scour collapse of Tadcaster Bridge. Left-side image is of the bridge in the summer before collapse (photograph provided by North Yorkshire County Council). The middle image and right-side image are taken from a video capturing the actual bridge collapse (video source: Press Association 2015)

to the bridge collapse on 29th December 2015. These image acquisitions were taken at 11-day intervals where possible. LIDAR data produced by the UK Environment Agency was then used in subsequent processing work as a Digital Elevation Model of 2 m resolution.

Both PSI and SBAS processing techniques were considered for this study to investigate whether a deformation signal in the area of failure could be observed over the bridge prior to its failure. Suitability and application of InSAR stacking techniques to bridges are heavily influenced by the form and geometry of the structure. As a masonry bridge this structure did not provide persistent scatterers for PSI techniques like the other bridges studied in this Chapter and in the literature. In this example, there was only one metal lamp post which would likely act as a point target reflector. In contrast, the SBAS technique was found to be more appropriate for this case, and 7 different distributed scatterer locations across the bridge were detected. The standard SBAS processing chain (Berardino et al. 2002b) was implemented within the SARscape software package (Sarmap SA 2014).

Time Series Motion Results

Deformation velocity maps are commonly plotted as outputs of InSAR analysis to show deformation over time (line of sight displacements as mm per year). This is useful for effects such as the study of deformation in cities after tunnelling or in the steady settlement of structures over time (Perissin et al. 2012), but in the case of bridge movements resulting from scour such with Tadcaster Bridge, this representation is misleading. The scatterers on the bridge using this form of plotting are all marked as 'steady' (indicated by a line of sight velocity of close to 0 mm/year) with no general trend of rising or falling, due to the assumption that points on the ground will move in a linear trend over time. Simply viewing the deformation velocity map (Fig. 8.16) would suggest that there is little to no movement occurring on the bridge. Depending on the structural form and layout, bridges could oscillate in response to many different load conditions (such as temperature or vehicular loading) or remain largely steady over years with a sudden change in deformation (say in the case of flooding causing localised scour around a bridge pier). It would be more relevant to consider the plot of the scatterers in terms of their movement over time.

The time series motion of the 7 scatterers attributed to Tadcaster Bridge (those scatterers spanning the River Wharfe – the river itself being absent of scatterer as water surfaces are not able to provide coherent scatterers) was plotted (Fig. 8.17). It was not possible to discern which exact area of the bridge each scatterer area was coming from, as each pixel contains the reflective response from multiple scatterers, with the final imaging result of the cell reflecting the vector sum of the scatterers. This becomes even less clear when accounting for processing effects (such as pixel averaging processes like 'multi-looking'). However, in the cluster of scatterers attributed to the bridge, scatterer 'b' in Fig. 8.17 was positioned on the upstream

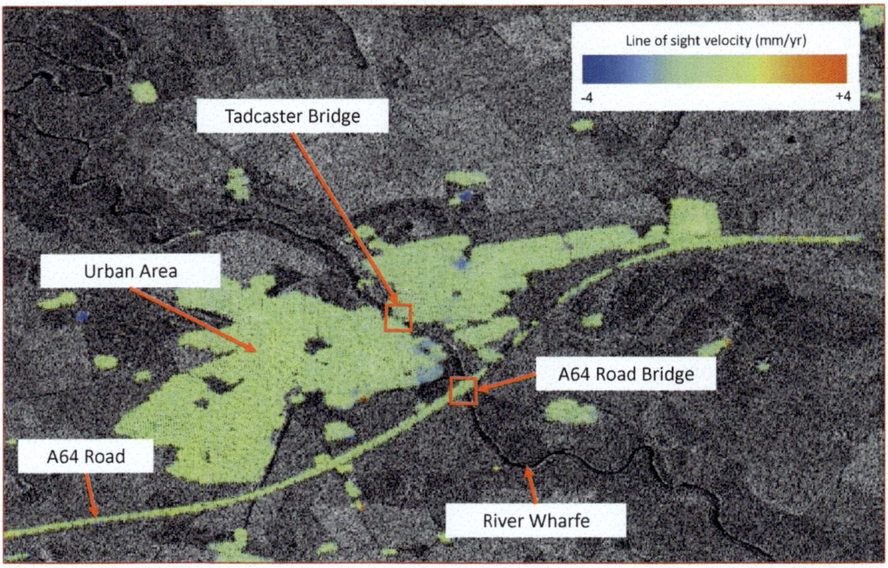

Fig. 8.16 SBAS velocity results over Tadcaster visualised over the mean SAR amplitude image of the site

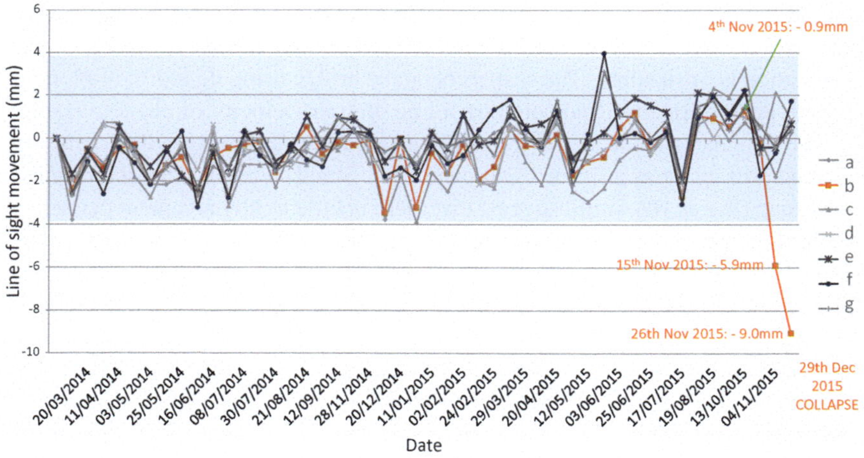

Fig. 8.17 Movement of scatterers attributed to the bridge plotted over time. The collection of 7 points plotted at a specific time on the x-axis corresponds to a series of SAR images acquired over the site. The y-axis marks movement in the line of sight of the SAR satellite. Blue shading denotes statistically assigned 'normal' behaviour of the bridge

side of the bridge deck in the middle region of the length of the bridge which directly correlated with the region of the failed pier.

The movement detected was the displacement in the line of sight of the SAR satellite over time, and was plotted as movement relative to the position of the bridge at the first acquisition, taken on 9th March 2014. The general variation in movement can be attributed to a combination of some real movement of parts of the bridge, and uncertainty within the measurements. Generally, a masonry arch bridge of this kind would be expected to remain roughly stable over time. The final two movement measurements plotted for scatterer ‘b’ on the 15th November 2015 and the 26th November 2015 were considered as outliers to all other measured bridge behaviour (the statistically ‘normal’ region being defined as three standard deviations away from the expected normal). These dates are the final measurements collected prior to the collapse on 29th December 2015. The significance of these results is that movement suggesting a precursor to failure could be seen in the data one month prior to collapse. As the abnormal movements identified are less than 1 cm, they would have been virtually impossible to detect by eye in a bridge during any visual inspection undertaken on the same dates.

The final acquisition was taken in the month prior to the collapse, with no further acquisitions available between November 26th 2015 and the collapse on 29th December 2015, where the majority of deformation would have occurred. The reliability would be improved if further images during this period were available to confirm the deformation trend. However, the results do demonstrate the potential of InSAR X-band data to detect unusual deformations in masonry arch bridge structures and as such, provide the potential capability to give early warning of scour failure (Selvakumaran et al. 2018). The use of InSAR as a potential method for early warning of bridge collapse is also illustrated in other studies, such as (Sousa and Bastos 2013).

8.6 Conclusions in the Context of Bridge Engineering

This Chapter provides an overview of a number of studies using InSAR to monitor bridges. Considering the learnings from each of these case studies, a short summary of ‘best practice’ for use of InSAR in bridge monitoring could be presented to end users to better understand how to make use of InSAR for bridge monitoring activities (Selvakumaran 2019):

1. Understanding whether InSAR is suitable and what the imaging or measurement requirements are.

    - Consider the SAR reflection response of the bridge of interest, and whether there will be measurement points will be available at points of interest. This might be done with radar simulations to better understand SAR viewing geometry or by looking at the SAR images and making an assessment. If it does not provide a suitable radar response, there is a possibility of considering

the installation of corner reflectors to augment the reflection response (but this is not suitable and/or practical in all cases).
- Understand what kind of bridge movements are expected, including direction and maximum movements—what can the satellite realistically capture given its wavelength, spatial resolution and pixel size).
- Understand what other in-situ datasets are available—whether the SAR data can be used to augment an existing sensor network and whether there is data to tie the relative SAR movements into real-world grids (e.g. GNSS measurements).
- Understand the number and resolution of images required (and which constellation(s) of satellites would achieve that, along with an understanding of how to weigh the benefits and limitations, e.g. better controlled orbits or shorter revisit periods); these can be tasked as new acquisitions, taken from the archive or a combination of both.

2. Processing of SAR data should be undertaken by someone with suitable specialist InSAR processing skills. Despite several simple-to-use software packages being developed in recent years, there is a risk in allowing users to plug in parameters without understanding the influence or impact of such parameters, or the assumptions built within the packages; this can impact the quality and reliability of the results presented.
3. Understanding SAR viewing geometry, line-of-sight measurements, etc. to know what is being measured (along with the nuances in signal processing that mean that the location of the persistent scatterer 'point' may not be where it appears to be projected on a map).
4. Understanding the precision and uncertainty of the measurements taken, especially when trying to convert line-of-sight measurements into directions and coordinate systems as desired by bridge engineers (e.g. converting to longitudinal displacement in a bridge has limitations, adds measurement uncertainty and has less accuracy in directions along the path of the satellite orbit).
5. Application of InSAR to bridge and civil engineering applications more generally should be a multidisciplinary endeavour. The best use of such tools going forward would be for InSAR specialists to work with educated and informed civil engineering end users, to apply the current state-of-the-art and develop new methods and automated tools for asset owners (ideally leveraging advancements in digital engineering and bridge engineering to inform the InSAR processing).

For successful implementation of InSAR in the monitoring of bridges, it is important to understand the limitations and opportunities of InSAR. The use of InSAR for bridge monitoring could potentially be improved as SAR data and processing specifications (resolution, frequency, processing assumptions) improve. At the current stage of technological development, it should be considered as a tool for specific bridges and failure mechanisms rather than a full bridge monitoring solution. In summary, InSAR monitoring does have the scale and precision of measurements to support bridge monitoring activities, but only for specific bridge cases, and for monitoring particular types of bridge deformation and behaviour.

The amount of data becoming available for civil applications is rapidly opening up at more frequent acquisitions and finer resolutions. Through understanding the potential to augment existing information and inspection methods, whilst acknowledging the limitations and applicability to monitoring different types of bridge assets, this form of monitoring can be used to complement existing methods and systems and provide valuable insights.

References

American Society Of Civil Engineers (ASCE). 2017. Bridge Infrastructure | Structurally Deficient Bridges | ASCE's 2017 Infrastructure Report Card.

Auer, S., S. Hinz, and R. Bamler. 2010. Ray Tracing Simulation Techniques for Understanding High Resolution SAR Images. *IEEE Transactions on Geoscience and Remote Sensing* 48 (3): 1445–1456.

Bamler, R., and P. Hartl. 1998. Synthetic Aperture Radar Interferometry. *Inverse Problems* 14 (4): 1–54.

Bamler, R., and B. Schättler. 1993. SAR Data Acquisition and Image Formation. In *SAR Geocoding: Data and Systems*, ed. G. Schreier, 53–102. Karlsruhe, Germany: Wichmann.

Berardino, P., G. Fornaro, R. Lanari, and E. Sansosti. 2002a. A New Algorithm for Surface Deformation Monitoring Based on Small Baseline Differential SAR Interferograms. *IEEE Transactions on Geoscience and Remote Sensing* 40 (11): 2375–2383.

———. 2002b. A New Algorithm for Surface Deformation Monitoring Based on Small Baseline Differential SAR Interferograms. *IEEE Transactions on Geoscience and Remote Sensing* 40 (11): 2375–2383.

Bovenga, F., A. Belmonte, A. Refice, G. Pasquariello, R. Nutricato, D.O. Nitti, and M.T. Chiaradia. 2018. Performance Analysis of Satellite Missions for Multi-Temporal SAR Interferometry. *Sensors (Switzerland)* 18 (5): 1–16.

Crosetto, M., O. Monserrat, M. Cuevas-González, N. Devanthéry, and B. Crippa. 2016. Persistent Scatterer Interferometry: A Review. *ISPRS Journal of Photogrammetry and Remote Sensing* 115: 78–89.

Cumming, I., and F. Wong. 2005. *Digital Processing of Synthetic Aperture Radar Data: Algorithms and Implementation*. Norwood, OH: Artech House Publishers.

Cusson, D., and I. Ozkan. 2019. Satellite-based InSAR Monitoring –Validation on Victoria Bridge in Montreal, Canada. In *5th Conference in Smart Monitoring, Assessment and Rehabilitation of Civil Structures (SMAR 2019)*, Potsdam, Germany, August 27–29, 2019.

Cusson, D., K. Trischuk, G. Hewus, M. Gara, and P. Ghuman. 2017. Satellite-Based Monitoring of a Highway Bridge in Canada - Challenges, Solutions and Value. In *ISHMII 2017*, no. December.

Davila, M., L. J. Butler, N. Gibbons, I. Brilakis, M. Z. E. B. Elshafie, and C. Middleton. 2016. Management of structural monitoring data of bridges using BIM. ICE Proceedings: Bridge Engineering, no. November, pp. 1–15.

Dheenathayalan, P., D. Small, and R.F. Hanssen. 2018. 3-D Positioning and Target Association for Medium-Resolution SAR Sensors. *IEEE Transactions on Geoscience and Remote Sensing* 56 (11): 6841–6853.

Doerry, A. W. 2008. *Reflectors for SAR Performance Testing* (Tech. Rep.) January. Albuquerque: Sandia National Laboratories.

Ferretti, A., C. Prati, and F. Rocca. 2000. Nonlinear Subsidence Rate Estimation Using Permanent Scatterers in Differential SAR Interferometry. *IEEE Transactions on Geoscience and Remote Sensing* 38 (5): 2202–2212.

———. 2001. Permanent Scatters in SAR Interferometry. *IEEE Transactions on Geoscience and Remote Sensing* 39 (1): 8–20.

Ferretti, A., A. Monti-Guarnieri, and C. Prati. 2007a. *InSAR Principles-Guidelines for SAR Interferometry Processing and Interpretation (TM-19)*. ESA, no. February, p. 48.

———. 2007b. *InSAR Principles: Guidelines for SAR Interferometry Processing and Interpretation (TM-19), Part C*. ESA.

Ferretti, A., A. Monti-guarnieri, C. Prati, and F. Rocca. 2007c. *InSAR Principles: Guidelines for SAR Interferometry Processing and Interpretation (TM-19), Part B*, 71. ESA.

Ferretti, A., G. Savio, R. Barzaghi, A. Borghi, S. Musazzi, F. Novali, C. Prati, and F. Rocca. 2007d. Submillimeter Accuracy of InSAR Time Series: Experimental Validation. *IEEE Transactions on Geoscience and Remote Sensing* 45 (5): 1142–1153.

Gabriel, A.K., R.M. Goldstein, and H.A. Zebker. 1989. Mapping Small Elevation Changes Over Large Areas - Differential Radar Interferometry. *Journal of Geophysical Research* 94 (B7): 9183–9191.

Gernhardt, S., and S. Hinz. 2008. Advanced Displacement Estimation for PSI Using High Resolution SAR Data. *International Geoscience and Remote Sensing Symposium (IGARSS)* 3 (1): 1276–1279.

Gisinger, C., U. Balss, R. Pail, X.X. Zhu, S. Montazeri, S. Gernhardt, and M. Eineder. 2015. Precise Three-Dimensional Stereo Localization of Corner Reflectors and Persistent Scatterers with TerraSAR-X. *IEEE Transactions on Geoscience and Remote Sensing* 53 (4): 1782–1802.

Hamill, L. 1998. *Bridge Hydraulics*. 1st ed. London: CRC Press (Taylor & Francis Group).

Hanssen, R.F. 2001. *Radar Interferometry: Data Interpretation and Error Analysis*. 1st ed. Netherlands: Springer.

Highways Agency. 2012. The Assessment of Scour and Other Hydraulic Actions at High-way Structures (BD97/12), (Tech. rep.). The Highways Agency, Transport Scotland, Welsh Government and The Department for Regional Development Northern Ireland.

Hu, J., Z.W. Li, X.L. Ding, J.J. Zhu, L. Zhang, and Q. Sun. 2014. Resolving Three-Dimensional Surface Displacements from InSAR Measurements: A Review. *Earth-Science Reviews* 133: 1–17.

Kirby, A., A. Kitchen, M. Escarameia, and O. Chesterton. 2015. *Manual on Scour at Bridges and Other Hydraulic Structures (C742)*. 2nd ed. London: CIRIA.

Lazecky, M., D. Perissin, M. Bakon, J. M. De Sousa, I. Hlavacova, and N. Real. 2015. Potential of Satellite InSAR Techniques for Monitoring of Bridge Deformations. In *2015 Joint Urban Remote Sensing Event (JURSE)*, (Lausanne, Switzerland), pp. 4–7.

Lea, F., and C. R. Middleton. 2002. *Reliability of Visual Inspection for Highway Bridges* (Tech. Rep. CUED/D-STRUCT/TR. 201). Cambridge: Department of Engineering, University of Cambridge.

Milillo, P., G. Giardina, D. Perissin, G. Milillo, A. Coletta, and C. Terranova. 2019. Pre-Collapse Space Geodetic Observations of Critical Infrastructure: The Morandi Bridge, Genoa, Italy. *Remote Sensing* 11 (12): 1–14.

Monserrat, O., M. Crosetto, M. Cuevas, and B. Crippa. 2011. The Thermal Expansion Component of Persistent Scatterer Interferometry Observations. *IEEE Geoscience and Remote Sensing Letters* 8 (5): 864–868.

Montazeri, S., F.R. González, and X.X. Zhu. 2018. Geocoding Error Correction for InSAR Point Clouds. *Remote Sensing* 10 (10): 1–22.

Moore, M., B. Phares, B. Graybeal, R. Dennis, and G. Washer. 2001. *Reliability of Visual Inspection for Highway Bridges* (Volume I: Final Report, Tech. Rep. FHWA-RD-01-020). McLean, VA: U.S. Department of Transport.

Moreira, A., P. Prats-Iraola, M. Younis, G. Krieger, I. Hajnsek, and K.P. Papathanassiou. 2013. A Tutorial on Synthetic Aperture Radar. *IEEE Geoscience and Remote Sensing Magazine*: 6–43.

Perissin, D., Z. Wang, and H. Lin. 2012. Shanghai Subway Tunnels and Highways Monitoring Through Cosmo-SkyMed Persistent Scatterers. *ISPRS Journal of Photogrammetry and Remote Sensing* 73: 58–67. Additional reference from which some figures are also used (PhD thesis).

Press_Association. 2015. Tadcaster bridge over river Wharfe collapses during flooding – video (available at: https://www.theguardian.com/environment/video/2015/dec/30/tadcaster-bridge-river-wharfe-collapses-flooding-video).

Qin, X., M. Liao, M. Yang, and L. Zhang. 2017. Monitoring Structure Health of Urban Bridges with Advanced Multi-Temporal InSAR Analysis. *Annals of GIS* 23 (4): 293–302.

Qin, X., X. Ding, and M. Liao. 2018. Three-Dimensional Deformation Monitoring and Structural Risk Assessment of Bridges by Integrating Observations from Multiple SAR Sensors. In *IEEE International Geoscience and Remote Sensing Symposium (IGARSS), (Valencia, Spain)*, 1384–1387. IEEE.

Quin, G., and P. Loreaux. 2013. Submillimeter Accuracy of Multipass Corner Reflector Monitoring by PS Technique. *IEEE Transactions on Geoscience and Remote Sensing* 51 (3): 1775–1783.

RAC Foundation. 2019. *Bridge Maintenance Backlog Grows* (Tech. rep.) London, UK: RAC Foundation

Ruck, G., and C. Krichbaum. 1968. *Radar Cross Section Handbook Supplement* (Tech. Rep.). Washington, DC: Advanced Research Projects Agency.

Sarmap SA, SARscape®, 2014.

Selvakumaran, S. 2019. Interferometric Synthetic Aperture Radar for Remote Satellite Monitoring of Bridges. https://doi.org/10.17863/CAM.51027

Selvakumaran, S., S. Plank, C. Geiß, C. Rossi, and C. Middleton. 2018. Remote Monitoring to Predict Bridge Scour Failure Using Interferometric Synthetic Aperture Radar (InSAR) Stacking Techniques. *International Journal of Applied Earth Observation and Geoinformation* 73: 463–470. https://doi.org/10.1016/j.jag.2018.07.004.

Selvakumaran, S., C. Rossi, A. Marinoni, G. Webb, J. Bennetts, E. Barton, S. Plank, and C.R. Middleton. 2020. Combined InSAR and Terrestrial Structural Monitoring of Bridges. *IEEE Transactions on Geoscience and Remote Sensing*. https://doi.org/10.1109/tgrs.2020.2979961.

Soergel, U., H. Gross, A. Thiele, and U. Thoennessen. 2006. Extraction of Bridges Over Water in High-Resolution InSAR Data. *Photogrammetric Computer Vision* 36.

Sousa, J.J., and L. Bastos. 2013. Multi-Temporal SAR Interferometry Reveals Acceleration of Bridge Sinking Before Collapse. *Natural Hazards and Earth System Science* 13 (3): 659–667.

Chapter 9
Measuring Subsidence in California and Its Impact on Water Conveyance Infrastructure

Cathleen E. Jones, Tom G. Farr, Zhen Liu, and Megan M. Miller

9.1 Introduction

Subsidence caused by groundwater pumping in the Central Valley has been a problem for decades. During the 2012–2016 drought, the California Dept. of Water Resources (DWR) engaged scientists at the Jet Propulsion Laboratory, California Institute of Technology, to determine where and by how much land surface elevation in the California Central Valley was changing. We applied radar remote sensing methods that were developed to study geophysical processes that cause deformation, such as earthquakes and volcanoes, to measure regional subsidence across the Central Valley and identify the major subsidence bowls, and to locate where localized subsidence was directly impacting the California Aqueduct. The results were provided in two separate reports to DWR, one published in 2015 that reported subsidence through early 2015 (Farr et al. 2015), and another in 2017 that extended the subsidence measurements through mid-2016 (Farr et al. 2017). Here we describe the project as a case study for monitoring a network of water conveyance and flood-control structures at the synoptic scale.

C. E. Jones (✉) · T. G. Farr · Z. Liu · M. M. Miller
Jet Propulsion Laboratory, California Institute of Technology, Pasadena, CA, USA
e-mail: Cathleen.E.Jones@jpl.nasa.gov

V. Singhroy (ed.), *Advances in Remote Sensing for Infrastructure Monitoring*, Springer Remote Sensing/Photogrammetry,
https://doi.org/10.1007/978-3-030-59109-0_9

9.2 The Problem of Subsidence in the Central Valley Groundwater Basins

The aquifer system of the southern Central Valley, as with many other alluvial basins of California, has both unconfined and confined parts comprising alternating layers of coarse- and fine-grained sediments. Water in the shallow coarse-grained, unconfined, or water-table aquifers is extracted or recharged easily and causes only minor "elastic" deformation, which is observed as seasonal subsidence and rebound of water levels and the land surface. However, many water wells exploit the deeper semi-confined and confined aquifers, where withdrawal of water causes drainage of the fine-grained confining layers called aquitards. A significant amount of water is available from the aquitards; however, they can become permanently compacted if groundwater is overdrafted. In general, if water levels remain above previous lows, deformation is elastic. However, if water levels fall below the previous record low, an inelastic, irreversible compaction of the fine-grained aquitards occurs. There is a time-delay between extraction of water and compaction of the material in the aquitard as the pressure equilibrates between the aquitard and the surrounding material. The layer properties determine how slowly the delayed compaction occurs and compaction can continue for decades, even after groundwater levels rebound. Furthermore, because aquitard pore space is reduced during compaction, there is reduced storage capacity for water to recharge the aquitard layers in the event of groundwater level recovery (see Galloway et al. 1999, pp. 8–13; Bertoldi et al. 1991 for reviews). Measuring and understanding subsidence as a function of groundwater dynamics can greatly improve resource management by providing information about what level of pumping results in elastic deformation without breaching the inelastic deformation threshold.

Consequences of land subsidence extend beyond the loss of groundwater storage. Areas that have subsided can be more prone to flooding as the change in slope of the land alters drainage patterns. In this way, historically higher land can unexpectedly become the pathway for flood waters. Differential subsidence also affects infrastructure through earth fissures which can break roads and exhume pipelines, and changes to the land gradient that reduce flow capacity in aqueducts and other water conveyance channels. Subsidence of levees and other flood-control structures leave the land at increased risk from overtopping or failure. Subsidence impact has been observed in water conveyance structures in the Central Valley where canal linings have been repaired and sides raised in the U.S. Bureau of Reclamation's Delta–Mendota Canal, California Dept. of Water Resources' (DWR's) California Aqueduct, and local agency irrigation canals. In addition, the East Side Bypass, a canal used to control runoff from the Sierra Nevada mountain range, has subsided so much that major effort has been required to restore its capacity for flood control. Identifying the location and probable cause of subsidence is particularly important in areas where economic costs for infrastructure maintenance and repair are high and social impact from flooding is severe.

9.3 Airborne and Satellite Radar Remote Sensing of California Groundwater Basins

NASA and a number of other countries' space agencies launched missions to observe the Earth from space using synthetic aperture radar (SAR) instruments suitable for interferometry (InSAR) (Table 9.1). Although each satellite operated only over a span of a few years, when combined together by concatenating multitemporal time series, the instruments provide continuous coverage of the Central Valley since 1992.

In this case study, the California Dept. of Water Resources funded acquisition and analysis of satellite and airborne SAR data. The satellite SAR provided the overview of subsidence in the Valley and the higher spatial resolution airborne SAR was used to identify localized infrastructure impact. For the satellite SAR, PALSAR images spanning the period June 2007–December 2010, Radarsat-2 images for the period May 2014–January 2015, and Sentinel-1 images from May 2015–September 2016 were processed to independent time series of subsidence within the Central Valley. Archived and contemporary Sentinel-1 radar data of California is available and provided free of charge by the European Space Agency. Two polar-orbiting satellites comprise Sentinel-1 and operate continuously at C-band (5.6 cm wavelength). The dual-platform constellation provides good temporal coverage of California, especially when including ascending and descending passes. Figure 9.1

Table 9.1 Past, present, and future radar satellites. The resolution specified is the value for the instrument prior to spatial averaging. When a range of spatial resolutions is given, it indicates that different operating modes of the instrument can be used. Finer resolution corresponds to the narrower swath width. X-band, C-band, and L-band correspond to 3.1, 5.5, and 23.8 cm wavelength for the SARs, respectively. "H" ("V") refers to horizontal (vertical) polarization, with HH (VV) indicating that the instrument transmits and receives horizontal (vertical) polarized electromagnetic radiation. "Quad" refers to an instrument that transmits and receives in both polarizations

Satellite	Dates	Resolution (m)	Swath (km)	Incident angles	Minimum revisit (days)	Band*/ Pol
ERS 1,2	1991–2010	25	100	25°	35	CVV
Envisat	2002–2010	25	100	15–45°	35	CVV, CHH
PALSAR	2006–2011	10–100	40–350	10–60°	46	L-quad
Radarsat 1	1995–2013	10–100	45–500	20–49°	24	CHH
Radarsat 2	2008–	3–100	25–500	10–60°	24	C-quad
TerraSAR-X	2007–	1–16	5–100	15–60°	11	X-quad
Cosmo-skymed	2007–	1–100	10–200	20–60°	<1	X-quad
PALSAR-2	2014–	3–60	50–350	8–70°	14	L-quad
Sentinel-1	2014–	20	250	30–45°	6	C-dual
NJSAR	2022	35	350	15–60°	12	L-quad

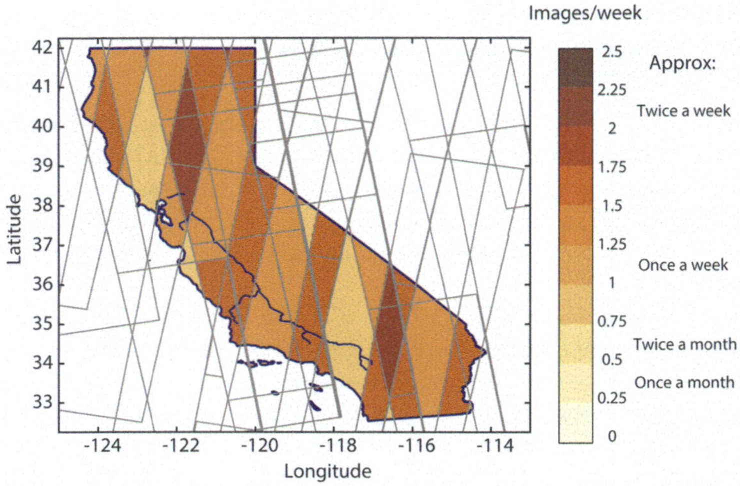

Fig. 9.1 Frequency of imaging of California by Sentinel-1 in 2018, in average number of images per week including both ascending (SE-NW) and descending (NE-SW) passes (gray outlines). The blue lines show aqueducts maintained by the California Dept. of Water Resources and used to transport water southward and westward

shows the Sentinel-1 imaging frequency for California in 2018. Every location within the Central Valley is imaged between 0.5 and 2 times per week. Given the regular repeated imaging, Sentinel-1 and other satellite SARs are well-suited to monitor regional subsidence.

NASA does not have a SAR in Earth-observing orbit that can provide InSAR data at present, but an instrument is currently being built for launch in the 2022–2023 timeframe as part of the NASA-ISRO SAR (NISAR) mission. In addition, the NASA Airborne Science Program operates an airborne SAR that is a prototype instrument for NISAR, the Uninhabited Aerial Vehicle SAR (UAVSAR) (https://uavsar.jpl.nasa.gov). This instrument is used for the airborne SAR component of this case study.

UAVSAR is flown on a piloted Gulfstream-3 aircraft and operates at 23.8 cm wavelength in the L-band part of the electromagnetic spectrum. The instrument is contained in a pod mounted below the fuselage (Fig. 9.2), suitable for porting to a large drone such as the Global Hawk, hence the name of the instrument. An airborne SAR, which operates from a much lower altitude, has a significantly higher signal-to-noise ratio than the satellite SARs, usually achieving a factor of 100 increase in signal through the use of a high-power instrument transmitting from 12.5 km altitude rather than from Earth orbit. UAVSAR temporal sampling is much less frequent than satellite SARs, particularly Sentinel-1, but it offers up to 1.7 × 1-m ground resolution, which allows significantly higher resolution. Spatial averaging is applied to all SAR data during processing to improve the accuracy of the

Fig. 9.2 The UAVSAR instrument is housed in a pod that mounts under the wings on the fuselage of a Gulfstream-3 aircraft. The synthetic aperture radar is a left side-looking instrument covering a ground swath 22 km wide at incidence angles extending from 22° in the near range to 67° in the far range (flat terrain values). Air ducts on the front of the pod provide ambient air cooling and a control system uses heaters to regulate the temperature of equipment in the pod. The aircraft uses a precision autopilot system to maintain its position within ±5-m of the planned track. Typical operating altitude for the aircraft is 12.5 km, above normal commercial air traffic

deformation measurement. UAVSAR products are processed to ~7-m resolution after averaging over 36 pixels, compared to the typical satellite SAR data processed to ~100-m resolution, so the effect of subsidence on infrastructure is better monitored using the airborne SAR.

Use of InSAR for measuring subsidence is not new to this study, although its application to measuring impact to the California Aqueduct is. Subsidence due to groundwater withdrawal in the western United States, evaluated based on satellite InSAR, has been done for Los Angeles (Bawden et al. 2001), the Antelope Valley (Galloway et al. 1998), Las Vegas (Hoffmann et al. 2001; Bell et al. 2008), Arizona (Miller and Shirzaei 2015; Conway 2016), Colorado (Reeves et al. 2011), the Santa Clara Valley (Sneed et al. 2003; Chaussard et al. 2017), the Coachella Valley (Sneed and Brandt 2007), and the southern Central Valley (Farr and Liu 2015; Sneed and Brandt 2013; Borchers and Carpenter 2014; Smith et al. 2017; Ojha et al. 2018, 2019; Murray and Lohman 2018). Previous InSAR results from UAVSAR include measurements of the Slumgullion landslide in Colorado (Delbridge et al. 2016), subsidence in New Orleans (Jones et al. 2016), fault slip in California (Donnellan et al. 2014), and landslides along the San Andreas fault (Scheingross et al. 2013).

9.4 Interferometric Synthetic Aperture Radar (InSAR)

InSAR is a technique whereby surface change occurring between two radar imaging passes can be measured and mapped to high precision (see Madsen and Zebker 1998; Massonnet 1997 for reviews). The ability to map surface deformation of less than 1 cm over large areas at spatial resolutions of 30 m or finer has opened up new possibilities for remote monitoring of groundwater resources.

The InSAR technique works by acquiring images from the same viewing geometry at two different times to determine whether a change in the surface position has occurred. The complex conjugate of the measured signals is generated to create an interferogram, from which the phase-difference can be used to create a deformation map that is precise to a small fraction of the radar wavelength. What is actually measured is not the change in surface elevation, but the change in distance along the line-of-sight (LOS) direction between the ground and the radar. In addition, the change is measured relative to a location in the scene which is selected during processing, and relative to the ground level on the date of the first acquisition.

In practice many acquisitions are used in the processing. After many pairs of radar images over an area are processed into interferograms, they can be further analyzed to create a time series of surface deformation. The InSAR time series analysis produces a history of LOS surface displacements similar to GPS time series observations but with much greater spatial coverage. InSAR time series recover both long-term mean velocities and the time varying components. This technique has been applied successfully for imaging non-steady-state deformation at volcanoes (Lundgren et al. 2004), deforming plate boundaries (Lundgren et al. 2009), aquifer dynamics (Farr et al. 2015; Lanari et al. 2004), and subsidence in the Sacramento–San Joaquin Delta (Sharma et al. 2016; Bekaert et al. 2018).

After the initial time series inversion, averaging in both space and time is applied to reduce random errors and noise and to smooth the deformation time series. A final step in post-processing is to project the LOS measurements to the vertical direction under the assumption that there is no significant horizontal surface movement gradient and all of the measured deformation comes from subsidence/uplift. This allows measurements from multiple satellites with different imaging geometries to be compared with each other. In this manner, the results can also be compared with vertical surface change measured with GPS and traditional surveying methods.

The estimated measurement precision for InSAR time series is generally on the order of mm/yr. to cm/yr., depending on the number of InSAR acquisitions and noise levels (Galloway et al. 1998; Ozawa and Ueda 2011; Chaussard et al. 2013). There are some potential error factors which must be considered when processing and interpreting InSAR because they can either mimic deformation signals or cause greater noise in the measurements. The most significant are orbital or track position error, atmospheric noise, topography-induced errors, and changes in the surface properties between image acquisitions (temporal phase decorrelation), the latter of which can arise from changes in soil moisture or grazing and agricultural activities. This latter effect is especially acute in agricultural areas like the Central Valley, where many areas regularly experience small-scale surface changes near the scale of the radar wavelength. Crop growth, trees blowing in the wind, or fields plowed between radar image acquisitions can cause loss of information. This effect can be ameliorated by using a longer wavelength radar, e.g., L-band as used for UAVSAR and PALSAR, and by selecting pairs of images for which there is small orbital track (baseline) separations and short time intervals between acquisitions. Orbital errors and topographic errors can be corrected during processing. Atmospheric water vapor and other variations in the Earth's troposphere and ionosphere remain a large error source because changes in meteorological conditions and total electron count in the ionosphere introduce phase delays that mimic surface displacement. Airborne SARs are not subject to the ionospheric error source. Tropospheric variations are usually mitigated by analyzing many interferometric pairs and averaging (stacking), under the assumption that ground deformation is steady and atmospheric phase delay is random in time. The longer the time series of images used in determining subsidence trends, the greater the reduction of atmospheric noise effects.

9.5 Broad-Scale Subsidence in the San Joaquin Valley

Maps of subsidence in the San Joaquin Valley were made from the Sentinel-1 time series for the period May 7, 2015–September 10, 2016 (Fig. 9.3) and are generally similar to the earlier PALSAR and Radarsat-2 results (Farr et al. 2015). Two main subsidence bowls can be seen in the maps of total subsidence: A southern one in the Tulare Basin about 40x100 km^2 in size, centered on Corcoran, and a northern one about 25x40 km^2 in size centered south of the town of El Nido. The highest magnitude cumulative subsidence during this time period in the Tulare Basin was ~55 cm near Corcoran. The maximum subsidence southeast of El Nido was ~40 cm. There is also continued development of a diffuse area northwest of Corcoran with subsidence of up to 30 cm. Subsidence of over 30 cm extends west to the California Aqueduct near Avenal; at the aqueduct there is a small patch of total subsidence of almost 45 cm, which occurred during two periods: May–Oct. 2015 and Apr.–Aug. 2016 (Fig. 9.4). This is better resolved in the UAVSAR results, discussed in the next section, because less spatial averaging was applied to the data.

The East Side Bypass flood-control canal runs through the main area of subsidence in the northern subsidence bowl southeast of El Nido, which contains some areas subsiding up to 50 cm. The large-scale subsidence map enables identification of a recently initiated area of subsidence about 11 km in diameter near Tranquility, which is 15 km southeast of Mendota. This area was first noted in late 2014 using the Radarsat-2 subsidence map, but the time series of deformation shows that subsidence in this particular area intensified. Its history of subsidence (Fig. 9.4) shows it subsided in 2 stages, first in early May–Oct. 2015 and then in Apr.–Aug. 2016.

The large-scale subsidence maps can also be used for isolated locations, albeit with less spatial accuracy than the airborne radar. The deformation histories of a few selected locations are plotted in Fig. 9.4. The large maximum subsidence in the

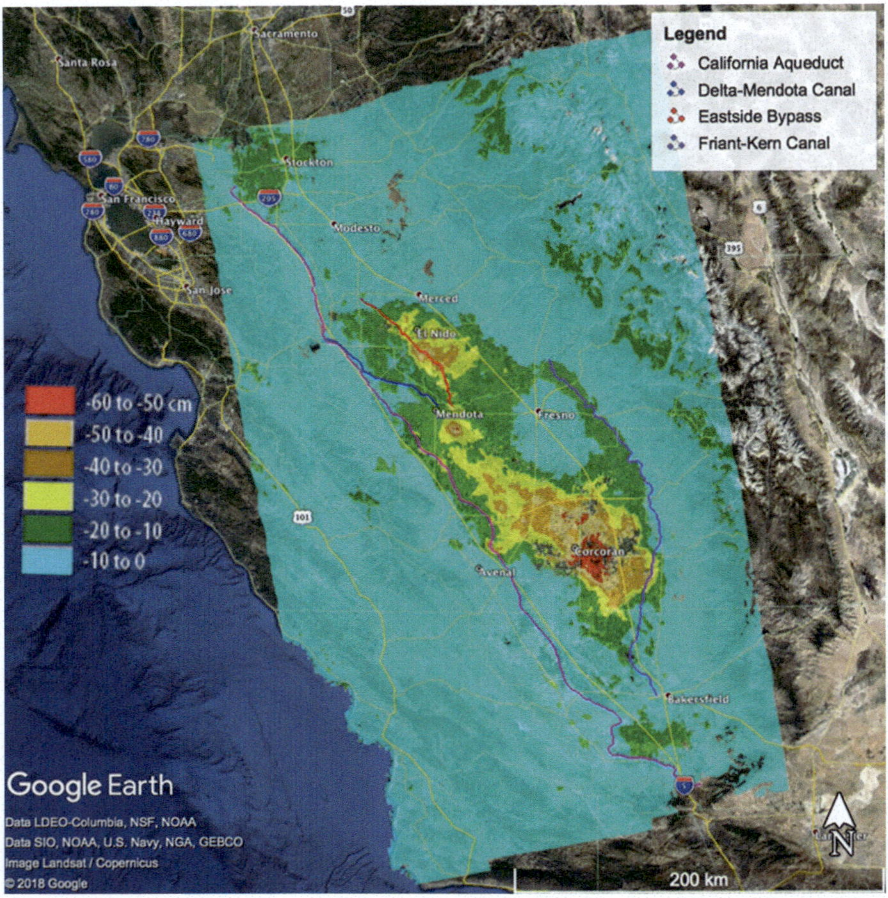

Fig. 9.3 Total subsidence in the San Joaquin Valley for the period May 7, 2015–Sept. 10, 2016, as measured by ESA's Sentinel-1, and processed at the Jet Propulsion Laboratory. Two large subsidence bowls are evident, (1) centered on Corcoran and (2) southeast of El Nido. There is also a small, newly developed subsidence feature between them, near Tranquility. An arm of the large Corcoran bowl extends to the California Aqueduct near Avenal

Corcoran area is clear and shows virtually no recovery at any time during the period of measurement. The maximum subsidence near El Nido shows some flattening between Oct. 2015 and Mar. 2016, probably related to relatively abundant rainfall that winter. The relatively new subsidence feature at Tranquility and the history of the aqueduct feature near Avenal also show a flattening between Oct. 2015 and Mar. 2016, but renewed subsidence after Apr. 2016. More recent data acquisitions show cessation of subsidence in many areas due to the heavy winter rains in 2017.

The transects shown in Fig. 9.5 give a more detailed picture of the total subsidence measured in the vicinity of the California Aqueduct and the East Side Bypass over the period of measurement. It is clear that the East Side Bypass has suffered

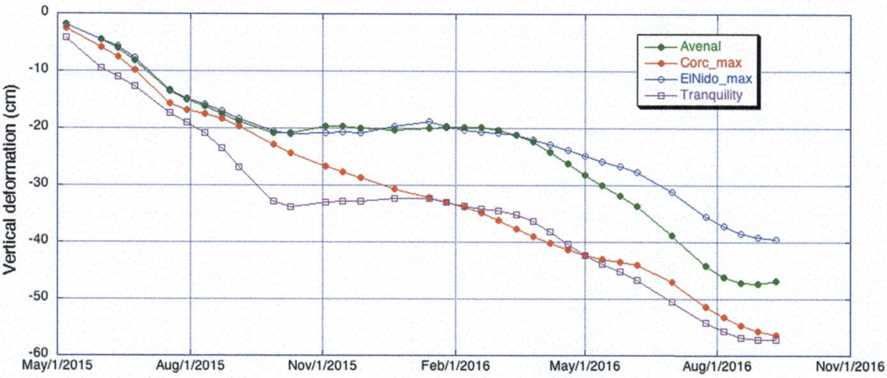

Fig. 9.4 Subsidence histories of a few locations in the San Joaquin Valley. "Avenal" is located at the maximum subsidence measured near the California aqueduct near the town of Avenal. This was also mapped by UAVSAR (Fig. 9.7). "Corcoran max" is located in the maximum subsidence pocket south of Corcoran. "El Nido max" is located in the pocket southeast of El Nido. "Tranquility" is located at the maximum subsidence 15 km southeast of Mendota, between the two main subsidence bowls

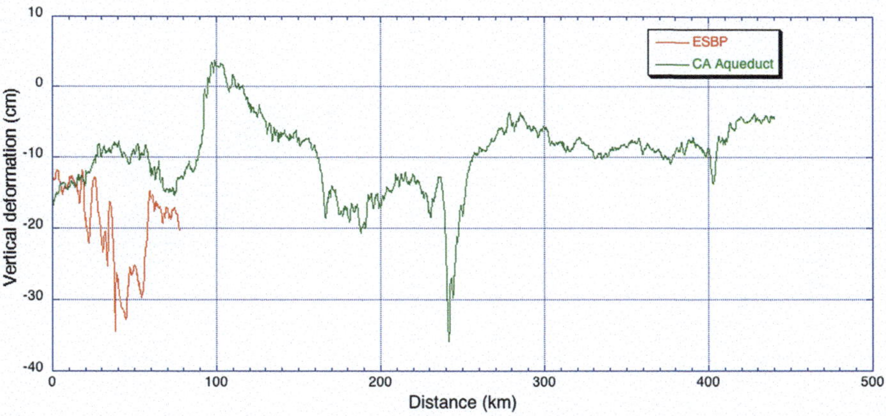

Fig. 9.5 Transects showing total subsidence along the California Aqueduct and the East Side Bypass (ESBP). The transects extend from N to S and represent total subsidence from May 7, 2015–May 25, 2016. Note the deep subsidence about midway along the East Side Bypass. That corresponds to the main subsidence bowl on the map (Fig. 9.3). The sharp pit near km 240 on the California Aqueduct corresponds to the feature near Avenal and shown in Fig. 9.4 and in Fig. 9.7

significant subsidence, concentrated in its central area. In contrast, most areas in the vicinity of the California Aqueduct experienced only a few cm of subsidence. The exception is the subsidence concentrated near Avenal and plotted near km 240. These measured amounts correspond to averages over the processed pixel (about 100 m), not the values on the aqueduct structure itself.

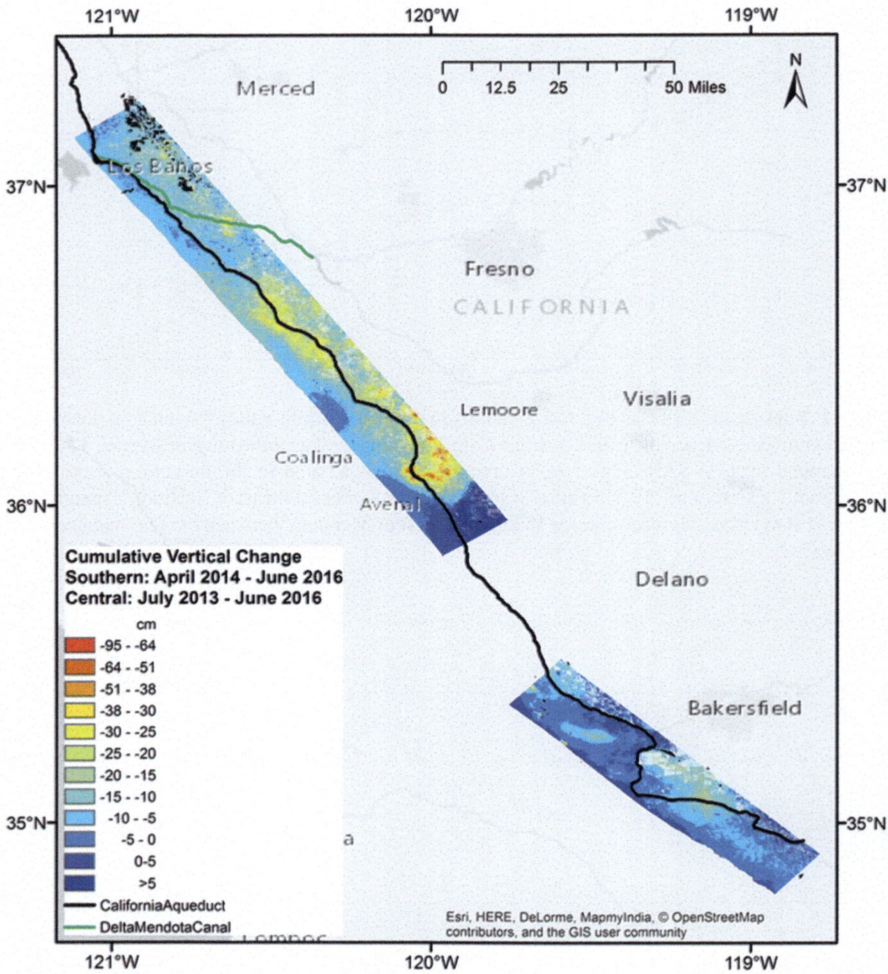

Fig. 9.6 Overview of subsidence measured in the two UAVSAR image swaths covering the California Aqueduct. Each swath is 22 km wide. The swaths were planned specifically to image the California Aqueduct in order to pinpoint impact to the structure from groundwater pumping, and, therefore, miss the large subsidence bowls to the east that are seen in the satellite SAR results. Part of the Delta–Mendota Canal is included in the northern swath

In the case of the Sentinel-1 data used for the subsidence measurements, uncertainties associated with the vertical displacement (subsidence/uplift) measurements were determined to be less than 2 cm and, in most areas, less than 1 cm.

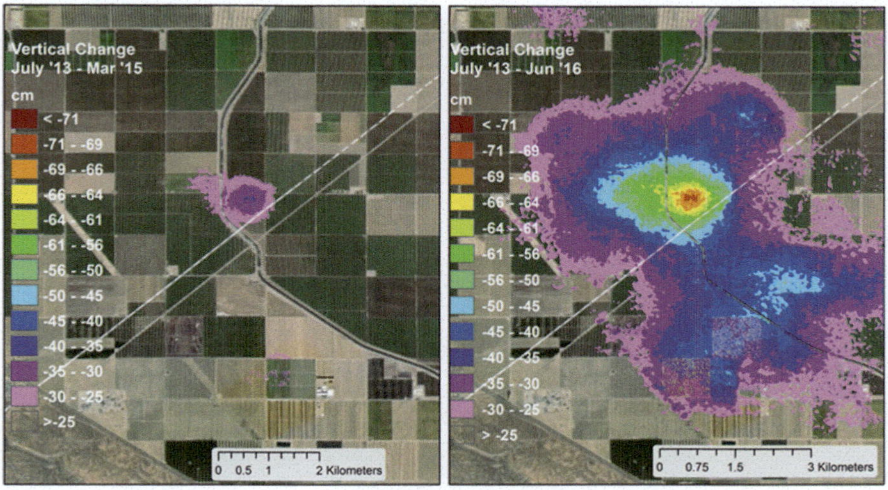

Fig. 9.7 Subsidence hot spot impacting the California Aqueduct (Avenal hot spot). The subsidence hot spot is centered <0.6 km east of the California Aqueduct. The two maps show the relative expansion of the subsidence bowl between March 2015 (left) and June 2016 (right). The same color scale is used for both, and only areas that subsided more than 25 cm during the interval are plotted. By June 2016, approximately 7.5 km of the aqueduct had subsided by more than 25 cm

9.6 Localized Subsidence Along the California Aqueduct and Delta–Mendota Canal

To evaluate subsidence on or near water conveyance structures, data were collected with the NASA UAVSAR airborne platform over two swaths roughly centered on the California Aqueduct (Fig. 9.6). The area also includes part of the Delta–Mendota Canal, a structure maintained by the federal government. The UAVSAR analysis for the State of California focused on small-scale subsidence directly impacting the aqueduct. Figure 9.6 shows cumulative change in ground elevation for the period of measurement, starting in mid-2013 (northern line) or spring-2014 (southern line) and extending through June 2016.

The northern imaged area has many locations experiencing significantly greater subsidence than is observed in the southern swath. The most rapid subsidence is observed along the northern stretch of the aqueduct at what is known as the Avenal hot spot. The rapid subsidence impacting the aqueduct at this location started in summer 2014, and both the amount of subsidence and the extent of aqueduct affected increased substantially by March 2015 (Fig. 9.7). The impact from the subsidence hot spot, which is centered east of the aqueduct, was such that approximately 7.5 km of the aqueduct experiencing 25 cm or greater subsidence after the measurements started in July 2013, with most of that occurring since summer 2014. The results from Sentinel-1 (Figs. 9.3, 9.4, and 9.5) and UAVSAR (Fig. 9.7) for the area around the Avenal hot spot show a similar rate of subsidence measured by both instruments.

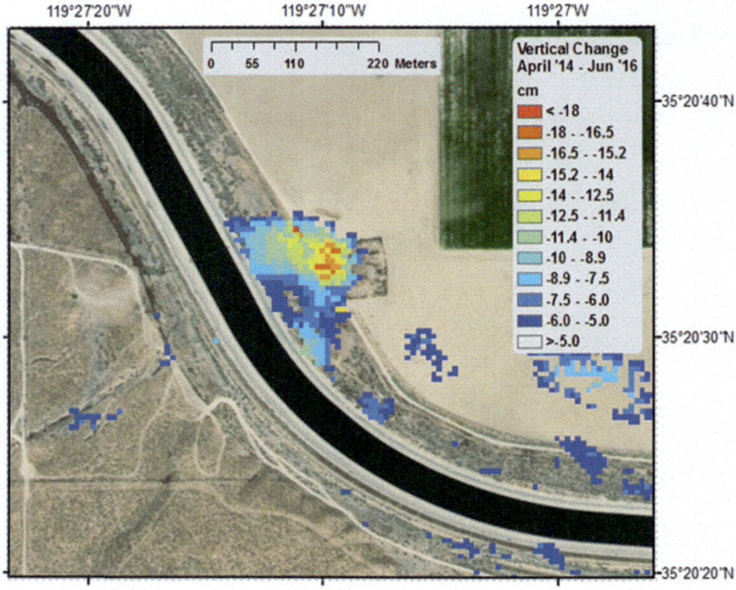

Fig. 9.8 Small subsidence feature in the shape of an arc, which is centered at the toe of the aqueduct's east slope and identified to be the location of a former seep. Maximum impact to the aqueduct is 14 cm for the period April 2014 to June 2016. This feature is not visible in the Sentinel-1 map

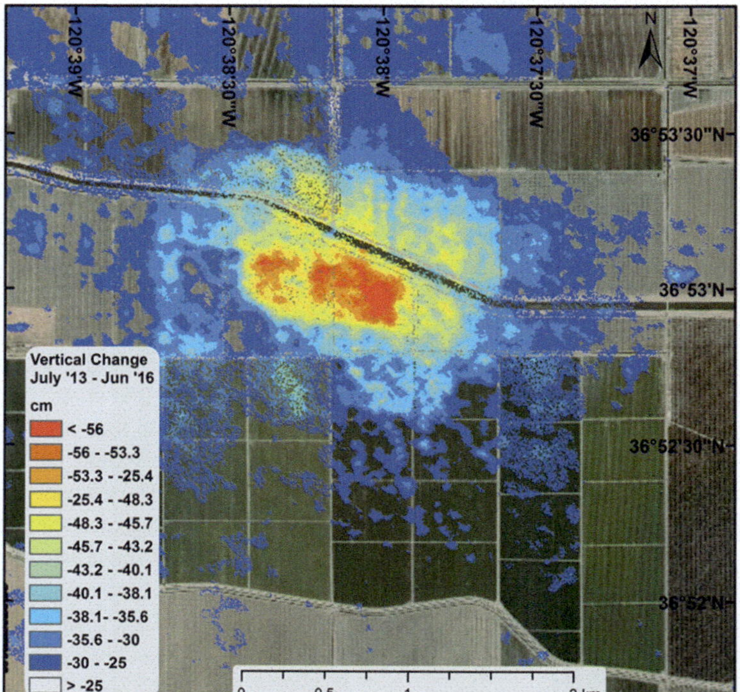

Fig. 9.9 Subsiding section of the Delta–Mendota Canal measured with UAVSAR

Several small areas of localized subsidence that did not show up in satellite data were detected with UAVSAR. Figure 9.8 shows an example of a feature of smaller size and magnitude, located very close to the aqueduct at the site of a former seep.

In addition to identifying changing elevation of the California Aqueduct, UAVSAR also identified a subsiding section of the Delta–Mendota Canal (Fig. 9.9).

9.7 Conclusions

InSAR has proven useful to the Ca. Dept. of Water Resources for monitoring subsidence from groundwater extraction and its impact on infrastructure. UAVSAR was used to measure subsidence of the California Aqueduct and to identify localized subsidence near or on the structure. Although the instrument also can map broader-scale subsidence, the image swath is substantially smaller than what one gets from a satellite (22 km vs. 200 km swath width). Sentinel-1, launched in April 2014, proved useful for making large-scale maps of subsidence in the Central Valley. Maps, as well as pixel histories of subsidence are obtained, and transects showing temporal and spatial details of subsidence can be produced from the InSAR data. The first acquisitions for California by Sentinel-1 were in late 2014 and are continuing, in general every 6–12 days. Early acquisitions over California were more sporadic, but have become more reliable as the coverage area has expanded along with ESA's capability to collect and process the data. We have begun downloading and processing Sentinel-1 data for other basins of California, including Antelope Valley, Coachella Valley, Borrego Valley, Imperial Valley, and the Salinas Valley, along with coastal areas. As the database expands, we expect to be able to present maps of those areas, as well as continuing monitoring of the Central Valley through the Advanced Rapid Imaging and Analysis (ARIA) project (Agram et al. 2018). In addition, UAVSAR will continue to provide high-resolution data along California aqueducts, as well as provide valuable prototype information for the upcoming NISAR instrument.

The cost of repairing and rebuilding California's critical infrastructure impacted by subsidence is high: The California DWR has spent 10s of millions of dollars in repairing the California Aqueduct over the last 40 years, and today a single repair can cost several million dollars (https://www.nbcbayarea.com/news/local/Damage-From-Sinking-Land-Costing-California-Billions-in-Infrastructure-Repairs-363597511.html). This study demonstrates that InSAR can provide information needed to target repairs to those areas most impacted and identify nascent localized subsidence so that it can be addressed before major loss of capacity or overtopping occurs.

Acknowledgements Processing of the spaceborne SAR subsidence measurements reported herein was funded by the California Department of Water Resources. The airborne UAVSAR acquisitions and analysis were funded by a combination of NASA and the California Department of Water Resources. This work was carried out at the Jet Propulsion Laboratory, California Institute of Technology, under contract with NASA and DWR. UAVSAR data are courtesy NASA/JPL-Caltech. The Alaska Satellite Facility (http://www.asf.alaska.edu/) archives and distributes the UAVSAR and Sentinel-1 data (which were acquired by the European Space Agency) used in this report.

References

Agram, P.S., D.P. Bekaert, S.E. Owen, and the ARIA team. 2018. *Development of standardized interferometric products and online processing capabilities, abs*. Washington, D.C.: American Geophysics Union Congress.

Bawden, G.W., W. Thatcher, R.S. Stein, K.W. Hudnut, and G. Peltzer. 2001. Tectonic contraction across Los Angeles after removal of groundwater pumping effects. *Nature* 412: 812–813.

Bekaert, D.P.S., C.E. Jones, K. An, and M.H. Huang. 2018. Exploiting UAVSAR for a comprehensive analysis of subsidence in the Sacramento Delta. *Remote Sensing Environment* 220: 124–134. https://doi.org/10.1016/j.rse.2018.10.023.

Bell, J.W., F. Amelung, A. Ferretti, M. Bianchi, and F. Novali. 2008. Permanent scatterer InSAR reveals seasonal an long-term aquifer-system response to groundwater pumping and artificial recharge. *Water Resources Research* 44: 2. https://doi.org/10.1029/2007WR006152.

Bertoldi, G.L., R.H. Johnston, and K.D. Evenson. 1991. *Ground water in the Central Valley, California: a summary report*. U.S. Geological Survey Professional Paper 1401-A, 44pp.

Borchers, J.W., and M. Carpenter. 2014. *Land Subsidence from Groundwater Use in California*. Report of findings to the California Water Foundation, http://www.californiawaterfoundation.org.

Chaussard, E., F. Amelung, H. Abidin, and S.-H. Hong. 2013. Sinking cities in Indonesia: ALOS PALSAR detects rapid subsidence due to groundwater and gas extraction. *Remote Sensing Environment* 128: 150–161.

Chaussard, E., P. Milillo, R. Bürgmann, D. Perissin, E.J. Fielding, and B. Baker. 2017. Remote sensing of ground deformation for monitoring groundwater management practices: application to the Santa Clara Valley during the 2012–2015 California drought. *Journal of Geophysical Research Solid Earth* 122. https://doi.org/10.1002/2017JB014676.

Conway, B.D. 2016. Land subsidence and earth fissures in south-central and southern Arizona, USA. *Hydrogeology Journal* 24: 649–655. https://doi.org/10.1007/s10040-015-1329-z.

Delbridge, B., R. Burgmann, E. Fielding, S. Hensley, and W. Schulz. 2016. Three-dimensional surface deformation derived from airborne interferometric UAVSAR: Application to the Slumgullion landslide. *Journal of Geophysical Research Solid Earth* 121: 3951–3977.

Donnellan, A., J. Parker, S. Hensley, M. Pierce, J. Wang, and J. Rundle. 2014. UAVSAR observations of triggered slip on the Imperial, Superstition Hills, and East Elmore ranch faults associated with the 2010 M 7.2 El mayor-Cucapah earthquake. *Geochemistry, Geophysics, Geosystems* 15: 815–829.

Farr, T.G., and Z. Liu. 2015. Monitoring subsidence associated with groundwater dynamics in the Central Valley of California using interferometric radar. In *Ch. 24 in remote sensing of the terrestrial water cycle, geophysical monograph*, ed. V. Lakshmi, vol. 206. Washington, DC: American Geophysical Union, Wiley.

Farr, T.G., C. Jones, and Z. Liu. 2015. *Progress report: subsidence in the Central Valley, California, submitted to CA DWR*. Available at: http://www.nasa.gov/jpl/nasa-california-drought-causing-valley-land-to-sink.

———. 2017. *Progress report: subsidence in the Central Valley, California, March 2015 – September 2016, submitted to CA DWR*. Available at: http://www.water.ca.gov/waterconditions/docs/2017/JPL%20subsidence%20report%20final%20for%20public%20dec%20 2016.pdf.

Galloway, D.L., K.W. Hudnut, S.E. Ingebritsen, S.P. Phillips, G. Peltzer, F. Rogez, and P.A. Rosen. 1998. Detection of aquifer system compaction and land subsidence using interferometric synthetic aperture radar, Antelope Valley, Mojave Desert, California. *Water Resources Research* 34: 2573–2585.

Galloway, D.L., D.R. Jones, and S.E. Ingebritsen. 1999. Land subsidence in the United States. In *U.S. Geological Survey Circular 1182*, 175pp.

Hoffmann, J., H.A. Zebker, D.L. Galloway, and F. Amelung. 2001. Seasonal subsidence and rebound in Las Vegas Valley, Nevada, observed by synthetic aperture radar interferometry. *Water Resources Research* 37: 1551–1566.

Jones, C.E., K. An, R.G. Blom, J.D. Kent, E. Ivins, and D. Bekaert. 2016. Anthropogenic and geologic influences on subsidence in the vicinity of New Orleans, Louisiana. *Journal of Geophysical Research Solid Earth* 121: 3867–3887.

Lanari, R., P. Lundgren, M. Manzo, and F. Casu. 2004. Satellite radar interferometry time series analysis of surface deformation for Los Angeles, California. *Geophysical Research Letters* 31. https://doi.org/10.1029/2004GL021294.

Lundgren, P., F. Casu, M. Manzo, A. Pepe, P. Berardino, E. Sansosti, and R. Lanari. 2004. Gravity and magma induced spreading of Mount Etna volcano revealed by satellite radar interferometry. *Geophysical Research Letters* 31: L04602.

Lundgren, P., E.A. Hetland, Z. Liu, and E.J. Fielding. 2009. Southern San Andreas-San Jacinto fault system slip rates estimated from earthquake cycle models constrained by GPS and interferometric synthetic aperture radar observations. *Journal of Geophysical Research* 114: B02403. https://doi.org/10.1029/2008JB005996.

Madsen, S.N., and H.A. Zebker. 1998. Imaging radar interferometry, ch. 6. In *Principles and Applications of Imaging Radar, Manual of Remote Sensing*, ed. F.M. Henderson and A.J. Lewis, vol. 2, 359, 866pp–380. New York: Wiley.

Massonnet, D. 1997. Satellite radar interferometry. *Scientific American* 276: 46–53.

Miller, M.M., and M. Shirzaei. 2015. Spatiotemporal characterization of land subsidence and uplift in Phoenix using InSAR time series and wavelet transforms. *Journal of Geophysical Research Solid Earth* 120. https://doi.org/10.1002/2015JB012017.

Murray, K.D., and R.B. Lohman. 2018. Short-lived pause in Central California subsidence after heavy winter precipitation of 2017. *Science Advances* 4: eaar8144. https://doi.org/10.1126/sciadv.aar8144.

Ojha, C., M. Shirzaei, S. Werth, D. Argus, and T. Farr. 2018. Sustained groundwater loss in California's Central Valley exacerbated by intense drought periods. *Water Resources Research* 54. https://doi.org/10.1029/2017WR022250.

Ojha, C., S. Werth, and M. Shirzaei. 2019. Groundwater loss and aquifer system compaction in San Joaquin Valley during 2012–2015 drought. *Journal of Geophysical Research Solid Earth* 124. https://doi.org/10.1029/2018JB016083.

Ozawa, T., and H. Ueda. 2011. Advanced interferometric synthetic aperture radar (InSAR) time series analysis suing interferograms of multiple-orbit tracks: a case study on Miyake-Jima. *Journal of Geophysical Research* 116: B12407. https://doi.org/10.1029/2011JB008489.

Reeves, J.A., R.J. Knight, H. Zebker, W.A. Schreüder, P.S. Agram, and T.R. Lauknes. 2011. High quality InSAR data linked to seasonal change in hydraulic head for an agricultural area in the San Luis Valley, Colorado. *Water Resources Research* 47. https://doi.org/10.1029/2010WR010312.

Scheingross, J., B. Minchew, B. Mackey, M. Simons, M. Lamb, and S. Hensley. 2013. Fault-zone controls on the spatial distribution of slow-moving landslides. *GSA Bulletin* 125: 473–489. https://doi.org/10.1130/B30719.1.

Sharma, P., C.E. Jones, J. Dudas, G. Bawden, and S. Deverel. 2016. Monitoring of subsidence with UAVSAR on Sherman Island in California's Sacramento-San Joaquin delta. *Remote Sensing of Environment* 181: 218–236. https://doi.org/10.1016/j.rse.2016.04.012.

Smith, R.G., R. Knight, J. Chen, J.A. Reeves, H.A. Zebker, T. Farr, and Z. Liu. 2017. Estimating the permanent loss of groundwater storage in the southern San Joaquin Valley, California. *Water Resources Research* 53: 2133–2148. https://doi.org/10.1002/2016WR019861.

Sneed, M., and J.T. Brandt. 2007. *Detection and Measurement of Land Subsidence Using Global Positioning System Surveying and interferometric Synthetic Aperture Radar, Coachella Valley, California, 1996–2005*: U.S. Geological Survey Scientific Investigations Report 2007–5251, 31pp.

———. 2013. *Detection and Measurement of Land Subsidence Using Global Positioning System Surveying and Interferometric Synthetic Aperture Radar, Coachella Valley, California, 1996–2005*: U.S. Geological Survey Scientific Investigations Report 2007–5251, v. 2.0, 31pp.

Sneed, M., S.V. Stork, and R.J. Laczniak. 2003. Aquifer-system characterization using InSAR. In *US Geological Survey Subsidence Interest Group Conference Proceedings, 2001*, ed. K.R. Prince and D.L. Galloway. USGS Open-File Rept. 03-308.

Part II
Urban Infrastructure

Chapter 10
Earth Observation Based Understanding of Canadian Urban Form

Ying Zhang

10.1 Introduction

Urbanization can be viewed as a complex dynamic process of land surface change, influenced by population pressure, employment availability, infrastructure development, economic activities and local topography (Bunting and Filion 2000; Batty 2005). Current urban land characteristics, e.g. cover, use, pattern and structure, reflect the impacts of accumulated human activities and responses to economic, demographic and topographic conditions. As footprints from intensive human activities, structure, pattern, composition and materials of the urban land are complex and differ significantly from rural footprints (for example, as shown in Anthropocene by Burtynsky at www.edwardburtynsky.com). Rapid urbanization is a global phenomenon (Seto et al. 2011, 2012; UN 2018) and Canada reflects this global urbanization trend. Canadian urban population percentage has increased to 81% in 2011 from about 70% in 1961 (StatsCan 2015) largely due to shifts in the country's economic base and the preferential settlement of new immigrants in large cities (Bunting and Filion 2000). The urban land structures of Canadian cities have also evolved during this urban growth process.

Information regarding current urban land surface conditions as well as their changes is essential in addressing issues such as urban energy efficiency, environment quality, disaster management, public health and environmentally sensitive land loss (Gueneralp et al. 2008; Johnson 2001; Poumanyvong et al. 2012). The knowledge gained from analyses of this information can lead to better understanding of the impacts of urbanization and urban land structure on urban development sustainability (Neptis 2013, 2014). Earth observation (EO) imagery and interpretative technologies

Y. Zhang (✉)
Canada Centre for Mapping and Earth Observation, Natural Resources Canada, Ottawa, ON, Canada
e-mail: ying.zhang@canada.ca

V. Singhroy (ed.), *Advances in Remote Sensing for Infrastructure Monitoring*, Springer Remote Sensing/Photogrammetry,
https://doi.org/10.1007/978-3-030-59109-0_10

can contribute significantly to the characterization of the urban forms of cities. The availability of remotely sensed images has facilitated detailed monitoring and mapping of urban land surfaces and thereby extended the role of remote sensing into operational management of urban development and urban activities (Carlson 2003). In addition to land cover and land use details, there is a recognition of the importance of urban land structure (Lynch and Rodwin 1958; Banai and Rapino 2009) that is the information at higher level than those of land cover and land use. Urban land structure, also referred to as urban form (Jabareen 2006; Williams 2014), includes the functional, morphological and demographic spatial distributions of built-up landscapes (Batty 2005; Clifton et al. 2008; Maantay and Maroko 2009; Aubrecht et al. 2009; Gerl et al. 2014; Downes and Storch 2013; Regnier and Legras 2018). The focus of this work is on spatial aspects of urban land structure.

Urban landscapes are complex with large diversity in characters of land covers, materials and morphology (Herold et al. 2004; Baldridge et al. 2009). The availability of image data at a broad range of spatial resolutions provides the opportunity to present urban land surface characteristics at a range of spatial levels. For extraction of detailed urban land surface information, especially in dense urban areas, imagery data with higher resolution are usually preferred. In using high or very high resolution remote sensing imagery, object-based, rather than pixel-based, image classification technologies have been found to be effective for extracting manmade objects (Herold et al. 2002; Darwish et al. 2003; Taubenböck et al. 2006; Weng and Quattrochi 2006; Navulur 2007; Khosravi et al. 2014; Georganos et al. 2018). Using multi-source data or information for urban land, information extraction can produce results with high accuracy. Land cover classification using multi-source information layers has been shown to improve accuracy by reducing confusion that arises from conventional spectral-based classification (Vogelmann et al. 1998). Data or information fusion technologies (Hall and Llinas 1997; Pohl and van Genderen 2016) have been developed as a multi-source data integration approach for more reliable results of information derivation and mapping. As the amount of urban data and information collected from earth observation and other sources (e.g. open source data) has grown significantly in recent years, integration technologies have played an increasingly important role for generation of new information layers describing the spatial distribution of land related characteristics.

Canada is a country with high automobile dependence in urban areas (Kenworthy and Laube 1999; Statistics Canada 2016a, b). Urban land use structure significantly influences intra-city commute activities (Cervero 1996; Crane 2000; Wheaton 2004; Shoaga and Muehleggerb 2015; Durantona and Turne 2018). For better understanding of these influences in Canadian cities, the information about urban land structures and changes can be derived based on land surface character maps generated from remote sensing imagery and information integration. This paper presents summarized results from research on Canadian urban land structure and changes, as well as their impacts, based on remote sensing data and technology. The scope of these studies includes urban land information extraction from image data, integration of information and spatial analysis and modeling for urban applications with focus on the impacts of urban land use structure on urban commuting effort. The data-information-knowledge processing framework is shown in Fig. 10.1. In the following section (Sect. 10.2), the

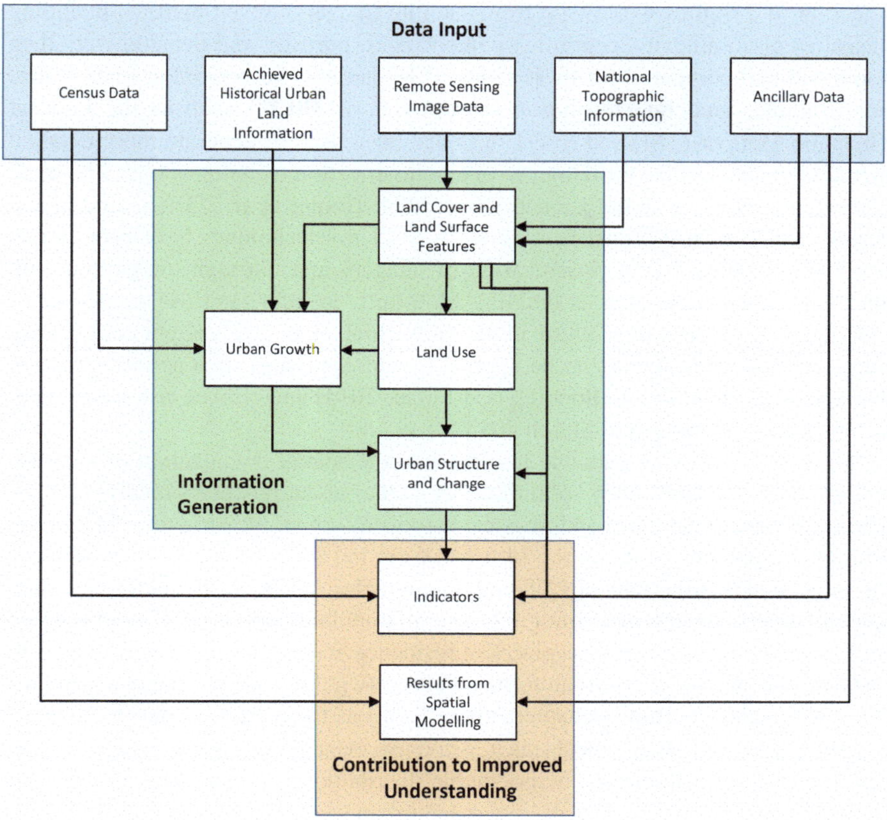

Fig. 10.1 Outline of processing from data to information and to the improved understanding in this work

results from characterization of Canadian land surfaces and the extraction of basic land surface information from remote sensing imagery are described. In Sect. 10.3, the generation and results of a time series of Canadian urban land growth involving integration of contemporary with archival land information are presented. Section 10.4 formulates the application of spatial analysis methods to understanding the impact of urban structure on intra-city work-related travel.

10.2 Characterization of Canadian Urban Land Surface

Urban Form

Traditionally, urban form has been defined by land and manmade characteristics that make up built-up areas. Overall, urban extent represents the first level of 'physical urban form' (PUF). Huang et al. (2007) derived overall urban extents of a selection of 77 cities worldwide from satellite imagery. Seven metrics, including measures of complexity, centrality, compactness, porosity and density, were then computed and compared for these cities. PUF can also be derived at finer scales. For example, moderate resolution sensors (i.e. 10–30 m) such as the Landsat Thematic Mapper (TM) and SPOT 4/5 have been used to generate more detailed thematic maps of cities utilizing per-pixel classification techniques (e.g. Zha et al. 2003; Guindon et al. 2004; Lu and Weng 2004; Homer et al. 2007; Guindon and Zhang 2009). At still higher resolutions (i.e. <2 m), techniques have been developed to use object-based classification of imagery and, through integration with ancillary information such as building footprints, to infer land use properties of individual city blocks (e.g. Wurm et al. 2009; Huck et al. 2011; Krehl et al. 2016). Such detailed information can be applied to address issues such as infrastructure damage assessment due to flooding (Gerl et al. 2014) and climate risk assessment in megacities (Downes and Storch 2013).

While PUFs provide valuable insights into the spatial organization of land use and physical structures, they are limited for studies about processes involving interactions of human activities and physical layouts of urban infrastructure. Therefore, a more comprehensive definition of urban form has been adopted which includes the integrated distributions of PUF and demographic attributes, the latter including spatially explicit mappings of not only overall population but also of more specific socio-economic factors such as places of residence and employment, age profiles in residential lands, etc. This urban form is defined (e.g. https://www.neptis.org/publications/growth-opportunity/chapters/urban-form) as the pattern of development in an urban area, including aspects such as the use of land (e.g. residential, commercial, industrial, institutional); demographic distributions on land. This urban form characterization is referred as an 'integrated urban form' (IUF).

Characteristics of Canadian Urban Form

The landscapes of Canadian cities share many similar attributes with urban centres in the USA. Historically, these cities included a high-density core both in terms of resident population and employment. This core is surrounded by lower density suburban areas dominated by low-rise single and double unit housing on land parcels with significant portions of vegetative cover. As North American cities are transitioning from manufacturing to service based economic activities, there is a gradual suburbanization of employment as well. Another key feature of these cities is the high level of segregation of land uses. Suburban employment is largely confined to large scale industrial parks, while commercial activities (e.g. shopping) occur in shopping malls with large land tracts dedicated to parking. As a result of this physical urban form, extensive use of private vehicles is made for intra-urban travel. This presents challenges for municipal governments to provide efficient and cost-effective public transit alternatives.

The urban landscape dominated by residential, single-family house communities is illustrated in Fig. 10.2, with maps of land use land cover, residential house type

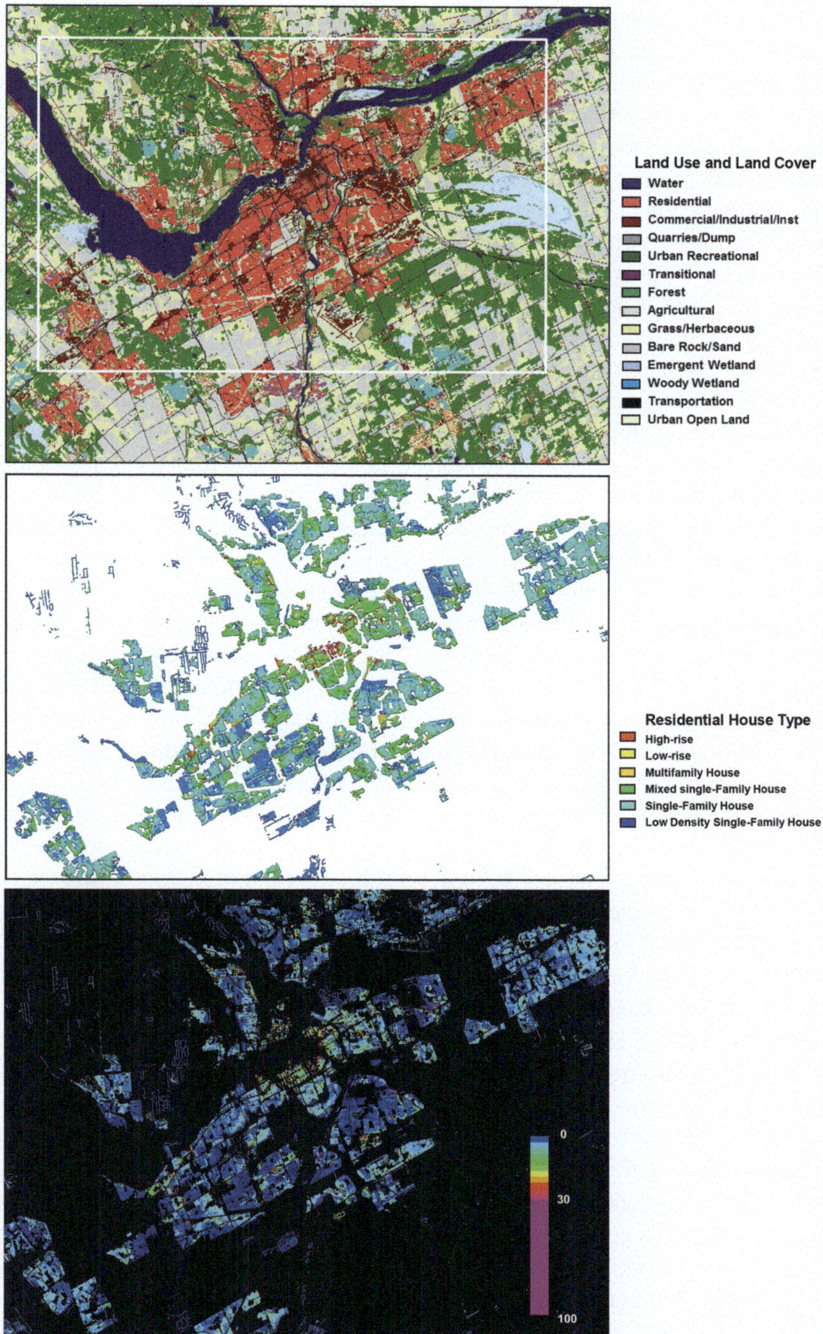

Fig. 10.2 Land use land cover (upper), residential house type (middle) and population density (lower) maps of Ottawa-Gatineau. The latter two maps were resulted from integration of the land cover land use map with census population and dwelling data from Statistics Canada

and population density in the Ottawa-Gatineau region circa 2001 (Zhang and Guindon 2009; Guindon and Zhang 2009; Zhang et al. 2010a, b; Zhang and Sun 2018). From these maps, the residential class (pink coloured in land use land cover maps) dominates the urbanized land of this region. The house type distribution illustrates the domination of single-family houses in residential lands where population density is low relative to the cores of the two cities, especially that of Ottawa. The maps of population density and dwelling density are generated from integration of the land cover's residential land layer at 30 m resolution and demographic data (2001 National Census from Statistics Canada) at the census block level. Furthermore, the distribution of residential house types was estimated by using dwelling density and validation with field collected information. As the input data for the integration process are pixel-based land use maps (those pixels classified as residential land use) and demographic data with polygons of census blocks, the integration process must address the fact that the spatial unit areas of the pixel-based land use map and census reporting units differ. According to Statistics Canada, the census block is the smallest reporting unit for accounting of population and dwelling data. The areas of census blocks vary and can contain not only residential lands but other uses as well. In the Ottawa-Gatineau region, the maximum percentage of residential land in any given census block is about 75%. To improve estimates of urban attributes such as population density, areas used are limited to the portions of census blocks classed as residential on satellite imagery. In Fig. 10.2 house types and population densities are presented at 30 m by 30 m pixel units. In the same way, employment-related census attributes such as employment density can be estimated using area estimates based on pixels assigned to the commercial-industrial-institutional land use class.

Urban land cover and land use information layers can be derived from satellite imagery data with different spatial resolutions. Figure 10.3 shows a land cover map derived from Quickbird PAN- sharpened image at 0.61 m resolution and a land use map based on Landsat imagery at 30 m resolution for a portion of Ottawa, Canada; the maps were generated using a pixel-based semi-supervised classification approach described in detail in Guindon et al. (2004). The land cover classes shown in Fig. 10.3a are impervious surface, bare soil, trees, grass and water. The impervious surface class, which includes building roofs, driveways, roads and parking lots, is an essential land cover class used to quantify impacts of urban development on the environment, hydrology and micro-climate. Numerous methods have recently been developed to map imperviousness from various sensors (see Weng 2007 for an overview of these methods). Land use classes, on the other hand, can encompass many land cover types. Effective production of land use information requires that land cover information derived from imagery be integrated with auxiliary GIS information. For example, a golf course would be classed as recreational land use shown in Fig. 10.3 but it includes various types of land covers (grass, tree, bare soil and impervious surface). Rather than relying solely on information from remote sensing imagery, urban recreation land information was obtained from the National Topographic Database (or alternatively, from a municipal information base). Similarly, the information layers of highway and secondary road networks, as well as that of wetland were integrated into the land use map.

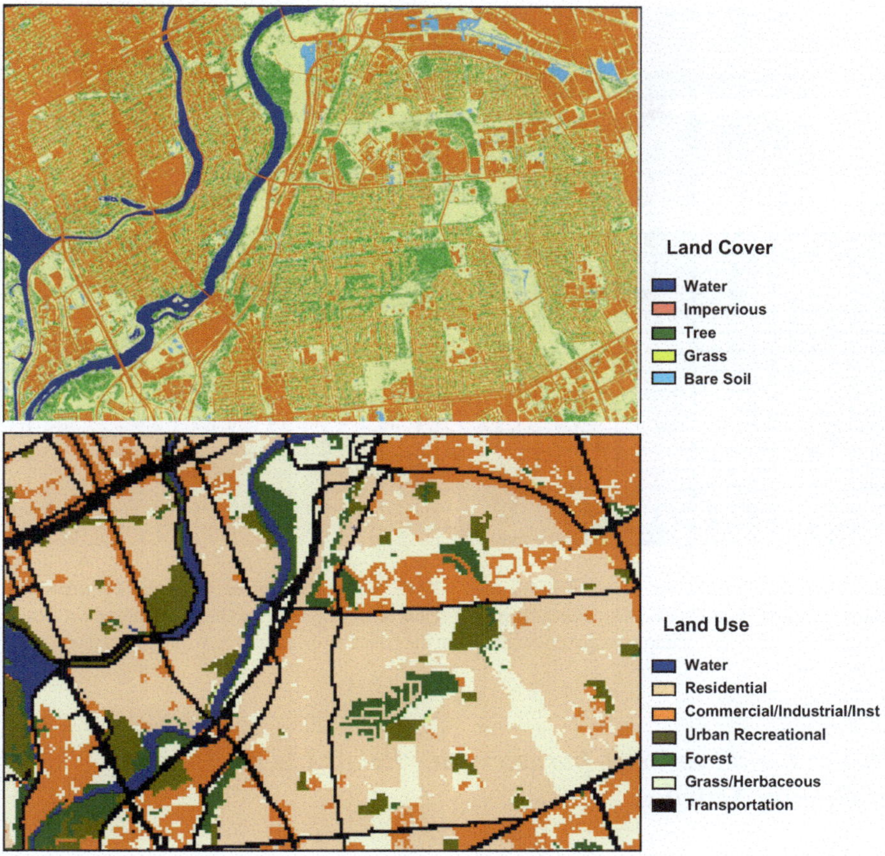

Fig. 10.3 Portions of (**a**) Basic land cover map (upper) derived from PAN-sharpened Quickbird (0.61 m resolution); (**b**) Land use map (lower) derived from Landsat image (30 m resolution) and with integrated information from diverse sources, covering a dense urban area of Ottawa

Residential and commercial-industrial-institutional, on the other hand, can be largely discerned from imagery alone based on spectral properties and spatial patterns of imperviousness and vegetation cover, as well as patterns of the road networks. Figure 10.4 shows high resolution (at 0.61 m) land cover samples in 30 m by 30 m windows sorted by impervious surface percentage in residential and commercial-industrial-institutional land use classes, respectively. The figure indicates that majority of the commercial-industrial-institutional class samples are dominated by high impervious land cover. However, the majority of residential samples exhibit mixtures of impervious surface, grass and trees with an average imperviousness of about 40%. In conclusion, the level of imperviousness surface land cover, readily captured by Landsat, provides good separability of these two urban land use classes (Zhang and Guindon 2009). These two classes not only contribute to the quantification of urban form but also provide spatially explicit information on commute travel origins and destinations.

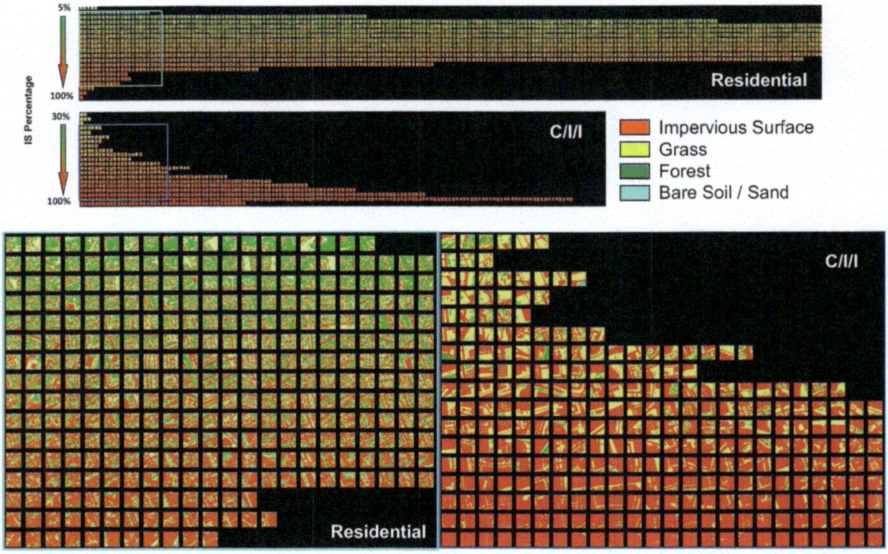

Fig. 10.4 The land cover samples in 30 m by 30 m windows of residential and commercial-industrial-institutional (C/I/I) land use classes at scale of impervious surface percentage (increasing downwardly). The samples were collected from the basic land cover map with 0.61 m resolution over Ottawa

Urban Land Surface Characteristics Related to the Intensity of Human Activities

The morphologic character of the urban landscape is another key attribute of urban form, which can reveal the density and intensity of human activities in Canadian urban areas. The downtown Central Business Districts (CBDs) or the city central areas in many cities exhibit high built-up density dominated by high-rise buildings, where are mainly of commercial-instructional land use. In recent years, through urban infill and renewal developments, there are some high-rise buildings with residential use in CBDs in major Canadian cities, e.g. those in the core of Toronto. Outside of the CBD core area, the lands in commercial and institutional use are concentrated with high percentage of impervious surface and with large size, boxy shaped low-rise buildings. The heights of buildings in the CBD can be related with the workplace office space; therefore, the distribution of the high- and low-rise office buildings in the commercial-industrial-institutional use lands reflects worker density. Therefore, the distribution in the urban built-up land morphology can be related with the worker density distribution pattern. Figure 10.5 shows the 3-D map (lower) of a portion of Calgary CBD and illustrates the transition of the morphology from high-rise building dominated zone to low-rise building zone in the CBD of Calgary, which is high in the core of CBD then sharply decreased with the distance to the core. This example of Calgary is representative of the abrupt change in the morphology from the urban central area outwards in major Canadian cities, especially for those with single core CBD (Montreal, Ottawa, Edmonton, etc.). The 3-D map of Digital Surface Model (DSM) was derived from LiDAR data with help of multi-spectral high resolution image. A LiDAR image data with a density of 10 points per square meter, Pleiades multi-spectral image with 50 cm resolution and building footprint polygons were registered together, processed and used for classification and grouping for generation of the building class. The irregular (non-flat) shapes of the building rooftops have been smoothed to remove the noise from the LiDAR data. The trees in the CBD area are sparse, the locations of which are validated with the land cover map from the Pleiades image. The map with building footprint and tree classes is shown in Fig. 10.5d.

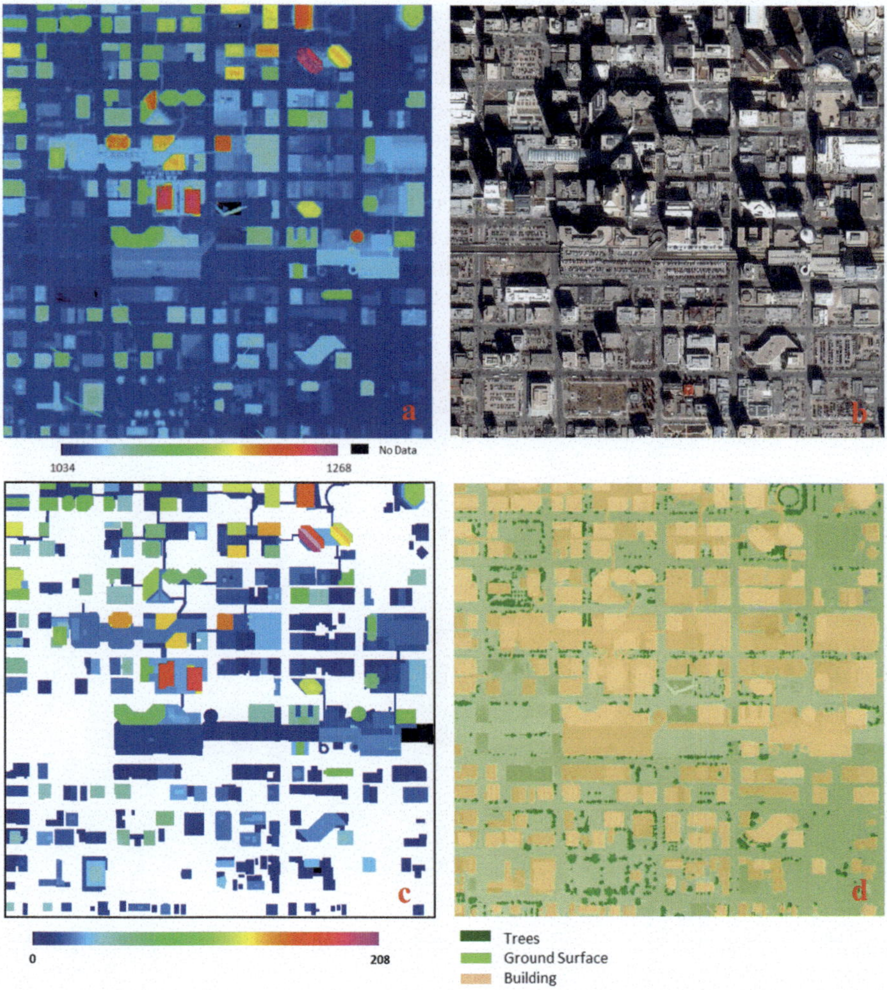

Fig. 10.5 A 3-D DSM map (lower) covering a portion of the Calgary CBD area (the vertical axis is in metres above sea level), derived from (**a**) LiDAR data and (**b**) Pleiades Pan-Sharpened image. (**c**) Building map derived from the LiDAR data and building footprints; (**d**) land cover map derived based on the LIDAR and Pleiades data

The spatial distributions of land surface feature targets, such as vehicles in parking lots, are also the essential information for the urban studies. Canadian intra-city transportation activities, e.g. commuting and shopping activities, are in high automobile dependency. The distribution of vehicles in parking in urban areas (except for the downtown core areas where some parking lots are underground) reveals the human activities such as those related with working and shopping, as well as the utilization of parking lots. In recent years, the high resolution (less than 1 m) satellite imagery (with PAN-sharpening) has been used widely for urban study and urban applications, which can provide the detailed information about the land surface. Figure 10.6 illustrates, as an example, the distribution of parking vehicles in commercial-industrial-institutional lands in a dense built-up urban area of Toronto.

In Fig. 10.6, the parking lots related with locations with different land use including (a) working place, (b) retailer, (c) car dealer and (d) park-and-ride parking lot, which indicate different purposes of the urban travel activities. The map shown in Fig. 10.6 reveals the intensity of vehicle use activities related with the land uses. The vehicles in parking were detected and extracted from PAN-sharpened multi-spectral WV3 image at 0.37 m resolution. Since the data spatial resolution is very high; the image pixel covering individual vehicle can be as pure pixel capturing the spectral and spatial behaviours that are different from those of other land surface targets. For the results shown in Fig. 10.6, both the image spectral and spatial information at segment level have been used in the extraction of vehicle targets.

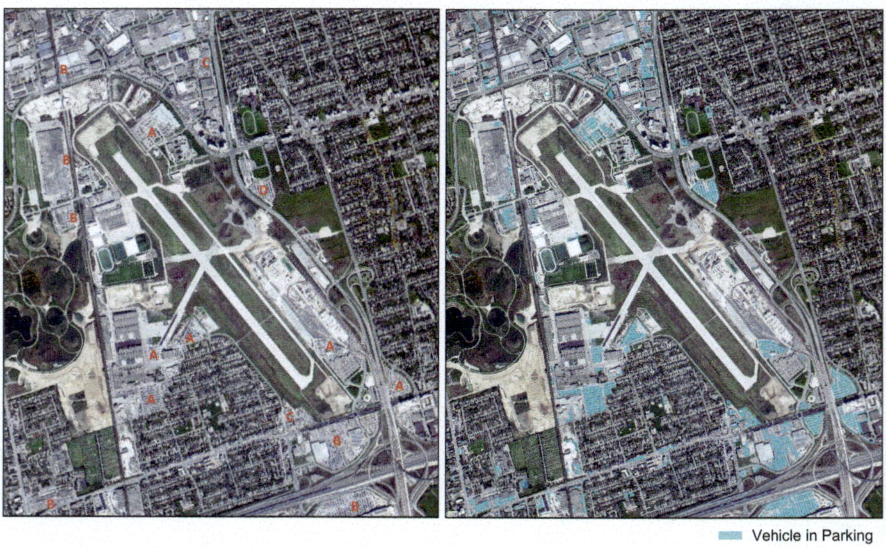

Fig. 10.6 PAN-sharpened WV3 RGB image (left) and the distribution of vehicles in parking (right) extracted from the image over a portion of Toronto. The parking lots are linked to (**a**) workplace; (**b**) retailer; (**c**) car dealer; (**d**) park-and-ride parking lot

10.3 Measuring Canadian Urban Land Expansion

The extensive archives of Landsat TM imagery can provide baseline data to portray urban land change at 30 m resolution from approximately 1984 to the present. However, the onset of rapid urbanization in many countries (such as in Canada) occurred earlier and therefore there is a need to create a consistent time series that dates back to at least the 1960s. Such long-term record information creation also requires further integration with diverse demographic datasets (Zhang et al. 2010a, b; StatsCan 2016a, b). For many built-up areas, there are archives of historical urban boundaries and land use, in both paper and/or digital form. Further, creation of a consistent time series of urban expansion involves fusion of information from contemporary land cover and land use (LCLU) maps with archived urban maps that may differ significantly in scale and thematic legend. An example of information fusion process for generating long-term (the 35-year period from 1966 to 2001) record of Canadian urban land growth is given below.

Our example involved assimilation of Landsat-based LCLU products with early federal land use program information (Cocklin and Smit 1982; Environment Canada 1976; Gierman 1981) covering the period 1966 to 1986. The baseline information for this period was collected as part of the Canadian Land Use Monitoring Program (CLUMP). The Landsat-based LCLU maps of 1986 and 2000 epochs of CUrLUS (Canadian Urban Land Use Survey, see Zhang et al. 2010a, b) were integrated with CLUMP maps of land use change. The CLUMP produced urban extent and land use information on a 5-year cycle, coincident with Canada's census years, and was generated through manual delineation using aerial photography. The CLUMP time series itself can be further subdivided into two self-consistent components hereafter referred to as CLUMP I (1966–1976) and CLUMP II (1981–1986). The former provides limited urban information, namely an urban 'mask' that delineated the physical extent of cities and was part of a broader program, the Canada Land Inventory. CLUMP-II involved a more comprehensive urban mapping program leading to detailed urban descriptions of both land cover and land use. Figure 10.7 illustrates CLUMP-I and CLUMP-II maps, as well as a CUrLUS map of Ottawa-Gatineau as example inputs for the information integration.

To generate a consistent time series, it was necessary to develop assimilation procedures to rationalize differences in thematic legends and spatial resolution among CLUMP-I, CLUMP-II and CUrLUS LCLU coverages. This process was facilitated by the fact that for each city of interest, a CUrLUS LCLU product could be generated from mid-1980s Landsat imagery and directly compared with its contemporary 1986 CLUMP product. The compatibility and consistency between CLUMP-II and CUrLUS information sets have been assessed at the overall urban mask level and the level of detailed built-up classes. There is good compatibility between CLUMP-II and CUrLUS in overall urban masks. Disagreement exists in the spatial detail of urban built-up classes and the class definitions between CLUMP-II and CUrLUS. Within the CLUMP information set, spatial and thematic detail exhibits improvement with time from the 1966 to the 1986. The differences in

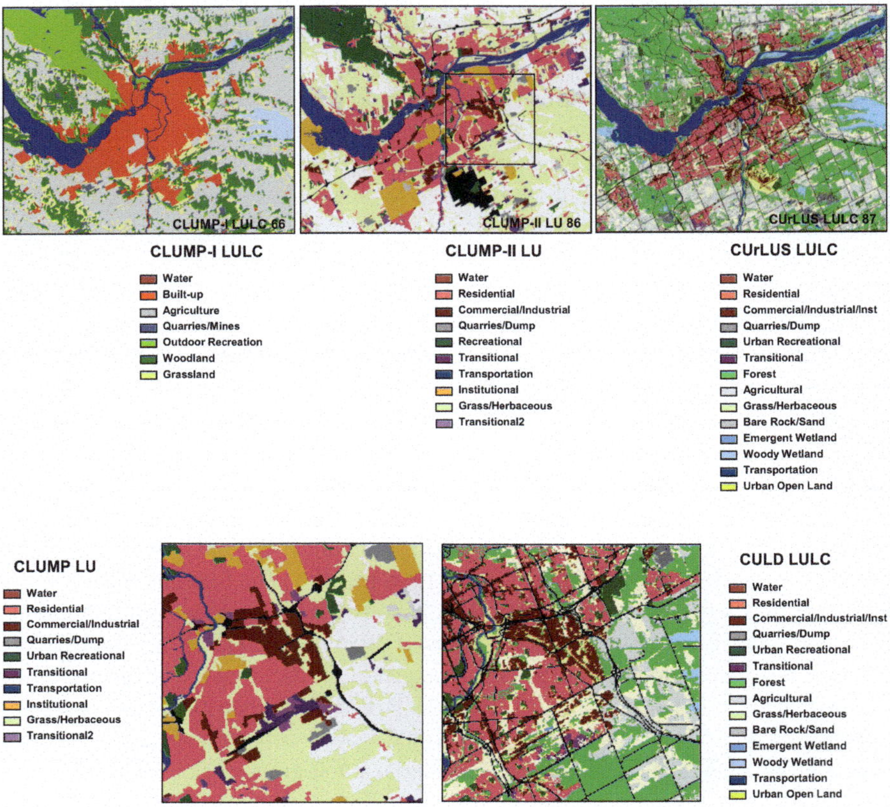

Fig. 10.7 Maps of CLUMP I (1966), CLUMP-II (1986) and CUrLUS (1987) of the Ottawa-Gatineau region

spatial detail can be overcome through the utilization of class assignment at the CUrLUS pixel level at 30 m. To retrieve the maximum level of detail for the early years, CUrLUS provides a set of reliable thematic spatial information cues that can be used within the constraints of two assumptions regarding the urbanization process, in particular its irreversible nature and the low probability of land use conversion once an area is built. While the first assumption is generally applicable, the second assumption should be used with caution (depending on the scales of the land conversion). For example, commercial or institutional or industrial land in city cores can be converted to residential use through a costly process involving, for example, 'brown-field' renewal and therefore has been limited in extent during this study period (from 1966 to 2001). The irreversibility of urbanization not only constrains the extent of a city's built-up land to remain the same or increase with time but must also be satisfied in the spatially explicit sense that if a land unit (e.g. pixel) is known to be not built at time $T2$ (later year), then it should also be not built at any time $T1$ (early year) where $T1 < T2$. It is also noted that the land cover and land use information of our latest epoch (2001) in CUrLUS is considered to be highly reliable due to the extensive field checking and the use of ancillary data to correct machine classification errors.

Urban land encompasses 'built-up' (e.g. residential and commercial/industrial land, transportation infrastructure) and herbaceous surface components. Another assumption that is invoked is that the transitions among these 'built-up' components tend to be rare, especially between residential (R) and commercial/industrial (C/I) over short-term time period (30–40 years). The primary exceptions to this temporal stability include urban renewal (usually in the core regions of cities) and brownfield redevelopment. However, in the case of large patches of urban openland or vegetated land, conversion to built-up lands can occur frequently in urbanization processes. For example, an industrial park is developed from urban openland over time through the construction of new businesses. Therefore, it is necessary to separate the variable component (vegetated land) from stable components (urban built-up land) within urban areas. Based on these constraints, a rule-based procedure for land cover and land use information fusion was developed.

The application of the assimilation methodology has been applied to retrieval and generation of consistent 35-year term urban land cover and land use change information for fourteen major Canadian urbanized areas encompassing approximately 90 municipalities with a total population of over 17 million (2001 census population counts). Using the temporally and spatially consistent urban land cover land use information, key variables quantifying urban expansion, such as urban land area, urban dwelling density, urban population density and natural land loss, as well as the influences of urban expansion on urban transportation and on the intensification of the urban heat island effect have been estimated.

Canadian cities experienced rapid urban land growth, especially in the decades immediately prior to 2001 (StatsCan 2016a, b). The majority of this urban growth has occurred in suburban regions and in satellite communities of major urban centres. The development of the Greater Toronto and Ottawa metropolitan areas are two prime examples of this expansion. Figure 10.8 illustrates the growth in four major urban centres (Ottawa-Gatineau, Halifax, Vancouver and Winnipeg) from 1966 to 2001. For example, the City of Ottawa has grown mainly in satellite towns as suburban residential developments in the east, west and south of the old city of Ottawa. These communities were formally amalgamated with the old city of Ottawa in 2001. The leapfrogging pattern of this expansion reflects the influence of the Greenbelt restrictions to development (ncc-ccn.gc.ca). Topographic conditions also influence the growth of cities. The growth patterns of Vancouver and Halifax are affected by the local geography such as water bodies and, in the case of Vancouver, local mountainous terrain. Winnipeg, on the other hand, a prairie city, exhibits more contiguous expansion pattern.

Urban land use and land structures have changed during this urban expansion. Figure 10.9 illustrates an example of urban land structure growth in the city of Calgary from 1966 to 2001. Residential and commercial-industrial-institutional land uses both have been grown during this period. However, Calgary exhibits growth in which added urban land has primarily been in the form of residential housing developments.

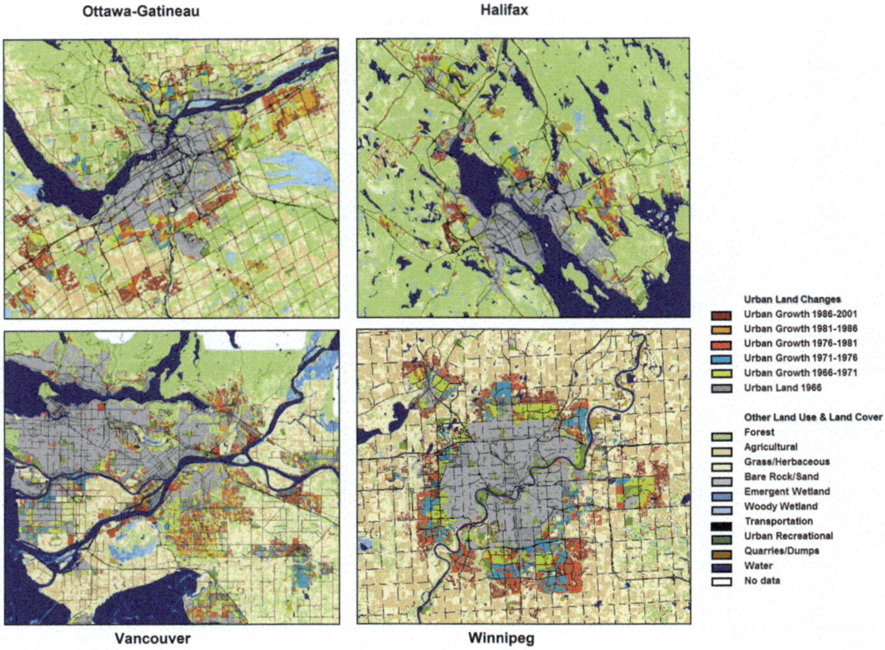

Fig. 10.8 The growth maps of four major Canadian urban areas from the original urban land in 1966 (in light grey)

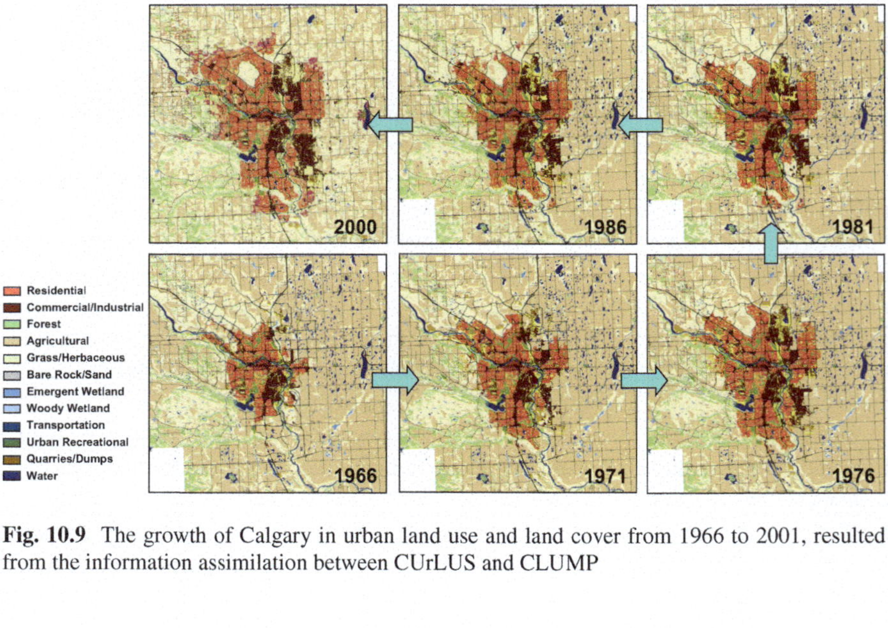

Fig. 10.9 The growth of Calgary in urban land use and land cover from 1966 to 2001, resulted from the information assimilation between CUrLUS and CLUMP

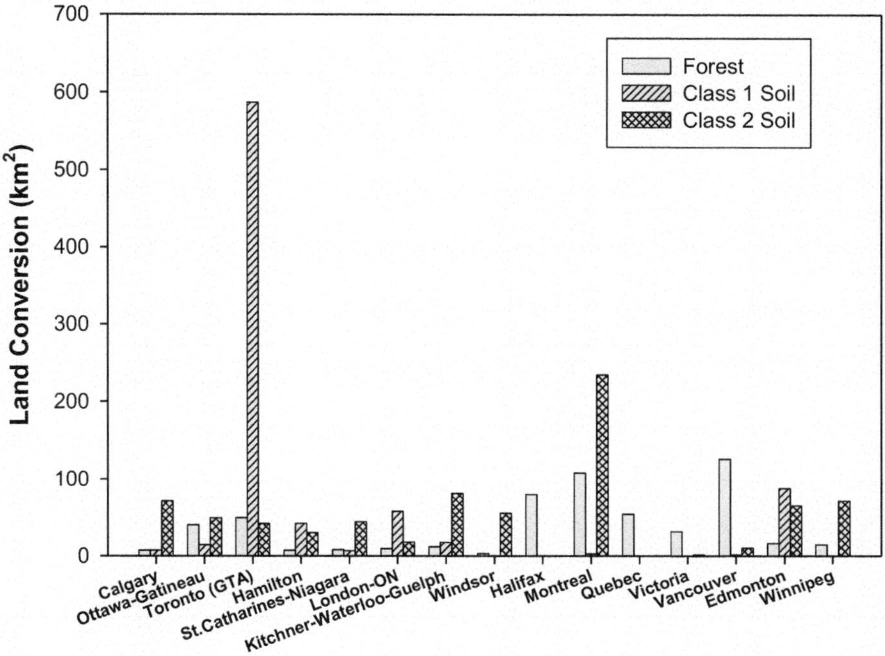

Fig. 10.10 The natural land (forest land and the land with good soil quality) conversion in major Canadian urban areas during 1966–2001 period

For major Canadian urban areas, the average urban built-up land area has increased about 123% from 1966 to 2001 corresponding to 3463 km^2. Correspondingly, rural land loss of 562 km^2 of forestland has occurred, while agricultural land of 822 km^2 and 767 km^2 of primary and secondary soil land, respectively, has been converted to urban land use. Details of these losses for individual urban areas are shown in Fig. 10.10. Dwelling densities in newly developed areas in all major cities has decreased over the time period in question. Figure 10.11 illustrates the changes of mean dwelling density in newly developed residential areas in major Canadian cities or urbanized regions. The dwelling density in 1966 was estimated using the total urban land. These trends show that the low-density residential suburbanization is the dominant urban expansion process in Canada (Zhang et al. 2014a, b). The trends of population density change vary among major Canadian cities. On the other hand, while population densities of large cities have remained approximately constant or increased, the densities of smaller urbanized areas are decreased.

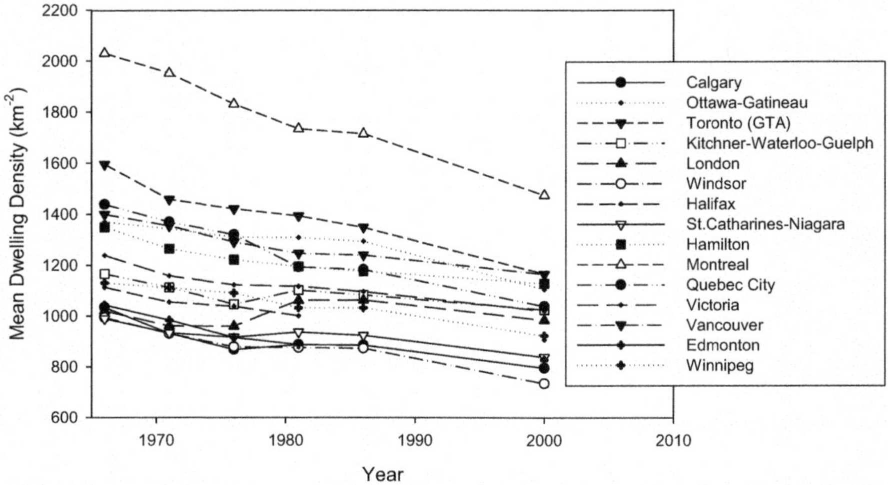

Fig. 10.11 The changes of mean dwelling density in newly developed residential areas in major Canadian cities or urbanized regions

10.4 Spatial Analysis of Urban Land Surface Information: Impacts on Intra-Urban Commuting

Integrated Urban Form (IUF) information can be used to address the impact of the physical structure of urban areas on a range of issues such as intra-urban transportation, energy use, human health and the local climate. Since EO-based urban land information is in raster format, spatial analyses, involving statistics and modeling of geospatial parameter relationships, can be conducted down to the sensor resolution level. Two approaches of spatial analysis are used in this paper, namely indicator quantification and spatial modeling. An indicator is a statistical measure derived directly or indirectly from data, as opposed to having been inferred from scenario-based model analysis. As such, it acts a surrogate or indirect measure of the state of some aspect of a complex system. Spatial modeling is a more sophisticated approach and can lead to a direct prediction of a quantity of interest. In addition, it can be employed to evaluate alternate scenarios (e.g. alternate urban physical structures or demographic distributions) to assess their impacts on quantities of interest or to track trends associated with urban growth.

This section is focused on utilizing IUFs to better understand the impact of Canadian urban land use structure on intra-urban commuting. In Canada, urban commuting is highly automobile-dependent. In 2011 on average 80% of Canadian workers used private vehicles to commute to work, while only 12% of workers used public transit (StatsCan, available at https://www12.statcan.gc.ca/nhs-enm/2011/as-sa/99-012-x/99-012-x2011003_1-eng.cfm). In individual major Canadian cities in that year, commuting via public transport varied considerably among major cities, i.e. 23.3% in Toronto, 22.2% in Montreal, 20.1% Ottawa-Gatineau, 19.7% in Vancouver and 15.9% in Calgary. The impact of urban form on urban transportation is widely recognized and has been the focus of previous studies (Duncan and Hartman 1998; Kenworthy and Laube 1999; Burchell et al. 2002; Song et al. 2017; Song and Knaap 2004; van Lierop et al. 2017; Zhang and Guindon 2006; Zhang et al. 2016). Urban travel patterns are inextricably linked to urban form. Population growth and suburbanization in Canada have led to dispersed urban commuting patterns that in turn have intensified the dependence on automobiles for intra-city travel (Raad 1998; Kenworthy and Laube 1999).

Indicators Related with Urban Commuting Activities

There are various levels of indicator quantification. Simple indicators, such as mean urban population density, can be obtained using rudimentary statistical methods and highly aggregated data. While attractive from the point of view of ease of understanding, simple indicators have a number of interpretation limitations. First, the connection between the indicator itself and the sought-after goals (such as commute activity attributes) is tenuous. Second, a simple indicator, such as scalar density, may be an inadequate measure of an attribute, such as spatially explicit urban form, that impacts the issue (such as transportation energy consumption) under study. As will be discussed later in an illustrative example, two cities can have similar overall population densities but differences in the distributions of the population within these cities can lead to significantly different urban travel and transportation patterns. On the other hand, a more detailed intra-urban indicator map visualizes the spatial distribution of the indicator and reveals the spatial behaviour of the information represented by the indicator. EO-based land information can be used as the baseline map. Through integration with other indicator information sets, a value-added information layer can be generated for spatial visualization and understanding improvement.

For more sophisticated indicators, EO-based land use information has been extensively used in spatial analysis methodologies such as 'Landscape Characterization', particularly in the landscape ecology arena (e.g. Guindon et al. 2004; Zhang et al. 2016). The essentials of landscape characterization are the description (core area, edge, shape, density, diversity, etc.) and analysis of spatial structures and pattern arrangements of land use land/cover elements such as forest patches, agricultural fields, etc. The technique is based on the premise that the spatial patterns of elements have linkages to their origins (e.g. forest patch characteristics are related to a priori de-forestation practices) and to their impacts on the environment, ecology and surrounding social/economic development. In the present study, the principles of landscape characterization are applied to urban indicator development. Spatial analysis methods are used to extract information about the spatial structure of urban form that is deemed relevant to transportation, for example, the spatial relationship between places of residence and places of work. While providing a closer linkage to transportation than those of the simple indicator category. Indicators of this category still are not based on an explicit model of travel but rather are linked to travel activity through an inference from spatial scale and pattern.

Population Density in Urban Land

A basic feature of urban 'sprawl' is inefficient land use that is reflected in low urban population density, or conversely, high urban land use per capita. In a city or urban area with low population density, public transportation is less efficient resulting in higher automobile dependence. Urban population density, UPD, can be simply expressed as,

$$\mathrm{UPD} = \frac{\mathrm{PU}}{\mathrm{LU}} \tag{10.1}$$

where PU is the population living in the urban area and LU is the total land used for urban activities. While this indicator is conceptually simple, there are three major challenges in quantifying it, namely (1) selection of land use classes that are considered 'urban', (2) accurate delineation of the urban-rural boundary and (3) linking census population counts to urban geographic extent. There is some ambiguity about what constitutes 'urban land'. Clearly, land occupied by buildings and parkland should be included, while water bodies and conservation lands should be excluded. On the other hand, land taken up by some forms of transportation infrastructure (e.g. airports) and vacant land within industrial parks are open to debate. For purposes of conducting a national urban survey that supports inter-city comparisons, we feel that the key issue is consistency of urban land definition across jurisdictions.

The total urban land LU includes all pixels labeled as 'urban' (i.e. either 'residential', 'commercial/industrial', 'recreational', 'transitional' or 'urban open land') in the land use land cover map, as well as all pixels with 'non-urban' labels that lie within a 3 pixel x 3 pixel buffer zone around each pixel with one of the above mentioned urban labels. Buffering is applied to account effectively for residential lots in low-density sprawl neighbourhoods. As an illustration, Fig. 10.12c, d shows the refined masks for the cities of Ottawa-Gatineau and Calgary. To link to census population information, the urban land is clipped to only include areas within the census sub-divisions (CSDs) of the census metropolitan area (CMA) or the census agglomeration (CA) in question. This interpolated population and the refined urban area are used to compute land use per capita. The total urban land LU has been estimated for each Canadian CMA with population over 200,000 (2001 census) as well as a selected sample of smaller cities (CMAs) (population > 100,000) located in southern Ontario. The urban mask derived from remotely sensed data has a number of advantages over traditional Census Urban Area (UA) delineations including improved inter-city consistency and the encapsulation of fuzzy urban boundaries.

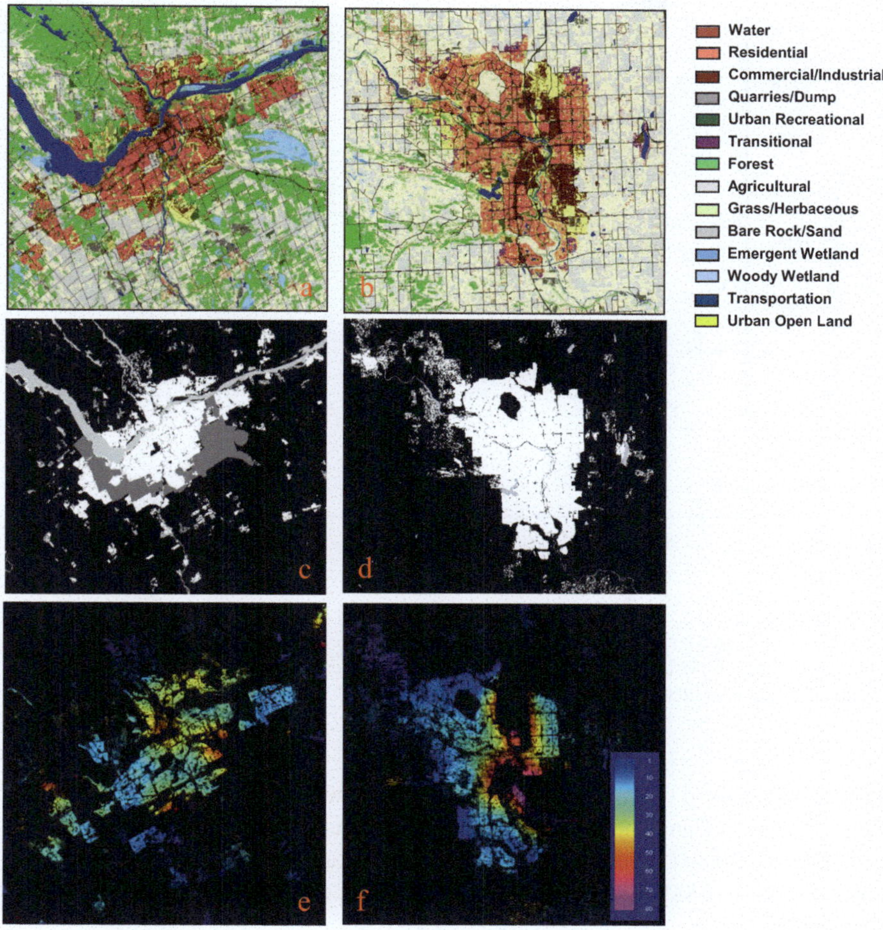

Fig. 10.12 Example maps of land use land cover (**a, b**), the urban land mask (**c, d**) with urban built-up land in white, water in light grey, the greenbelt conservation zone in dark grey and non-urban built-up land in black, the Mix Index with 3 km window (**e, f**) for Ottawa-Gatineau and Calgary

To illustrate the density range of typical Canadian cities, Fig. 10.13 illustrates plot of urban population density against urban use land area LU for 16 Canadian cities (CA) with population larger than 100,000. Densities range from approximately 1400 to 2900 peoples/km^2. In addition, there is a weak trend of increasing density with increasing city size.

While Canadian and US urban forms are similar in many aspects, e.g. the predominance of suburban single-family house types in terms of urban coverage, they

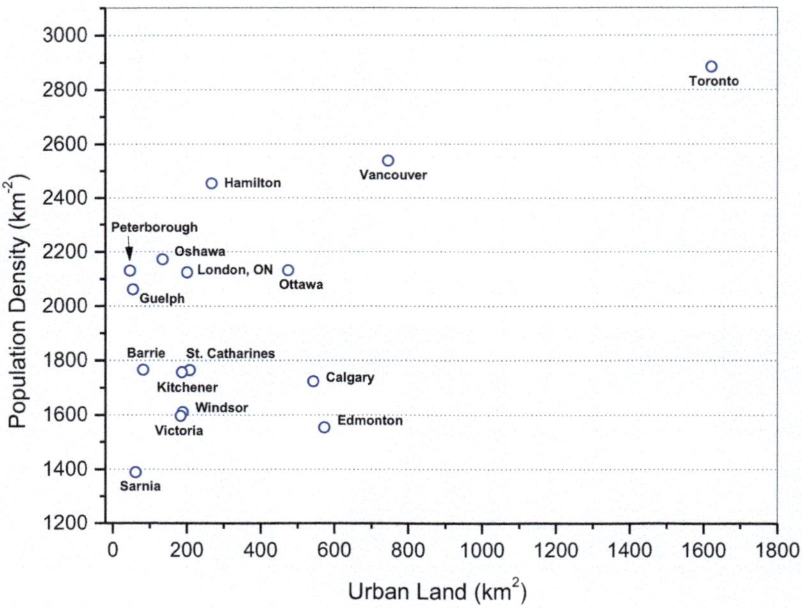

Fig. 10.13 Inter-city comparison of urban population density for 16 metropolitan areas with population over 100,000. Urban land masks for these cities were derived from LandSat data. Population census data was provided by StatsCan

do differ in terms of urban density. Figure 10.14 shows the systematic differences in density between Canadian and American urban areas or metropolitans in the Great Lake Region, based on the demographic data from Canadian census of 2001 and the US census of 2000 and land cover land use Landsat classifications (Zhang et al. 2016).

Mix Index

From a transportation point of view, the level of land use mix, for example, of residential areas and places of work or shopping, impacts travel activities, trip lengths and travel mode options. The *geographic scale* at which *mixed use* is appropriately measured depends on the *mode of transportation*. For example, to support a high degree of walking and cycling, residential and employment uses must be mixed on a very small scale. Current zoning practices in North

American cities typically separate urban land into large parcels by use (i.e. residential developments, industrial parks, shopping malls, etc.). This type of urban form results in longer urban trips than would be the case of a highly mixed land use form. It also increases the dependence on private cars resulting in more inner-

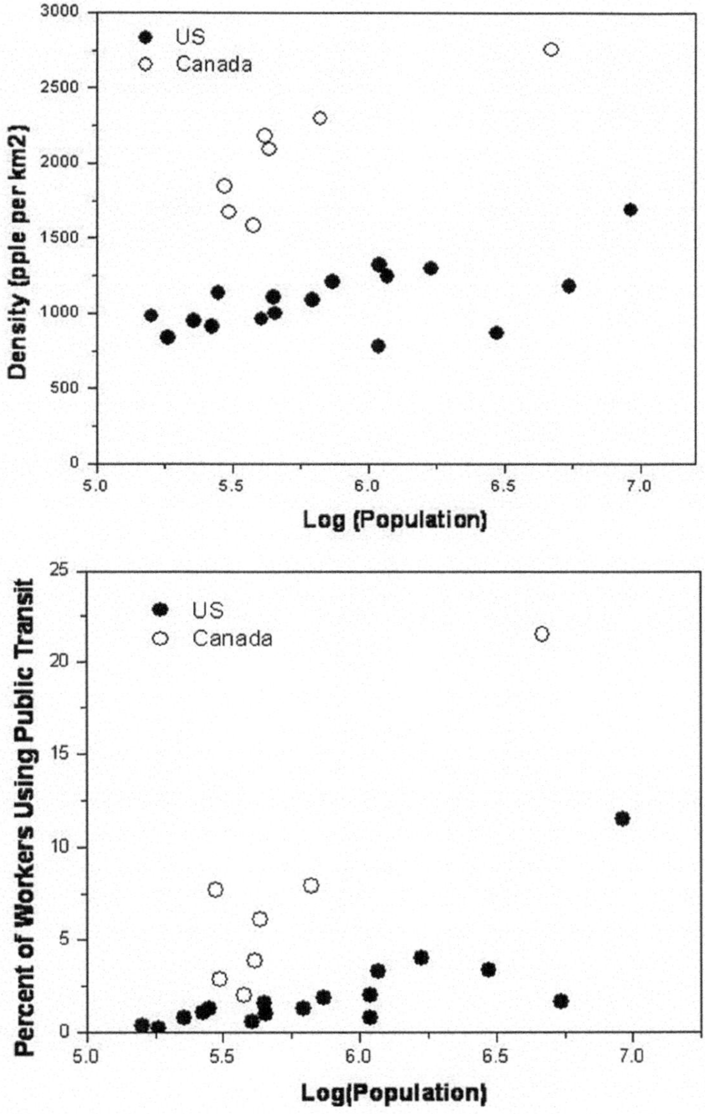

Fig. 10.14 The land density of urban areas in the Great Lakes Region

city traffic congestion. Figure 10.15 schematically illustrates this urban land use structure—travel distance relationship. The challenge, in this case, is then to translate the concepts expressed above by non-technical decision-makers into a measure, hereafter referred to as the 'Mix Index' (MI), that quantifies the impacts of

Fig. 10.15 Simplified diagram about the relationship between the urban land structure and the commuting travel distance

the land use mix and urban form structure on the feasibility of various modes of transportation.

For each residential land use pixel, the level of mixing of relevant land use classes within a circular distance of radius equal to the travel horizon is computed. For work-related commute travel, the 'mix index', MI, measures the level of residential (home) vs. commercial/industrial/institutional (workplace) land use mixing. For a circular window centred on a residential pixel located at (x,y), it is defined as

$$\mathrm{MI}(x,y)=\frac{\sum_i L_c}{\sum_i (L_c+L_r)} \tag{10.2}$$

where Lc and Lr are the areas of commercial-industrial-institutional land use and residential land use, respectively, within the circular window. MI can be computed from simple pixel counts (i) of the appropriate thematic classes. Three mixing radii corresponding to the horizons of three travel modes, namely 0.5 km for a walking scenario, less than 3 km for cycling, and greater than 3 km for motorized vehicle use were considered. For example, if a residential pixel exhibits a low MI value for radii less than 3 km, then this implies that for people living at that location, walking and cycling are unlikely to be feasible travel mode options and therefore residents must rely on some form of motorized transport.

Figure 10.12e, f shows the maps of MI (with 3-km window radius, respectively) for Ottawa-Gatineau and Calgary. The MI map is a portrayal of 'local workplace availability' based on the spatial distribution of the level of land use mixing across a city. Blue corresponds to low mix levels, i.e. residential areas that are highly segregated from potential work areas, while red corresponds to high levels of access to commercial/industrial services. Typically, these figures show that people living in satellite communities (e.g. Orleans and Barrhaven in east and south suburbs of Ottawa), and rural-residential areas (e.g. in the northwest of quadrant of the Calgary) have little opportunity for walking or bicycling to work.

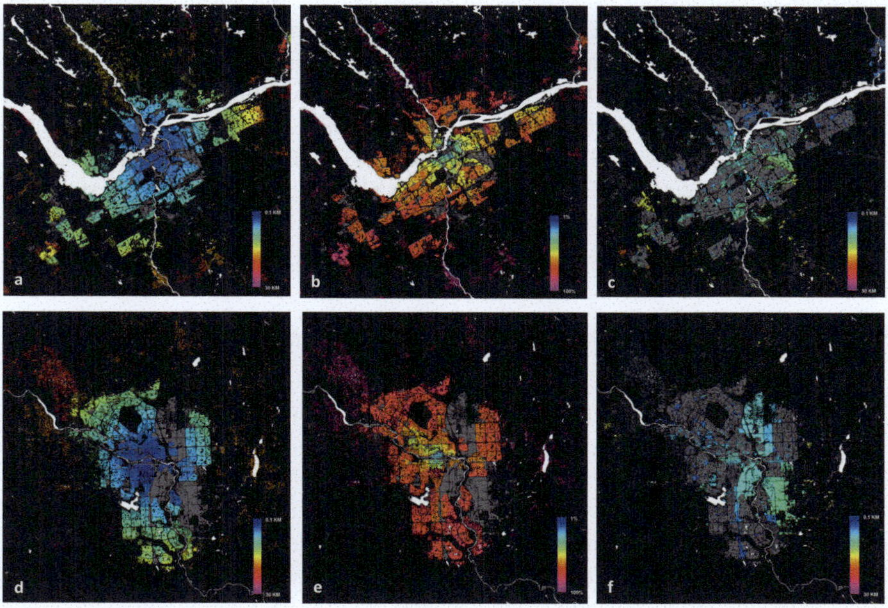

Fig. 10.16 Maps of averaged worker travel distance from residence (**a** and **d**), automobile use for commute (**b** and **e**) and averaged worker travel distance from workplace (**c** and **f**) for Ottawa-Gatineau and Calgary

Automobile Dependency and Travel Distance Maps

Integrated urban form information and commute statistics can be combined to generate value-added information products that improve understanding of commute-related issues. Figure 10.16 displays three such products in the form of maps of average travel distance to work from home, average travel distance to home from workplace and the percentage of automobile use for work-related travel, for Ottawa-Gatineau and Calgary. The commute statistics used were quantified at the census tract level. The residential land and commercial-industrial-institutional land use classes, derived from Landsat TM image classifications were used. Automobile use for work is representative of the level of automobile dependence on urban commuting. It is interesting that for both metropolitan centres, the commuting automobile dependence is low (about 30–40%) in residential areas close to downtown areas, while sharp increases in automobile use occur (approaching 100%) in outlying suburbs. Travel distance from home exhibits a similar pattern; distances are short in residential areas close to the downtown area. This indicates that both metropolitan centres include a core central business district (CBD) where the employment is highly concentrated. Automobile dependence is influenced by the distance of home to work. People living close to CBDs rely less on automobile use for travel to work. The suburban lands are mainly the 'bedroom' communities. Even where employment is available in some suburban areas (e.g. Kanata of Ottawa), travel distance from home to work and automobile dependence are still high.

There is a trend in North American cities toward migration of employment to outlying areas (e.g. industrial parks) suggesting that reduction in reliance on cars will be difficult to achieve in the future. Also, one of the causes of high travel distances seen for workers employed in areas outside the CBD is that suburban employment tends to be found in a few industrial parks of very large spatial dimension and low employment density compared to that found in CBDs.

Spatial Modeling of Urban Commute Distance

To assess the impacts of individual components in complex systems or characterizing a system with insufficient direct data, spatial modeling based on geospatial data (such as the land use maps) and realistic assumptions is a useful tool to isolate relevant components and to fill these data gaps. The next approach in terms of sophistication to understanding urban form–travel relationships is spatial modeling that combines urban form information and simulations of travel behaviour. Output from the model will then include, predicted travel effort for regions within an urban area (e.g. regions such as specific census tracts) and cumulative distance traveled, a parameter which is directly linked to energy consumption. Model quality can be assessed through comparison of predicted travel-related characteristics that are reported by independent travel surveys (e.g. by national census).

In our approach, each 'residential' pixel gives rise to a trip origin. On the other hand, the location of each pixel classed as 'commercial-industrial-institutional' becomes a potential trip destination. For each residential pixel a set of candidate 'commercial-industrial-institutional' destination pixels is randomly selected based upon a Monte Carlo modeling technique. From among these candidates, one is selected based upon a 'human decision' factor and employment density models. The factor involves favouring destination candidates that are closest to the residential pixel in question. Execution of the model produces trip origin-destinations for each residential pixel and a travel distance estimate based on minimum Euclidean distance. The employment model is based upon the fact that the CBD employment density is high and density decreases with increasing distance from the city centre. By combining this demographic information with LCLU information, the employment density can be expressed in terms of a weighting of numbers of jobs per pixel classed as 'commercial-industrial-institutional'. This form is suitable for our modeling effort since weighting can be used, for example, to favour CBD 'commercial-industrial-institutional' pixels in the candidate destination selection process. Statistics Canada provides an important source of census information to validate simulated travel patterns. This information is in the form of the proportion of a city's workforce whose employment-related travel distances fall into a specified set of distance intervals.

Figure 10.17 shows the probability of home-work travel as function of travel distance based on land use maps for Ottawa-Gatineau and Calgary. As comparison

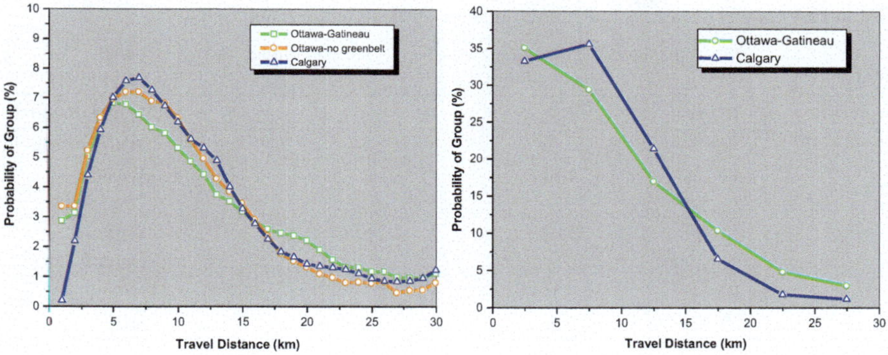

Fig. 10.17 Averaged commute travel distance simulated (left) and from the census 2001 (right) for Ottawa-Gatineau and Calgary

for validation, the probability curves of home-work travel derived from the census data are shown in Fig. 10.17 in larger sample group sizes than that in the simulation. The employment density model used in the simulation assumes a sharp decrease outwards from the CBD for both Ottawa-Gatineau and Calgary. For each residential unit (pixel), three trip destinations are randomly selected from among commercial pixel pools. The distances to each destination are computed and the shortest distance is then chosen. The travel distances for all residential pixels of each city have been histogrammed to obtain a commuting distance profile. These profiles can then be compared with equivalent information derived from information acquired during the 2001 Canadian census. The patterns of distance variation in the predicted curves agree well with the census data. Ottawa-Gatineau has more long-distance trips (i.e. >15 km) but fewer short-distance trips (<3 km) than Calgary. Since the estimation is mainly based on land use spatial distribution, this suggests that urban form is a major factor influencing trip distance. It should be noted that varying the 'human decision' algorithm (i.e. choosing the minimum distance from a greater number of randomly selected trips) does not change the basic patterns shown in Fig. 10.17, except that the peaks of both curves with more choices move towards shorter distances. However, the structure of city's urban form, i.e. the availability of working places that are close to homes, controls the minimum distance value that the curve peak can reach.

It should be noted that in the case of Ottawa-Gatineau, two features, namely the Ottawa River and the Greenbelt conservation zone (in white and light grey in Fig. 10.12c), play a significant role in increasing average travel distances for suburban communities. To gauge the importance of the Greenbelt, we have examined the selected trip for each residential pixel and subtracted any distance contribution that involved traversing this conservation area. These compensated trip profiles are shown in Fig. 10.17 and summarized in Table 10.1. Compared to the profiles taking into account full trip distance, it can be seen that the Greenbelt enhances the number of long trips as well as mean travel distance.

Table 10.1 Travel distance simulation results for Ottawa-Gatineau and Calgary

City	Mean distance (km)	Percentage (%) of trips >5 km	Human choice (no. of locations)	Other parameters
Ottawa-Gatineau	10.76	78.75	3	With Greenbelt
	9.49	70.49	5	With Greenbelt
	9.25	74.71	3	Without Greenbelt
	8.01	65.61	5	Without Greenbelt
Calgary	10.24	80.19	3	N/A
	9.00	72.56	5	N/A

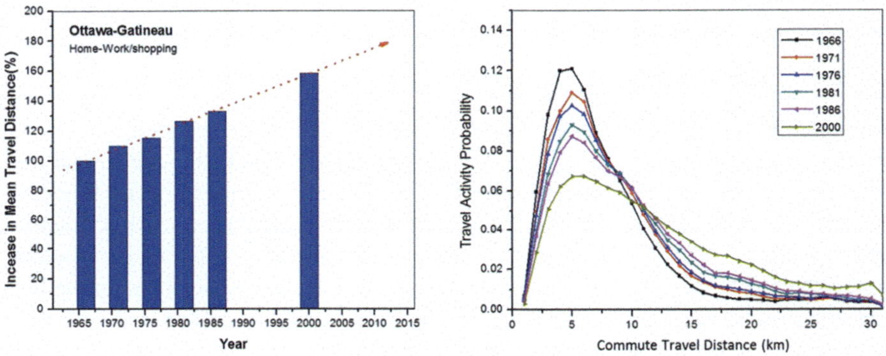

Fig. 10.18 Increase in averaged commute travel distance (left) and changes in the relationship between commute travel distance and travel activity probability (right) simulated as land growth in Ottawa-Gatineau from 1966 to 2000

Assessing the impact of the Greenbelt on transportation also highlights another aspect of sustainability, namely the need to balance what can appear to be conflicting sustainability goals. In the Ottawa-Gatineau case, the existence of the 'Greenbelt' zone increases commuting distances and hence transportation energy consumption. From the energy viewpoint, the Greenbelt is detrimental to urban sustainability. On the other hand, the Greenbelt has significant positive impacts on the urban environment, such as preserving biodiversity and urban green space. From the 'quality of life' viewpoint the Greenbelt can be viewed as contributing to sustainability.

The impact of urban growth on commute travel distance has been investigated through simulation for the Ottawa-Gatineau case. Figure 10.18 illustrates the changes in average travel distance (left) and in the relationship between travel distance and the travel activity for the Ottawa- Gatineau region. Overall, average commute travel distance has increased through years. This increase is significant and estimated to be 50% higher than that of 1966. This increase is mainly the result of increased long-distance travel as shown in Fig. 10.18 (right).

10.5 Summary and Discussion

This paper presents research on the utilization of earth observation imagery for urban land cover land use characterization and the integration of demographics leading to the creation of an improved quantification of urban form. It is argued that this integrated version of urban form is a key information source for understanding the impacts of urban landscapes and structures on human activities such as intra-urban travel, human health and local climatic alterations. Aspects of work-related travel in Canadian cities have been used to demonstrate these concepts including the development of targeted travel indicators to assess the feasibility of alternate modes of transportation and of spatial travel models to predict details of travel effort such as travel distance patterns within cities. Understanding historic patterns of urban growth and hence of changing urban form is critical in assessing parallel trends in issues such as growth sustainability.

While intra-urban travel is the focus of this paper, it is emphasized that similar analyses can be effectively applied to address a myriad of critical urban issues. Since many such issues involve landscape–population interactions, there is a need for the systematic creation of integrated urban form information consistent at both the national and inter-national level. Pioneer work in this area has been undertaken in the creation of the Canadian Urban Land Use Survey and the bi- national Great Lakes Urban Survey.

Acknowledgement The author thanks Dr. Bert Guindon for collaboration in the research works presented in this paper; Ms. Krista Sun for her preprocessing of the spatial and demographic data used in this paper; Ms. Nancy Hoffman of Statistics Canada and Mr. Peter Reilly-Roe (previously in Office of Energy Efficiency, Natural Resources Canada) for discussions and constructive advice, the City of Ottawa, the City of Toronto and GEOBase team members of Natural Resources Canada for providing related information or data, the Canadian Space Agency for partially funding aspects of the research presented in this paper, the anonymous reviewer and Ms. Lucia Huang for helpful comments that have improved the manuscript.

References

Aubrecht, C., K. Steinnocher, M. Hollaus, and W. Wolfgang. 2009. Integrating earth observation and GIScience for high resolution spatial and functional modeling of urban land use. *Computers Environment and Urban Systems* 33: 15–25.

Baldridge, A.M., S.J. Hook, C.I. Grove, and G. Rivera. 2009. The ASTER spectral library version 2.0. *Remote Sensing of Environment* 113 (4): 711–715.

Banai, R., and M.A. Rapino. 2009. Urban theory since a theory of good city form (1981)—A progress review. *Journal of Urbanism: International Research on Placemaking and Urban Sustainability* 2 (3): 259–276.

Batty, M. 2005. *Cities and complexity*. Cambridge: MIT Press.

Bunting, T., and P. Filion. 2000. *Canadian cities in transition: The twenty-first century*. 2nd ed, 576. Oxford: Oxford University Press.

Burchell, R.W., G. Lowenstein, W.R. Dolphin, C.C. Galley, A. Downs, S. Seskin, K.G. Still, T. Moore. 2002. Costs of sprawl – 2000. Transit Cooperative Research Program (TCRP) Report 74, published by Transportation Research Board, Washington.

Carlson, T. 2003. Applications of remote sensing to urban problems. *Remote Sensing of Environment* 86 (3): 273–274.

Cervero, R. 1996. Mixed land uses and commuting: Evidence from the American housing survey. *Transportation Research* 30 (5): 361–377.

Clifton, K., P. Ewing, Y. Song. 2008. Quantitive analysis of Urban form: A multidisciplanry review. *Journal of Urbanisum International Research on Placemaking and Urban Sustainability*, 1 (1): 17–45.

Cocklin, C., and B. Smit, 1982. A method to assess the implications of future urban expansion on rural land. Environment Canada, Lands Directorate, CLI working paper No. 23, 40.

Crane, R. 2000. The influence of urban form on travel: An interpretive review. *Urban Planning Literature* 15 (1): 3–23.

Duncan, B, J. Hartman, 1998. Toward sustainable urban transportation in Canada: Proc. Conf. Transportation Land Use Air Quality, Reston, VA, (USA), 1998, 11–20.

Darwish, A., K. Leukert, W. Reinhardt, 2003. Urban land-cover classification: An object based perspective. In *2nd GRSS/ISPRS Joint Workshop on Remote Sensing and Data Fusion over Urban Areas*, Berlin, Germany, 2003, pp. 278–282.

Downes, N., and H. Storch. 2013. Understanding urban structures—An approach for assessing climate risk in emerging megacities. In *Proceedings of EnviroInfo 2013: Environmental Informatics and Renewable Energies*, p. 9.

Durantona, G.M., and A. Turne. 2018. Urban form and driving: Evidence from US cities. *Journal of Urban Economics* 108: 170–191.

Environment Canada. 1976. Canada land inventory: Land capability for agriculture. Environment Canada, Lands Directorate, Report No. 10, p. 26.

Georganos, S., T. Grippa, S. Vanhuysse, M. Lennert, M. Shimoni, and E. Wolff. 2018. Very high resolution object-based land use–land cover urban classification using extreme gradient boosting. *IEEE Geoscience and Remote Sensing Letters* 15 (4): 607–611.

Gerl, T., M. Bochow, and H. Kreibich. 2014. Flood damage modeling on the basis of urban structure mapping using high-resolution remote sensing data. *Water* 6: 2367–2393.

Gierman, D. M. 1981. Land use classification for land use monitoring. Environment Canada, Lands Directorate, CLI Working Paper No. 17, p. 40.

Gueneralp, B., and K.C. Seto. 2008. Environmental impacts of urban growth from an integrated dynamic perspective: A case study of Shenzhen, South China. *Global Environmental Change* 18 (4): 720–735.

Guindon, B., and Y. Zhang. 2009. Automated urban delineation from Landsat imagery based on spatial information processing. *PE&RS* 75: 845–858.

Guindon, B., Y. Zhang, and C. Dillabaugh. 2004. Landsat urban mapping based on a combined spectral-spatial methodology. *Remote Sensing of Environment* 92: 218–232.

Hall, D.L., and J. Llinas. 1997. An introduction to multisensor data fusion. *Proceedings of the IEEE* 85 (1): 6–23.

Herold M., J. Scepan, A. Müller, S. Günther. 2002. Object-oriented mapping and analysis of urban land use/cover using IKONOS data. In *Proceedings of 22nd EARSEL Symposium Geoinformation for European-Wide Integration*, Prague, June 2002.

Herold, M., D.A. Roberts, M.E. Gardner, and P.E. Dennison. 2004. Spectrometry for urban area remote sensing—Development and analysis of a spectral library from 350 to 2400 nm. *Remote Sensing of Environment* 91 (3-4): 304–319.

Homer, C., J. Dewitz, J. Fry, M. Coan, N. Hossain, C. Larson, N. Herold, A. McKerrow, J.N. Van Driel, and J. Wickham. 2007. Completion of the 2001 National Land Cover Database for the Conterminous United States. *Photogrammetric Engineering and Remote Sensing* 73: 337–341.

Huang, J., X.X. Lu, and J.M. Sellers. 2007. A global comparative analysis of urban form: Applying spatial metrics and remote sensing. *Landscape and Urban Planning* 82: 184–197.

Huck, A., S. Hese, and E. Banzhaf. 2011. Delineating parameters for object-based urban structure mapping in Santiago de Chile using Quickbird data. *International Archives of the Photogrammetry, Remote Sensing and Spatial Information Sciences* XXXVIII-4/w-19: 131–136.

Jabareen, Y.R. 2006. Sustainable urban forms. *Journal of Planning Education and Research* 26: 38–52.

Johnson, M.P. 2001. Environmental impacts of urban sprawl: A survey of the literature and proposed research agenda. *Environment and Planning A: Economy and Space* 33 (4): 717–735.

Kenworthy, J.R., and F.B. Laube. 1999. *An international sourcebook of automobile dependence in cities 1960-1990*, 704. Boulder., ISBN 0-87081-532-7: University Press of Colorado.

Khosravi, I., M. Momeni, and M. Rahnemoonfar. 2014. Performance evaluation of object-based and Pixel-based building detection algorithms from very high spatial resolution imagery. *Photogrammetric Engineering & Remote Sensing* 80 (5): 519–528.

Krehl, A., S. Siedentop, H. Taubenbock, and M. Wurm. 2016. A comprehensive view on urban spatial structure: Urban density patterns of German city regions. *International Journal of Geo-Information* 5: 21.

Lu, D., and Q. Weng. 2004. Spectral mixture analysis of the urban landscape in Indianapolis with Landsat ETM+ imagery. *Photogrammetric Engineering and Remote Sensing* 70: 1053–1062.

Lynch, K., and L. Rodwin. 1958. A theory of urban form. *Journal of the American Institute of Planners* 24 (4): 201–214.

Maantay, J., and A. Maroko. 2009. Mapping urban risk: Flood hazards, race, & environmental justice in New York. *Applied Geography* 29 (1): 111–124.

Navulur, K. 2007. *Multispectral image analysis using the object-oriented paradigm*, 163. New York., ISBN-13:978-1-4200-4306-8: CRC Press.

Neptis Foundation. 2013. The big picture about land use and why it matters. www.nettis.org.

———. 2014. Rethinking Sprawl. www.nettis.org.

Pohl, C., and J. van Genderen. 2016. *Remote sensing image fusion: A practical guide*, 266. New York., ISBN 9781498730020: CRC Press.

Poumanyvong, P., S. Kaneko, and S. Dhakal. 2012. Impacts of urbanization on national transport and road energy use: Evidence from low, middle and high income countries. *Energy Policy* 46: 268–277.

Raad, T. 1998. The car in Canada: A study of factors influencing automobile dependence in Canada's seven largest cities, 1961–1991. *Thesis*, University of British Columbia, p. 253.

Regnier, C., and S. Legras. 2018. Urban structure and environmental externalities. *Environmental and Resource Economics* 70: 31–52.

Seto, K.C., M. Fragkias, B. Güneralp, and M.K. Reilly. 2011. A meta-analysis of global urban land expansion. *PLoS ONE* 6 (8): e23777. https://doi.org/10.1371/journal.pone.0023777.

Seto, K.C., B. Güneralp, and L.R. Hutyra. 2012. Global forecasts of urban expansion to 2030 and direct impacts on biodiversity and carbon pools. *PNAS* 109 (40): 16083–16088.

Shoaga, D., and E. Muehleggerb. 2015. Commuting times and land use regulations. *Procedia Engineering* 107: 488–493.

Song, Y., G.J. Knaap. 2004. Measuring Urban Form: Is Portland winning the war on sprawl? *Journal of the American Planning Association* 70 (2): 210–225. https://doi.org/10.1080/01944360408976371

Song, Y., G. Shao, X. Song, Y. Liu, L. Pan, and H. Ye. 2017. The relationships between urban form and urban, commuting: An empirical study in China. *Sustainability, MDPI* 9: 1150.

Statistics Canada. 2015. Canada goes urban. *The Daily*, achieved document, www.150.statcan.gc.ca.

———. 2016a. The changing landscape of Canadian metropolitan areas, 1971 to 2011. *The Daily*, March 22, 2016, www.150.statcan.gc.ca.

———. 2016b. Commutes using sustainable transportation in census metropolitan areas, Census of Population, 2016, No. 98-200-X2016029, ISBN 978-0-660-23763-3. https://www12.statcan.gc.ca/census-recensement/2016/as-sa/98-200-x/2016029/98- 200-x2016029-eng.pdf.

Taubenböck, H., T. Esch, A. Roth. 2006. An urban classification approach based on an object-oriented analysis of high resolution satellite imagery for a special structuring within urban areas. In *1st EARSeL Workshop of the SIG Urban Remote Sensing*, Humboldt-Universität zu Berlin, 2–3 March 2006.
United States. 2018. World urbanization prospects: The 2018 revision. population.un.org/wup/publications.
van Lierop, D., G. Boisjoly, E. Grisé, and A. El-Geneidy. 2017. Evolution in land use and transport research. In *Planning knowledge and research*, ed. T. Sanchez. New York: Routledge.
Vogelmann, J., T.L. Sohl, P.V. Campbell, and D.M. Shaw. 1998. Regional land cover characterization using Landsat TM data and ancillary data sources. *Environmental Monitoring and Assessment* 51 (1): 415–428.
Weng, Q. 2007. *Remote sensing of impervious surfaces*, 488. Boca Raton: CRC Press.
Weng, Q., and D.A. Quattrochi. 2006. *Urban remote sensing*, 3–10. Boca Raton: CRC Press.
Wheaton, W.C. 2004. Commuting, congestion, and employment dispersal in cities with mixed land use. *Journal of Urban Economics* 55 (3): 417–438.
Williams, K. 2014. Urban form and infrastructure: A morphological review. Foresight, Government Office for Science, URN GS/14/808. www.gov.uk/go-science.
Wurm, M., Taubenbock, H., Roth, A. and Dech, S. 2009. Urban structuring using multisensoral remote sensing data: By the example of German cities Cologne and Dresden. 2009 Urban Remote Sensing Event, p. 8.
Zha, Y., J. Gao, and S. Ni. 2003. Use of normalized difference built-up index in automatically mapping urban areas from TM imagery. *International Journal of Remote Sensing* 24: 583–594.
———. 2009. Multi-resolution integration of land cover for urban subpixel density mapping. *International Journal of Digital Earth.* 2: 89–108.
Zhang, Y., and B. Guindon. 2006. Using satellite remote sensing to survey transportation-related urban sustainability. Part I: Methodologies for indicator quantification. *Applied Earth Observation and Geoinformation.* 8: 149–164.
Zhang, Y., and L. Sun. 2018. Spatial-temporal impacts of urban land use land cover on land surface temperature: Case studies of two Canadian urban areas. *International Journal of Applied Earth Observation and Geoinformation* 39 (5): 1491–1509.
Zhang, Y., B. Guindon, and K. Sun. 2010a. Measuring the trajectories of urban land use change: Assessment for major Canadian urban areas in 1966–2001. *Journal of Land Use Science* 5: 217–235.
———. 2010b. The concepts and application of the Canadian urban land use survey. *Canadian Journal of Remote Sensing.* 36 (3): 224–235.
———. 2014a. Spatial-temporal-thematic assimilation of Landsat-based and archived historical information for measuring urbanization processes. *Journal of Land Use Science* 10 (4): 369–387.
Zhang, Y., X. Li, B. Guindon, N. Lantz, and Z. Sun. 2014b. Target-driven extraction of urban land surface information from high-resolution imagery. *Journal of Applied Remote Sensing* 8 (1): 084594.
Zhang, Y., B. Guindon, and K. Sun. 2016. Exploring the link between urban form and work related transportation using combined satellite image and census information: Case of the Great Lakes region. *International Journal of Applied Earth Observation and Geoinformation* 47: 139–150.

Chapter 11
Extraction of Building Footprints from LiDAR: An Assessment of Classification and Point Density Requirements

Jean-Samuel Proulx-Bourque, Heather McGrath, Denis Bergeron, and Charles Fortin

Plain Language Summary
LiDAR data is increasingly being used for creating datasets other than terrain. In this study we look at the minimum requirements of LiDAR point cloud data (classified and un-classified) in order to successfully extract a building footprint. Our findings indicate that vendor classified data with a minimum density of 4 pts/m² is sufficient to accurately extract building footprints with >75% confidence. However, vendor classification comes at a significant cost. For those with limited budgets or access to existing un-classified LiDAR data, we present a fully automated open-source re-classification technique that requires a point density of at least 8 pts/m² to meet this same confidence level.

11.1 Introduction

Extraction of elevation data to create digital terrain and surface models is the most popular use of Light Detection and Ranging (LiDAR) data. However, the LiDAR point cloud data is valuable for a variety of other uses, including extracting vegetation, critical infrastructure, hydrography, and buildings. Specifically, extracting building footprints from LiDAR data is desirable for a number of reasons, including estimation of urban population, energy demand, and/or quality of life (Zhang et al. 2006). As well, building footprints are useful in the determination of property taxes, planning, public works/public safety, and 3D city models (Zhao and Wang 2014). Beyond these uses, extracted building footprints may be used in urban landscape

J.-S. Proulx-Bourque (✉) · H. McGrath · D. Bergeron · C. Fortin
Natural Resources Canada, Canada Centre for Mapping and Earth Observation, Sherbrooke, Québec, Canada
e-mail: jean-samuel.proulx-bourque@canada.ca

V. Singhroy (ed.), *Advances in Remote Sensing for Infrastructure Monitoring*, Springer Remote Sensing/Photogrammetry,
https://doi.org/10.1007/978-3-030-59109-0_11

models for the assessment of urban heat islands, as the base elevation for flood insurance models, and in natural hazard risk assessment models.

Deriving building footprints from optical sensors is not new. In the earliest stages, a manual process was followed to generate building footprints from aerial photography (Zhang et al. 2006). However, in recent years semi-automated and automated processes have become more popular (Brooks et al. 2015). With technological advancements and new sensors becoming available, there has been a shift from aerial photography to satellite imagery, and now LiDAR to map building footprints. While satellite imagery provides global coverage, historically such systems (e.g., Landsat, MODIS, and AVHRR) provide insufficient spatial resolution for detailed building detection (Zhao and Wang 2014). Higher resolution systems (e.g., IKONOS and Quickbird) are more successful in detecting building footprints, however, the limitations to these systems, namely, sun shadows, relief displacement, and occlusion, hinder the ability to extract and geocode building footprints (Zhao and Wang 2014) and they are only two-dimensional.

LiDAR data is collected as reflected laser beams, which have intensity data associated with a three-dimensional position. Thus, these data points are free from the limitations of the previous, pixel-based methods. In addition, LiDAR data collected from airborne systems have the advantage of being highly accurate in the vertical dimension (Aguilar et al. 2010). While some methods of building extraction have combined LiDAR and high-resolution optical imagery, errors may be present in the derived dataset due to differences related to collection time, displacement introduced by differing resolution, shadows, and relief displacement. As well, this dependence on such a secondary dataset can be limiting.

Existing Techniques

Researchers, including Zhang et al. (2003, 2006), Zhao and Wang (2014), and Pérez-García et al. (2012) have, therefore, tested and evaluated many different LiDAR only building footprint extraction techniques, which range from semi-automatic to automatic. Building footprint extraction can be categorized into three steps: (i) isolation of non-ground points, (ii) segmentation of building points, and (iii) building outline extraction.

Previous research into isolation of non-ground points can be summarized into two broad categories: separation of non-ground objects from the point cloud and a non-hierarchical method. In the former category, many types of filters have been tested. Such filtering techniques include: (i) morphological filters (filter based on acceptable height differences between neighboring points) (Kilian et al. 1996; Zhang et al. 2003, 2006), (ii) progressive densification filters (which use a triangulated surface between ground seed points, and the ground is densified by finding the offset and distance angles of each point to the triangulated surface) (Pérez-García et al. 2012), (iii) surface based filters (initially sets all points to ground then progressively removes non-ground points based on weights applied to the points) (Pfeifer 2005), and (iv) segmentation based filtering (grouping points into segments using the local normal of each point, using the local neighborhood, and then segments are modified using region growing techniques) (Filin and Pfeifer 2006). Each of these filtering techniques performs reasonably well in flat terrain; however, their performance is poorer in undulating terrain (Zhao and Wang 2014). In the non-hierarchical method, all land use types are separated simultaneously. Often, a pixel-based classification method is utilized in this category. Some researchers have added point intensity to this technique, however, as the intensity values of objects such as roads and parking lots are similar to that of building rooftops, this addition does not generally improve the results (Zhao and Wang 2014).

Once the non-ground points are isolated, the next step is segmentation, as the non-ground points typically contain a mixture of trees, utilities, buildings, etc. A traditional way to separate these is using thresholds on a digital surface model (DSM) or a normalized DSM (difference of DTM and DSM), which may be globally or locally applied. This is not always successful since trees and buildings often have similar heights and may be spatially close together (Zhao and Wang 2014). Other methods that have been applied to separate trees from building points include morphological filters, texture analysis, mathematical algorithms such as plane-fitting techniques, and hierarchical object-oriented classification (Zhang et al. 2006; Awrangjeb and Fraser 2014). Depending on the complexity of the study area, the accuracy of these results vary (Zhao and Wang 2014).

Finally, once the building points have been isolated, a building outline extraction process is required to generate a polygon or footprint, from the noisy and irregularly spaced boundary points. As with the previous steps, multiple techniques have been proposed. One of the most common is to determine dominant directions of the structure. Tested techniques for determining dominant directions include a least squares algorithm (Sampath and Shan 2003), non-linear least squares (Lokhat and Touya 2016), angle histogram of boundary points (Alharthy and Bethel 2002), Douglas–Peucker (Zhang et al. 2006), weighted line segmentation (Zhang et al. 2006), and invariant parameters based on known roof type (Maas and Vosselman 1999).

The quality of the derived building footprint can vary based on the selected extraction technique, point density, building shape, and building density. Maas and Vosselman (1999) found 1 pt/m^2 or greater can produce precision building dimensions within 0.1–0.2 m. Rutzinger et al. (2009) and Zhang et al. (2006) also found 1 pt/m^2 was able to successfully detect buildings, while Awrangjeb (2016) reported 90% success rate for building detection in urban area with point density as low as 4 pts/m^2. Alternatively, Ramiya et al. (2017) found that point density was not a factor in the performance of their algorithm for building detection.

Canadian LiDAR Collection

As part of the Government of Canada's High-Resolution Digital Elevation Model (HRDEM)—CanElevation series, airborne LiDAR data is being collected across the country to support the National Elevation Strategy (Natural Resources Canada

(NRCan) 2017). Currently, 51 of Canada's 100 most populated cities are covered by LiDAR data (Canada Centre for Mapping and Earth Observation (CCMEO) 2017). This LiDAR data has been collected since 2009 by different companies without uniform specifications; thus the point density and the details of the point data classification (fully classified, ground/non-ground, and un-classified) in the available datasets differ by locale and over time. Generally, LiDAR data collection vendors use semi-automated methods to process and classify point data. This semi-automated process involves manual editing in addition to automated classification techniques and can be costly, especially when multiple point classes are required.

Objective

In this paper, we consider LiDAR data collected in a Canadian city to explore the capability to automatically extract building footprints from vendor-classified and re-classified (classified using open-source tools) point cloud data at varying densities to evaluate the quality and completeness of the derived building footprints. Re-classification is included in these tests in order to determine the suitability of extracting building footprints from existing un-classified LiDAR, and given the cost of vendor-classification when requisitioning LiDAR, evaluate its ability for building footprint extraction. Through a comparative analysis of the ability to detect buildings and the quality of the match of the derived footprint shape, we make recommendations for the desired minimal point density for building footprint extraction from airborne LiDAR for urban communities. The methods applied to derive building footprint (isolation of non-ground points, segmentation of building points, and building outline extraction) are discussed in Sect. 11.3. These tools were selected based on their availability in open-source software solutions, ability to be exploited in a high-performance computing environment, and based on the relative performance of those methods. In Sects. 11.4 and 11.5, a comparative analysis of this technique with known building footprints is presented and discussed, finally the key findings and recommendations are summarized in the conclusion.

11.2 Study Area and Data

The study area is located in Québec City, Canada. Québec City is the capital of the province of Québec and it is the second largest city. Québec City is a slightly atypical Canadian city and represents a more complex example than other Canadian cities considering it has been developed over hundreds of years, and contains a large mix of construction styles, materials, and building layouts. The original French settlers built in an architectural style similar to the French style, adapted for the colder Quebec climate. In Old Quebec, many colonial buildings can be found. In the areas of Saint-Jean-Baptiste the colorful houses are built into the steep hillside. Whereas in the borough of Limoilou, the majority of residences are condos or apartments in 3-storey buildings which are either built into one another or built very close together, such that an entire block can appear to be one building.

This location was chosen due to the importance of its metropolitan area and to the availability of high-density, recently collected (2017) LiDAR point cloud data. The study area is composed of 34 tiles of 1 km^2 each with a density of 13.25 points/m^2, Fig. 11.1. The term "vendor-classified" is used subsequently to represent the semi-automated method of point cloud classification obtained directly from the vendor. To complete the semi-automatic classification involved considerable manual effort that comes with significant cost. Thus, in Sect. 11.3.1, a fully automated process is described wherein classifications from the vendor are removed and the point cloud is re-classified using a series of open-source tools.

In order to validate the building footprint extraction results, high-resolution imagery available from ESRI Imagery basemap and Google Earth, with acquisition dates consistent with the LiDAR, was used. The coverage of the ESRI Imagery basemap in the selected study area is high-resolution, 50 cm, or better with an accuracy of 4.23 m (Esri 2019). The resolution of Google Earth imagery ranges across the globe, from 15 m to 15 cm, with little indication of the exact accuracy for any specific location (Wikipedia 2019). The Google Earth imagery is a combination of satellite data, airborne imagery, and StreetView data from a variety of vendors and is stitched together. Throughout the Québec City study area, the imagery is high resolution and details such as rooftop features (e.g., air conditioning vents) and vehicle side-view mirrors are discernable.

11.3 Methods

Section 11.3.1 describes how the building footprints were extracted from LiDAR point clouds using a variety of open-source scripts while Sect. 11.3.2 provides details about the method used to adjust the density of the LiDAR point clouds.

Extraction of Building Footprints from LiDAR

The method used to extract the building footprints from the LiDAR point clouds consists of four steps: classification, building footprint extraction, building splitting, and squaring, Fig. 11.2.

The LiDAR data provided for this project was re-classified using three tools of the LAStools software suite (LAStools 2019): lasground (a variant of Zhang et al. 2003), lasheight, and lasclassify. The first step is to differentiate ground from non-ground points via a progressive morphological filter applied to the point cloud data. Next, a height value was given to all non-ground points by subtracting the surface of a TIN of the ground points from the non-ground point's elevation value. Finally,

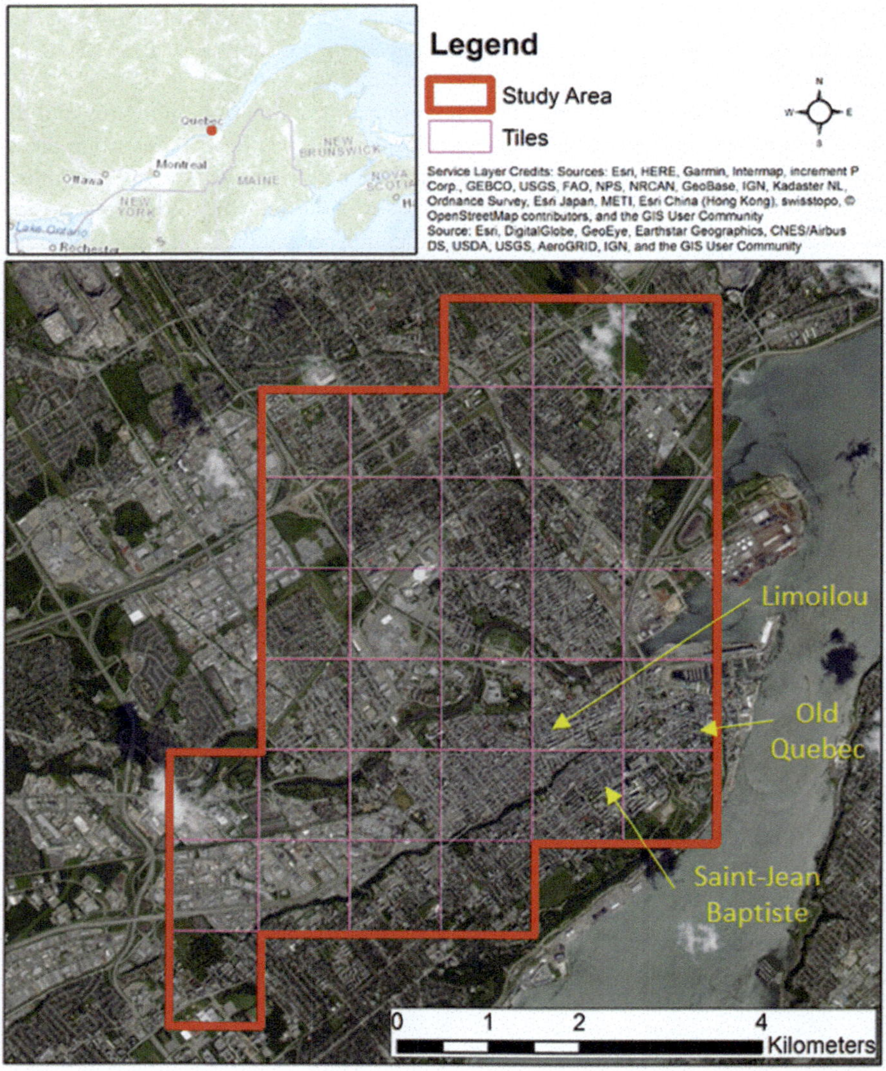

Fig. 11.1 Extent of the study area, Québec City, QC, Canada

the building points were classified from the other non-ground heights, e.g. trees, bridges, electrical lines, etc., using texture analysis. In the final classification step, neighboring points are analyzed and clustered based on height (>2 m) and the tolerance on standard deviation (0.1 m). This methods works well for rooftops that have planar surfaces.

A preliminary building footprint was generated in Step 2 by converting the classified building point clusters into vector geometry by computing a non-convex

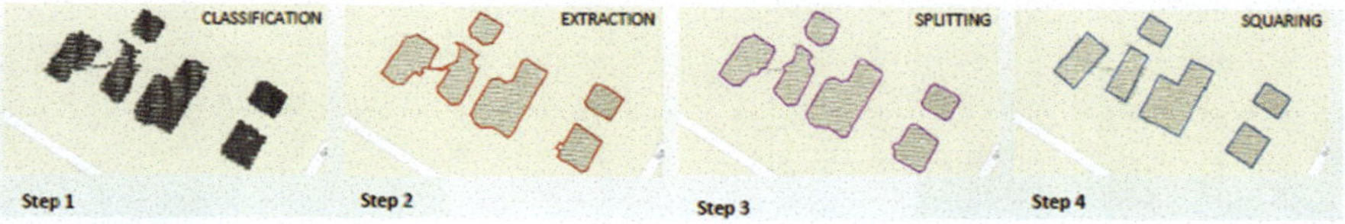

Fig. 11.2 Step-by-step method to extract building footprints from LiDAR

envelope around each individual building cluster, using the lasbounday function in LAStools. During this step, a minimal dimension of 10 m^2 was applied.

Several buildings were linked together as a result misclassified building points. Thus, step 3 involved a splitting process to separate these artificially merged polygons. The geometry splitting was a two-step process: (i) application of a negative buffer on the polygons to eliminate narrow geometries and (ii) a positive, or standard buffer, to return the geometries unaffected by the negative buffer back to their original size, values of −0.5 m and 0.5 m, respectively, were applied.

Finally, a squaring method was applied based on the non-linear least squares method developed by Lokhat and Touya (2016). The model is based on five constraints that, given a tolerance, are aimed at minimizing vertex displacement and forcing 90°, 45°, and 135° angles, as well as forcing parallelism and straight lines. A tolerance of 40° was applied to force right angles and allow for a displacement up to 1 meter in order to straighten line segments. Buildings with a circularity index value above 0.85 were not squared using this method, which allowed irregularly shaped buildings to be captured in this process. For example, in the bottom left of Fig. 11.3, the round shaped building is successfully captured.

Point Clouds Density Adjustment

The initial point cloud had a density of 13.25 pts/m^2. In order to evaluate the impact of density on the successful detection of buildings and the geometric representation of the buildings, supplemental points clouds were generated to, respectively, match densities of 10, 8, 6, 4, 2, and 1 pts/m^2.

For each of the 34 tiles, the total number of returns and the total number of first returns were calculated. The ratio of first returns to total returns was then calculated for each tile. Using this first-total ratio as a baseline, columns of points were removed to match the required point density and maintain uniform point spacing over the surface of the tile. The PointCloudThinner from Feature Manipulation Engine (FME) was used to perform the density reduction while the lasinfo method from LASTools was applied to validate the resulting density.

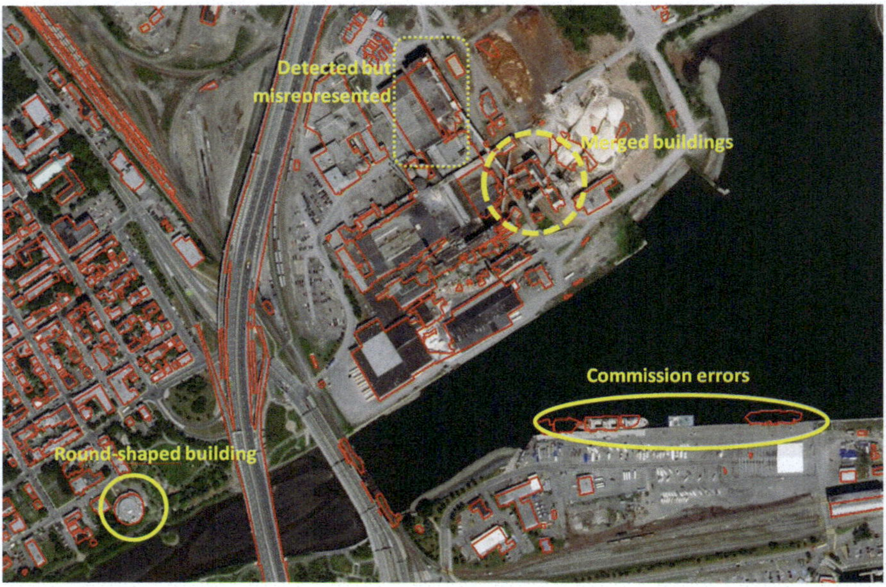

Fig. 11.3 Example of buildings from re-classified LiDAR data, including (i) detected and correctly represented, (ii) detected but misrepresented, (iii) merged buildings, and (iv) commission errors

11.4 Results

The results were evaluated quantitatively through statistical sampling and qualitative evaluation using Esri and Google Earth imagery for validation. In total, 338 samples across the study area were generated for each point density tested and for the vendor-classified and re-classified datasets. The number of samples was calculated with an expected error of 2% and a maximum error of 5%, to ensure that they were representative of, and could be used to characterize the dataset as a whole (Pfaffenberger and Patterson 1981). The samples were selected randomly and evenly distributed over the entire study site.

All samples were checked manually at a scale of 1:400 by the same GIS technician. Every building within each window was checked. This qualitative analysis involved classifying the building geometry as (i) detected and correctly represented, (ii) detected but misrepresented, (iii) merged, and if there were (iv) commission errors or missing buildings, Fig. 11.3. The criteria used to identify if a building was detected but misrepresented included identifying if the building was missing sections and/or the footprint did not match the corresponding imagery. This was done through visual analysis. If these two criteria were not present for a detected building, the building shape was considered to be correctly represented and corresponded to ground truth. Commission errors included polygons representing items such as boats or viaducts, which were above the minimal dimension threshold and were incorrectly detected as buildings.

For the datasets that are vendor-classified, the results for the quality of the representation of the buildings are nearly stable when the density of points is greater than 4 pts/m^2. Between 77% and 80% of the buildings were detected and correctly represented, Fig. 11.4, and greater than 97.8% of the buildings being detected, when the point density was greater than or equal to 4 pts/m^2, Table 11.1. These results are consistent with the findings of Awrangjeb (2016). However, as the point density decreases to 1 pts/m^2 or 2 pts/m^2, the ability to correctly detect buildings is greatly diminished. In these circumstances less than half of the buildings are correctly represented, Fig. 11.4.

Comparing the results of vendor-classified and re-classified methods of building extraction in the most dense sample, 13.25 pts/m^2, finds the fully automated open-source re-classification technique a bit poorer (−2.5%) at correctly detecting and representing buildings. This trend continues with decreasing point density, with the smallest difference of 1.2% at 8 pts/m^2 and largest, 10.4%, at 6 pts/m^2. Thus, the vendor classification is marginally better at correctly detecting and representing the building footprints. It was also found that a point density between 4 pts/m^2 and 13.25 pts/m^2, the "detected but misrepresented" buildings were between 4.5 and 7% of the sample in the vendor-classified result, and between 7.4 and 10.4% in the re-classified result. The results show that the ability to correctly represent a building decreases gradually when reducing the density and are systematically above the 5% error threshold when the density is below 8 pts/m^2. The results of the re-classified 10 pts/m^2 (73.9%) are lower than the results for the 8 pts/m^2 (76.9%), showing that density is probably not the only factor affecting the results.

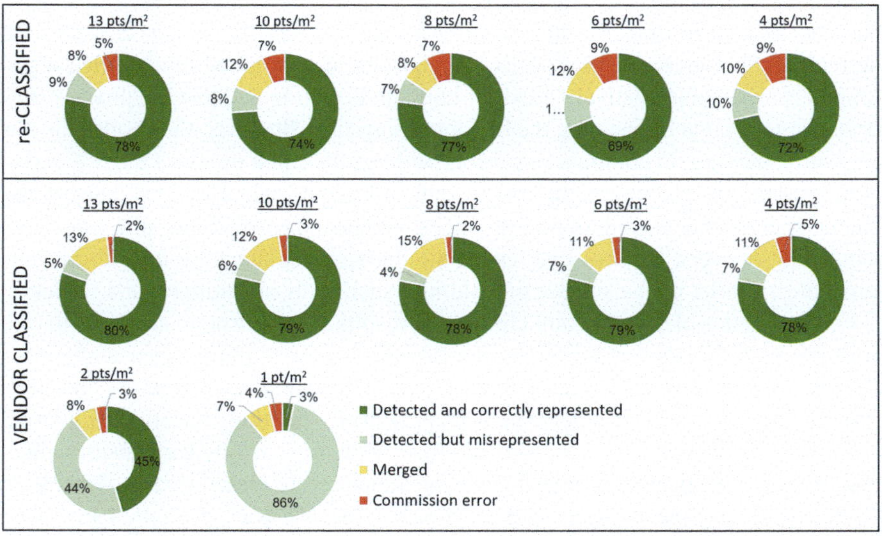

Fig. 11.4 Results from user analysis and classification of samples from classified and un-classified building detection at different point densities

Table 11.1 Percent of buildings successfully detected or omitted

	Vendor-classified (pts/m^2) (%)							Re-classified (pts/m^2) (%)				
	13	10	8	6	4	2	1	13	10	8	6	4
Detection	99.4	98.5	97.8	98.0	98.3	97.3	96.1	98.5	98.6	99.0	98.6	98.4
Omission	0.6	1.5	2.2	2.0	1.7	2.7	3.9	1.5	1.4	1.0	1.4	1.6

The vendor-classified dataset contains lower commission errors than the re-classified point cloud, with a difference over 3% at the reference density. The results for commissions errors are comparable from one density to another with differences between 3% and 6% for a given density, between the vendor-classified and the re-classified datasets. Except for the reference dataset at density of 13 pts/m^2, the vendor-classified datasets have slightly more omissions errors than the re-classified datasets, Table 11.1.

11.5 Discussion

The quality of the vendor classification is directly responsible for the detection performance. This explains why detection level is so high, given that their classification was corrected semi-automatically. In the re-classified dataset, omissions errors were misclassified in the point cloud. The higher rate of omission error seems to indicate that the semi-automatic correction resulted in the deletion of actual buildings. This is possibly due to the minimal dimension criteria for buildings used by the vendor when doing the classification or the re-classification method used.

The results in Table 11.1, indicating detection/omission rate for buildings, show that detection performance is not greatly affected whether the vendor-classified or the re-classified dataset is used. This is a sign that the designed re-classified algorithm performance is relatively good for building detection and that the process performance is good at delineating the building footprints. However, the results show a substantial increase in commissions errors with results equal or above the 5% maximum error threshold between the vendor and re-classified data, Fig. 11.4. This shows that the open-source algorithm performance could be improved, especially when distinguishing between features that have a similar spectral signature to rooftops and constitute a similar planar surface to buildings, such as: boats, bridges, and viaducts.

For the vendor-classified point clouds, the results show a drop in ability to detect buildings with 2 pts/m^2 and less. Successful building detection is consistently possible at 4 pts/m^2 and above. This finding may help guide specifications and help governments save on costs when preparing projects for building detection. However, it should be noted that these results could have been impacted by the density reduction. The original dataset acquired from the vender were classified at a density of 13.25 pts/m^2 and the density of the classified LiDAR was resampled to lower densities. If the vendor classification had been done at 2 pts/m^2, this may have had a different impact on the results. The resulting point pattern using the aforementioned PointCloudThinner technique is likely dissimilar to acquisitions patterns.

Thus, testing an alternative point density reduction technique, such as sub-sampling the flight-line data, could be considered and tested to evaluate the impact of density reduction on the results.

Manual intervention at lower density would have likely produced better results than the results produced through the automatic density degradation. Furthermore, the processing steps used to generate the geometries could have enhanced this phenomenon by shifting segments and eliminating building sections, thus making the geometric representation inaccurate.

For the re-classified method, there is good performance where density is 8 pts/m^2 and above. However, because of commission errors above the maximum error threshold we would not recommend pursuing this solution as-is. However, for organizations with the necessary resources and expertise to edit the extraction results, this re-classification method could provide an interesting alterative if buildings are desired. This finding could be helpful for governments that already have important LIDAR data holdings that were not already classified for buildings.

The results in Fig. 11.3 indicate a large variability with commission and omission errors that is not necessarily consistent with an increase or decrease in the LiDAR density. These are possibly linked to the location of the sampling window. In cases where these windows include motorways or ports, or any other type of location which is favorable to include false positives. The random sampling approach was meant to reduce bias in the analysis, however, this may have impacted the results. In the re-classified approach, there are impacts on the relative percentage of omissions and commission errors when the sampling window brought the evaluator over areas with important infrastructures, such as yards, ports, or motorways. This could also explain why results for re-classified project at 10 pts/m^2 are uncharacteristically poor. Further testing which includes alternatives such as increasing sample size and proportional sampling by land use may provide additional data to support this hypothesis. Further testing should also be done to validate the impact of building density on the results.

11.6 Conclusion

LiDAR data collected in a Canadian city was used to automatically extract building footprints from semi-automatic vendor-classified and re-classified (via open-source tools) point cloud data at varying densities in order to evaluate the quality and completeness of the derived building footprints. Through a comparative analysis of the ability to detect buildings and the quality of the match, recommendations are made for the desired minimal point density for building footprint extraction from airborne LiDAR for urban communities. The methods applied to derive building footprints included: classification based on textural analysis of the non-ground point cloud data, building footprint extraction through generation of non-convex envelope, building splitting (as needed), and squaring using non-linear least squares.

The findings indicate that building footprints are correctly represented by the semi-automatic vendor-classification method at point densities of 4 pts/m^2 and above. Below the 4 pts/m^2 threshold the quality of the representation becomes unacceptable, with respect to a 5% error threshold. The baseline for quality, used in this context, is at or above 75% detection rate of buildings. Up to 15% of the buildings were incorrectly merged with adjacent buildings in the vendor-classified data with densities of 4 pts/m^2 or higher. At point densities of 1 or 2 pts/m^2, the ability to correctly detect and represent buildings from the vendor-classified point data is 3% and 45%, respectively, thus, the footprint extraction method described here is not recommended at these lower densities.

With respect to the open source, fully automatic, re-classification method, the representation of building footprints is within acceptable tolerance, given a point density of 8 pts/m^2 and higher. However, the quantity and type of omission error (5–9%) associated with this re-classification method is deemed unacceptable. The remaining 20% is split, fairly equally, between merged buildings and detected but misrepresented buildings. Either manual intervention or improvement in the tested algorithms would be required to reach an acceptable quality threshold. Thus, while acquiring un-classified data and processing it via the described open-source tools would be an economic savings, there would be additional costs and/or expertise needed to meet the same level of quality and completeness as the semi-automatic vendor-classified data—given the point density is 8 pts/m^2 or higher.

An option, which may be considered for future studies, would be the inclusion of supplemental datasets, such as roads or waterbodies to eliminate some of the false positives in the re-classified method. In addition, improvements in building classification algorithms using machine learning approaches such as Deep Neural Network's could provide a solution to improve the performance of the algorithms used to classify buildings in LiDAR point clouds. Finally, it is recommended to further test this method in order to determine if the suggested minimal point density is also optimal for rural communities.

References

Aguilar, F.J., J.P. Mills, J. Delgado, M.A. Aguilar, J.G. Negreiros, and J.L. Pérez. 2010. Modelling vertical error in LiDAR-derived digital elevation models. *ISPRS Journal of Photogrammetry and Remote Sensing* 65 (1): 103–110. https://doi.org/10.1016/j.isprsjprs.2009.09.003.

Alharthy, A., and J. Bethel. 2002. Heuristic filtering and 3D feature extraction from LIDAR data. *International Archives of Photogrammetry Remote Sensing and Spatial Information Sciences* 34 (3/A): 29–34.

Awrangjeb, M. 2016. Using point cloud data to identify, trace, and regularize the outlines of buildings. *International Journal of Remote Sensing* 37 (3): 551–579.

Awrangjeb, M., and C.S. Fraser. 2014. Automatic segmentation of raw LiDAR data for extraction of building roofs. *Remote Sensing* 6 (5): 3716–3751.

Brooks, R., T. Nelson, K. Amolins, and G.B. Hall. 2015. Semi-automated building footprint extraction from orthophotos. *Geomatica* 69 (2): 231–244. https://doi.org/10.5623/cig2015-206.

Canada Centre for Mapping and Earth Observation. 2017. Synthesis report: study on the potential of lidar data for the extraction of cartographic objects other than elevation, Geomatics Canada, Technical Note 5. https://doi.org/10.4095/300243.

Esri. 2019. World Imagery Service Layer in ArcMap. World Boundaries and Places - Esri, HERE, Garmin, (c) OpenStreetMap contributors, and the GIS user community, World Imagery - Source: Esri, DigitalGlobe, GeoEye, Earthstar Geographics, CNES/Airbus DS, USDA, USGS, AeroGRID, IGN, and the GIS User Community.

Filin, S., and S. Pfeifer. 2006. Segmentation of airborne laser scanning data using a slope adaptive neighborhood. *ISPRS Journal of Photogrammetry and Remote Sensing* 60 (2): 71–80.

Kilian, J., N. Haala, and M. Englich. 1996. Capture and evaluation of airborne laser scanner data. *International Archives of Photogrammetry and Remote Sensing* 31 (Part B3): 383–388.

LAStools. 2019. LAStools. GitHub, https://github.com/LAStools/LAStools. Accessed 26 April 2019.

Lokhat, I., and G. Touya. 2016. Enhancing building footprints with squaring operations. *Journal of Spatial Information Science* 13 (13): 33–60. https://doi.org/10.5311/JOSIS.2016.13.276.

Maas, H.G., and G. Vosselman. 1999. Two algorithms for extracting building models from raw laser altimetry data. *ISPRS Journal of Photogrammetry and Remote Sensing* 54 (2-3): 153–163. https://doi.org/10.1016/S0924-2716(99)00004-0.

Natural Resources Canada. 2017. High-resolution digital elevation model generated from LiDAR data – new data available. Government of Canada. http://www.nrcan.gc.ca/earth-sciences/geography/topographic-information/free-data-geogratis/whats-new/20044. Accessed 12 April 2019.

Pérez-García, J.L., J. Delgado, J. Cardenal, C. Colomo, and M.A. Ureña. 2012. Progressive densification and region growing methods for LIDAR data classification. *International Archives of the Photogrammetry, Remote Sensing and Spatial Information Sciences* 39 (B3): 155–160. https://doi.org/10.5194/isprsarchives-XXXIX-B3-155-2012.

Pfaffenberger, R.C., and J.H. Patterson. 1981. *Statistical Methods*. Illinois: Homewood.

Pfeifer, N. 2005. A subdivision algorithm for smooth 3D terrain models. *ISPRS Journal of Photogrammetry and Remote Sensing*. 59 (3): 115–127.

Ramiya, A.M., R.R. Nidamanuri, and R. Krishnan. 2017. Segmentation based building detection approach from LiDAR point cloud. *The Egyptian Journal of Remote Sensing and Space Science* 20 (1): 71–77. https://doi.org/10.1016/j.ejrs.2016.04.001.

Rutzinger, M., F. Rottensteiner, and N. Pfeifer. 2009. A comparison of evaluation techniques for building extraction from airborne laser scanning. *IEEE Journal of Selected Topics in Applied Earth Observations and Remote Sensing* 2 (1): 11–20. https://doi.org/10.1109/JSTARS.2009.2012488.

Sampath, A., and J. Shan. 2003. Building Segmentation and Regularization from Raw Lidar Data, https://engineering.purdue.edu/~jshan/publications/2004/ASPRS_2004_p5798_Lidar_Building.pdf. Accessed 12 Feb 2019.

Wikipedia. 2019. Google Earth, https://en.wikipedia.org/wiki/Google_Earth#Resolution_and_accuracy. Accessed 22 May 2019.

Zhang, K., S.C. Chen, D. Whitman, M.L. Shyu, J. Yan, and C. Zhang. 2003. A progressive morphological filter for removing nonground measurements from airborne LIDAR data. *IEEE Transactions on Geoscience and Remote Sensing* 41 (4): 872–882. https://doi.org/10.1109/TGRS.2003.810682.

Zhang, K., J. Yan, and S.C. Chen. 2006. Automatic construction of building footprints from airborne LIDAR data. *IEEE Transactions on Geoscience and Remote Sensing* 44 (9): 2523–2533. https://doi.org/10.1109/TGRS.2006.874137.

Zhao, T., and J. Wang. 2014. Use of lidar-derived NDTI and intensity for rule-based object-oriented extraction of building footprints. *International Journal of Remote Sensing* 35 (2): 578–597. https://doi.org/10.1080/01431161.2013.871394.

Part III
Infrastructure Damage

Chapter 12
Flood Mapping from Multi-Sensor EO Data for Near Real-Time Infrastructure Impact Assessment: Lessons Learned from the 2017 Spring Flood in Eastern Canada

Ian Olthof, Simon Tolszczuk-Leclerc, Brad Lehrbass, Victor Neufeld, and Vincent Decker

12.1 Introduction

The Emergency Geomatics Services (EGS) is a section within the Canada Centre for Mapping and Earth Observation, Natural Resources Canada, responsible for providing geospatial intelligence during natural disasters including flooding. EGS maps floods to help mitigate impacts on people and infrastructure by providing an overview of current flood extents that allow emergency responders including the military to prioritize and deploy help where it is needed most. Mitigation measures include shutting down power and gas transmission lines, erecting sandbag walls to protect buildings and other infrastructure, as well as mass evacuations to safeguard people from flooding.

On 23 April 2017 EGS was activated by Public Safety Canada for flooding caused by snowmelt and heavy precipitation in communities located on the Ottawa River between the Quebec–Ontario provincial border and Lake of Two Mountains, Canada. As record amounts of precipitation continued to fall into May 2017, the activation area expanded to include the Ottawa River from west of Ottawa to east of Montreal as far as Lac St-Pierre further downstream (Fig. 12.1).

At the outset of the 2017 flood, EGS was in the midst of changing its operations; from the way, it extracted surface water extents from satellite imagery to the incorporation of citizen geographic information to improve its flood map products. This chapter describes lessons learned amid this transition during EGSs' 2017 springtime flood activation in Eastern Ontario and Western Quebec. Data received by EGS

I. Olthof (✉) · S. Tolszczuk-Leclerc · B. Lehrbass · V. Neufeld · V. Decker
Emergency Geomatics Services, Canada Centre for Mapping and Earth Observation, Natural Resources Canada, Ottawa, ON, Canada
e-mail: ian.olthof@canada.ca

V. Singhroy (ed.), *Advances in Remote Sensing for Infrastructure Monitoring*, Springer Remote Sensing/Photogrammetry,
https://doi.org/10.1007/978-3-030-59109-0_12

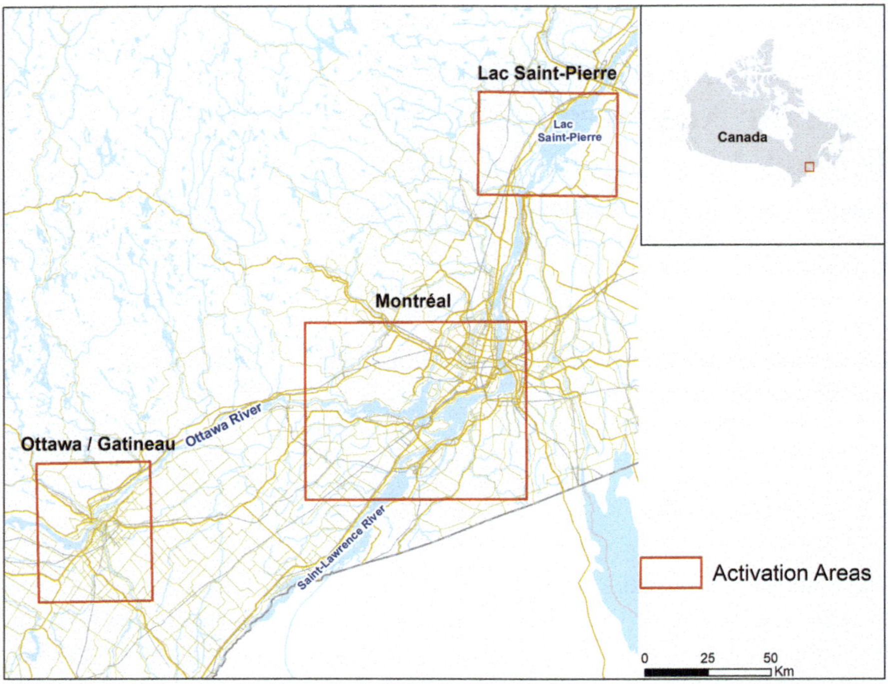

Fig. 12.1 2017 flood activation area from west of Ottawa to Lac St-Pierre

used to generate accurate and consistent near real-time flood information is listed, including satellite imagery, as well as airborne and ground-based reference data used to improve and validate flood maps. The transition that was taking place within EGS to update surface water mapping methods is described as this coincided with the 2017 activation. Due to the challenges faced by processing vast quantities of satellite imagery received during the event, EGS decided to adopt the new methods due to improved mapping efficiency and consistency. The rationale for this transition including advantages of the new methods are explained.

Background

In the year prior to the 2017 activation, EGS had made significant progress on automated surface water mapping to extract both open water and flooded vegetation extents from optical and radar satellite data. This work was ongoing within Public Safety's floodplain characterization programme with the goal of generating useful information for EGS, including historical surface water maps to help precisely delineate floodplains as well as calibrate satellite flood models for operational mapping. This new methodology for automated flood mapping was demonstrated using historical optical satellite imagery from the Landsat mission obtained through the USGS in Olthof (2017).

This methodology was subsequently adapted to single and dual-polarization radar imagery from RADARSAT-1 and 2 and work began to generate historical surface water maps over the Saint John River floodplain to compare with optical data with the eventual goal of seamlessly integrating information derived from both sensors (Olthof and Tolszczuk-Leclerc 2018). The overarching objective of this work was to produce a robust, accurate, automated and efficient methodology to map surface water from a range of sensors for both operational and historical flood mapping with minimal user intervention. The large data volumes of satellite imagery that exist in historical archives necessitated the transition from manual methods to full automation, while efficiency is needed since EGS has set a goal to publish maps within 2–4 h of satellite data reception to provide current emergency situational awareness. The 2–4 h timeframe includes several time-consuming steps from data download, processing, quality control including editing when necessary, to map production and publishing.

At the request of Public Safety Canada, the International Charter on Space and Major Disasters was activated on 6 May 2017 for flooding in Southern Quebec. The Charter members consist of 17 space agencies from around the world who provide free imagery from 34 satellites to relief organizations working in areas affected by natural or manmade disasters. Upon activation, the methodology was applied to data from different sensors received through the Charter in addition to Canada's own RADARSAT-2. The majority of data were from radar satellites, providing the ability to detect water through cloud during near continuous cloud cover that was present in the weeks of flooding. Fortunately, the methodology proved to be sufficiently reliable to be adapted to other sensors that include different radar wavelengths, polarizations and spatial resolutions, requiring specification of only three sensor-specific parameters to account for these differences.

Data

Satellite, airborne, ground-based and ancillary data layers were integrated to provide near-real-time flood maps that were quality checked on a best effort basis prior to release to government agencies and the public.

Satellite Imagery

EGS tasks Canada's RADARSAT-2 satellite to image flood-prone regions through acquisition plans prior to the springtime flood season and gets priority tasking during an emergency activation. EGS tasks RADARSAT-2 dual-polarization imagery in standard or wide-mode to ensure sufficient coverage over the affected region. Through the Charter, EGS received both optical and radar imagery from a number of different satellite sensors with detail ranging from sub-metre to 250 m resolution and coverage from tens of square kilometres to thousands (Table 12.1).

Table 12.1 Satellite provided by the International Charter on Space and Major Disasters to the Emergency Geomatics Services during the 2017 flood.

Sensor	Bands	Agency/Company	Country
KOMPSAT5	SAR (X-band)	KARI	Korea
ALOS-2 PALSAR-2	SAR (L-band)	JAXA	Japan
RADARSAT-2	SAR (C-band)	CSA/MDA	Canada
Sentinel-1A, 1B	SAR (C-band)	ESA	European Union
TerraSAR-X	SAR (X-band)	DLR	Germany
MODIS	Optical	NASA	USA
Pleiades-1A, 1B	Optical	CNES	France
VIIRS	Optical	NOAA/NASA	USA
ASTER	Optical	NASA/JAXA	USA/Japan
Resourcesat-1/AWiFS	Optical	ISRO	India
TanDEM-X	SAR (X-band)	DLR	Germany
WorldView-2, -3	Optical	DigitalGlobe	USA

Priority needed to be given to the best available scenes due to the volume of data received through the Charter. From April 23 to May 24, 112 scenes were evaluated for map production and publishing, the majority of which were discarded due to excessive cloud cover in optical, high wind in radar or some other quality issue. Of the 112 scenes, 41 were processed to some degree and 22 were fully processed and published as maps simultaneously on the Federal Geospatial Platform (FGP) used to disseminate flood products internally to government and Open Maps to the public.

NASP

From May 7–16, 2017, Transport Canada's National Aerial Surveillance Program (NASP) aircraft were tasked to acquire oblique pictures of the flooding along approximately 600 km of the waterfront from Pembroke, Ontario eastward to Quebec City. Four separate flight zones were defined, with flights rotating through each flight zone over the 10-day period. Pictures captured every 5 s during flights totaled nearly 14,000 images taken during the 11-day period. These pictures were used to populate an online interactive web map through ArcGIS Online to distribute information and help improve and validate flood maps as they were generated from satellite imagery (Fig. 12.2).

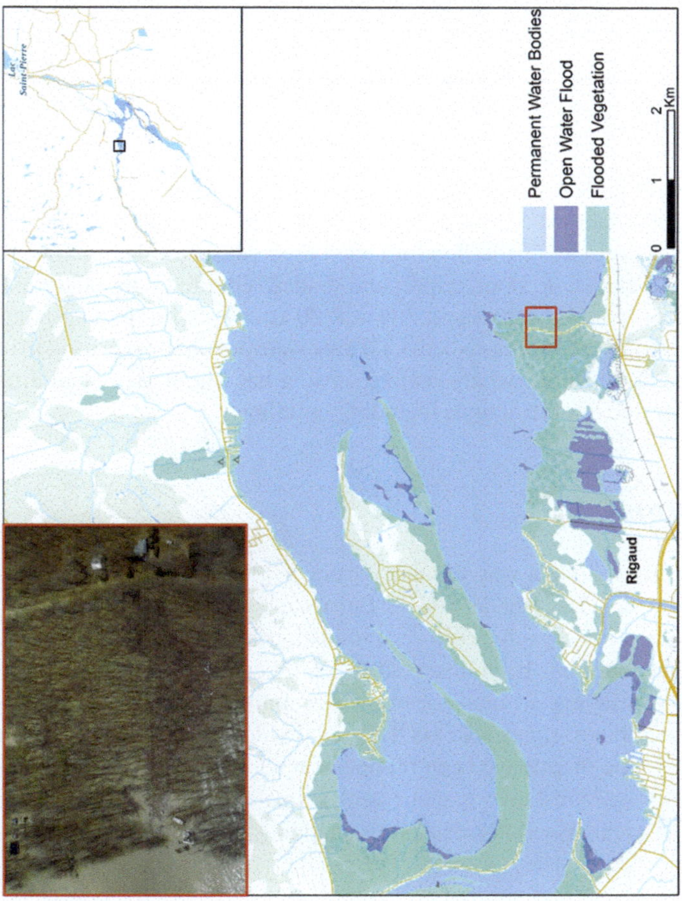

Fig. 12.2 Open water and flooded vegetation mapped from RADARSAT-2 near Rigaud, Quebec, confirmed by NASP oblique airphotos

CGI

Prior to the activation, a beta version Citizen Geographic Information (CGI) application was released to select trusted users to acquire near real-time in-situ flood information. The application notifies users of a satellite overhead about to acquire an image, at which time the user is requested to take a geotagged picture of flooding and upload it to a central geodatabase for analysis. The image metadata also includes azimuth, enabling co-registration of features in CGI pictures with features in the satellite image. When available, this information is used to verify flood extents in satellite products and better tailor products by correcting obvious errors before release on the FGP and Open Maps.

LIDAR HRDEM

Canada's new High-Resolution Digital Elevation Data Model (HRDEM) LIDAR-derived elevation data were made available to EGS prior to their official release to the public over portions of the flood-affected area where available. The complete dataset includes elevation, slope, aspect and shaded relief for a digital terrain model available at 2 m spatial resolution in 20 km × 20 km tiles with floating-point precision. Tiles covering urban areas where CGI data were acquired were used to improve flood mapping in locations where remote sensing has traditionally had difficulties mapping water due to obstructions including buildings.

Water Occurrence

A water occurrence map was obtained from a global product generated from 32 years of historical Landsat data by the European Commission (Pekel et al. 2016). Water occurrence is also referred to as inundation frequency because it maps the percentage that water has been observed at each location in satellite imagery. It is generated by overlaying a time-series of water maps and for each pixel location, counting the number of times water is detected divided by the total number of valid observations. These maps have been produced at a global scale to support applications including water resource management, climate modelling, biodiversity conservation, food security and flood response mapping.

12.2 Emergency Geomatics Services New Operational Flood Mapping Methods

Open Water Mapping

The new EGS flood mapping methodology performs multi-channel supervised machine learning classification of available radar polarizations to map open water, and then bright threshold region growing from open water to map flooded vegeta-

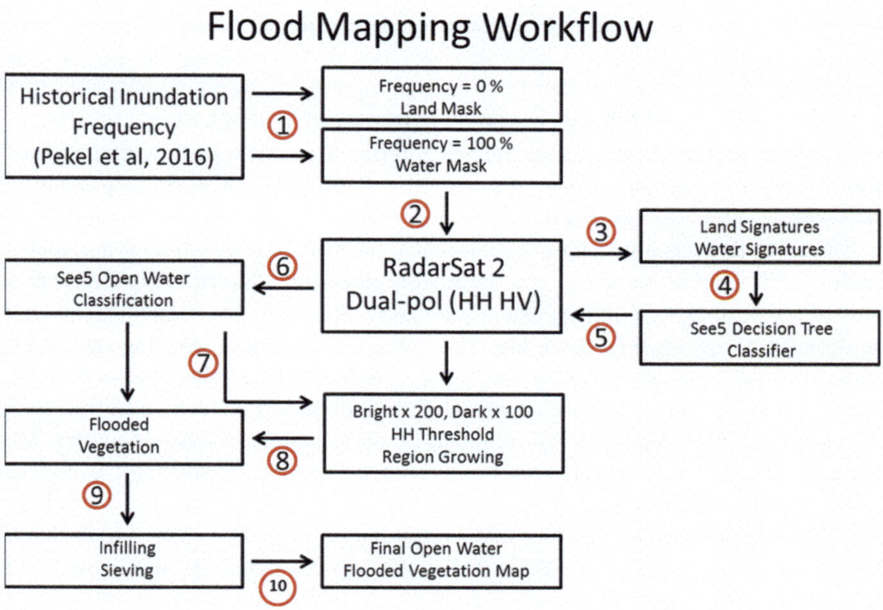

Fig. 12.3 New methodology put into operations for the 2017 flood activation

tion (Fig. 12.3). Smooth open water is a specular surface that produces a dark appearance in radar imagery caused by single bounce of incident radiation away from the sensor. Previously, EGS mapped open water using the traditionally accepted method of single polarization interactive dark image thresholding even when multiple polarizations were available (Bolanos et al. 2016; Li and Wang 2015). While thresholding has been shown to perform well for open water under ideal conditions, water surface roughness due to ice that is sometimes present during the spring flood season as well as waves can increase backscatter to a level where a single threshold value cannot reliably separate water from land (Henry et al. 2006). Automated methods used to determine optimal threshold values are compromised by these factors, while manual thresholding can better tune the value to minimize errors of omission and commission. Even still, a significant amount of post-processing is often required to reduce errors to an acceptable level (White et al. 2014).

Information contained in multiple radar polarizations can help to reduce errors in open water extraction; however making use of this information requires an approach other than single band image thresholding. The supervised multispectral classification has long been used in terrestrial remote sensing applications, but has seen limited use for surface water mapping from radar. One challenge is that supervised classification approaches require spectral signatures for each class to train a classifier, for example, signatures representing land and water in the case of water extraction. The use of standard spectra to classify land and water will not achieve an optimal classification result for several reasons. First, the spectral variability of water in radar due to wind and ice can cause confusion between water and land. Second, spectral signatures of land also vary due to the presence of several land cover types that change through time because of vegetation phenology, moisture and atmosphere. An additional complicating factor is that in order to perform traditional multispectral classification such as minimum distance or maximum likelihood, separate signatures must be sampled for all land cover types present in the scene.

To deal with these limitations, an automated open surface water extraction methodology that is an extension of one already developed in Olthof (2017) was implemented. The approach makes use of recently available inundation frequency products from historical Landsat data (Pekel et al. 2016; Olthof 2017) to sample all scene-specific signatures representing every land and water class present in the image. Land signatures are extracted where inundation has never occurred (0% occurrence) based on historical inundation maps, while water signatures are extracted where water has always been observed (100% occurrence). Sampled, scene-specific signatures are used to train the C50 (See5) decision tree machine learning algorithm that classifies the entire image into the water versus land. Machine learning is used because it does not assume a statistical distribution and can handle vast amounts of training data to precisely tune the classifier.

Flooded Vegetation Mapping

A significant amount of flooding in Canada occurs in vegetated areas, which has lead to considerable research to improve flooded vegetation mapping in recent years. Flooded vegetation presents a challenge to remote sensing because the signal must penetrate the vegetation layer to sense water below. While optical remote sensing has a limited capability to detect flooding beneath vegetation and can do so reliably only during leaf-off in the early spring or late fall, radar is able to detect water beneath leaf-on canopies under many conditions. Radar is an active sensor that is side-looking, causing double bounce of the incident beam first off horizontal water surfaces beneath the canopy, and second off vertical trunks and stems acting as corner reflectors before returning to the sensor (Hess et al. 1990). Leaf size, wavelength and incidence angle are all factors that affect water detection beneath vegetation from radar, with longer wavelengths relative to leaf size (Townsend and Walsh 1998) and shallower incidence angles (White et al. 2015) generally providing greater signal penetration through the canopy.

Double bounce off flooded vegetation causes a high-intensity return to the sensor, producing a bright signal in radar imagery particularly in like-polarization channels (e.g. HH). Bright target thresholding has been used in the past to map flooded vegetation; however, this approach was never operationalized by EGS because it leads to significant commission error caused by the presence of other corner reflectors such as buildings and rock outcrops in the scene. Post-classification editing in a GIS environment can be performed to help reduce commission error, but this relies on ancillary data layers that may be out of date or contain errors. Manual editing has also been done to improve flood products but is both time-consuming and subjective in the absence of reliable ancillary information. A robust flooded vegetation extraction methodology independent of ancillary data requirements is preferred for emergency flood mapping for product quality, improved processing speed and efficiency.

To minimize processing time and commission error, the new methodology maps flooded vegetation by iteratively region growing from open water into adjacent bright intensity areas characteristic of double bounce using a bright threshold value criterion. Region growing has the advantage over thresholding the entire image by assigning only adjacent pixels to the flooded vegetation class, thereby reducing commission errors caused by double bounce elsewhere in the scene. Double bounce off buildings adjacent to open water was corrected using the urban class from a recent Landsat land cover of North America (Latifovic et al. 2017), which was generated including information from Canada's road network layer. Once the first region growing using a bright threshold value is complete, a second region growing is performed using a dark threshold value from open water and flooded vegetation to help infill dark areas contained within bright flooded vegetation and better connect nearby areas of open water. Both instances of region growing stop when either no additional adjacent pixels meet the threshold criterion or after a set number of iterations. Open water and flooded vegetation are subsequently improved using morphological operators to infill small land areas entirely surrounded by water, followed by filtering and sieving to reduce both errors of omission and commission.

Because the methodology relies on signatures extracted from each scene to classify open water, and only user specification of radar polarization, bright and dark threshold values are needed to map flooded vegetation, the approach is sensor-independent. As new charter data was received during the activation, tools were modified to accept the number of polarizations available to work on single polarization (e.g. TerraSAR-X), dual-polarization (e.g. Sentinel-1) and quad-polarization (e.g. RADARSAT-2) (Fig. 12.4). Where multiple polarizations are available, specification of the polarization to use for flooded vegetation region growing must be specified. Generally, the polarization with the best contrast between flooded vegetation and surrounding water and land is selected. Where available, like-polarization provide the best contrast, for example, HH in RADARSAT-2.

Urban Flood Mapping

A major shortcoming of the approach applied to radar data was its inability to accurately map open water in urban environments. The presence of buildings causing shadow made dark water target detection impossible, while confidence that any detected dark targets were water was low. Further, urban areas located on the water were often mapped as flooded vegetation due to the presence of corner reflectors

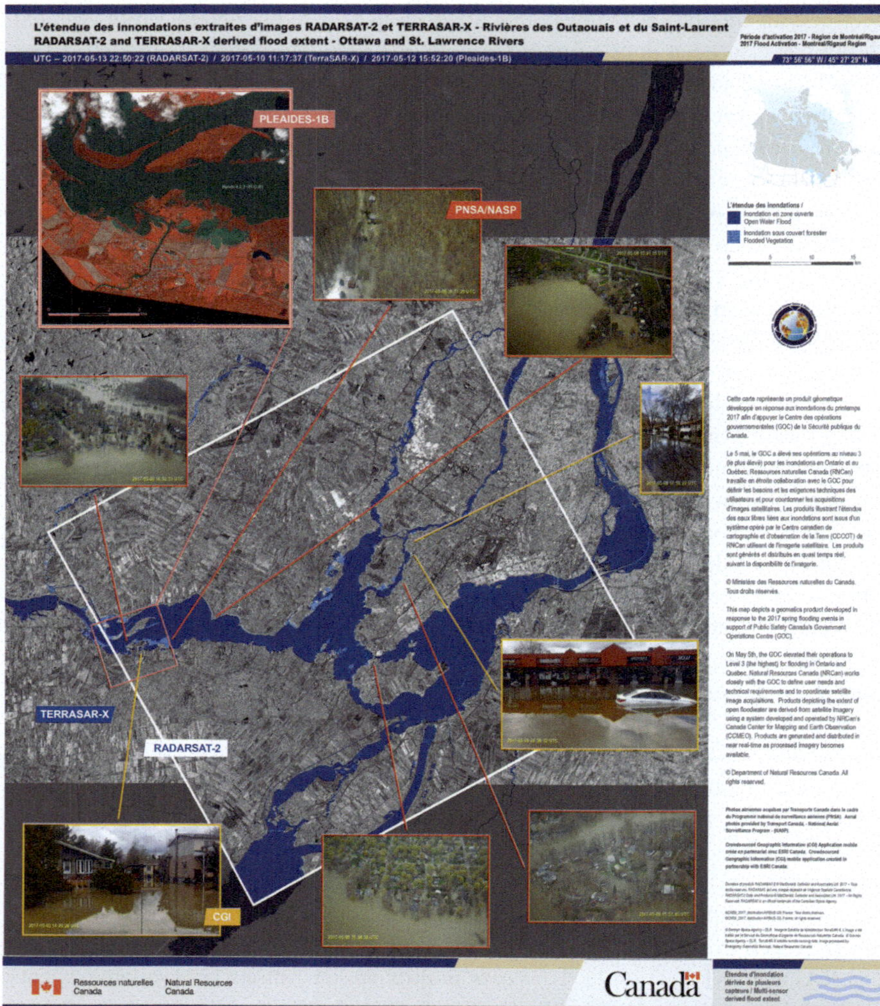

Fig. 12.4 Example of multi-sensor flood mapping using the new operational methodology from RADARSAT-2 (interior white box), and TerraSar-X (outside), showing the consistency in mapping open water and flooded vegetation between sensors

such as road and buildings producing a high-intensity return. Post-classification improvements were made with land cover data to remove flooded vegetation in urban settings; however, this also removed any real water that was detected.

As Citizen Geographic Information (CGI) data submitted during the event became available during the activation, tests began on using it in combination with newly available LIDAR HRDEM data to improve urban flood mapping. A total of 394 observations were submitted across three flood-affected regions including Lac St-Pierre, Montreal and Lake of Two Mountains, and Ottawa-Gatineau between

March 29 and April 14, 2017. The use of LIDAR data for flood mapping requires a local water depth measurement, which can be estimated from CGI data assuming the observer is taking pictures at or near the flood perimeter and can be verified by looking at the pictures themselves. Once an elevation along the flood perimeter has been established from CGI, the full local flood extent can be mapped by filling adjacent pixels below the established flood elevation at the flood perimeter, as shown in Fig. 12.5 using 80 observations submitted in Gatineau 6 April 2017. The extent to which infilling can be done varies with the degree of local variation in the geoide. For example, if a single elevation threshold were applied to a LIDAR DEM from Ottawa to Montreal, flooding would be under predicted in Ottawa and simultaneously over predicted in Montreal due to the slope of the geoide from Ottawa to Montreal that causes water to flow in that direction. While the extent of urban flooding is considerably less than the distance between Ottawa and Montreal, CGI data acquired in Montreal's West Island should not be used to estimate urban flooding in the East End. Therefore, local CGI data can be used to estimate local residential/urban flooding from inland to the nearest adjacent permanent water body, but not between separate, distant locations.

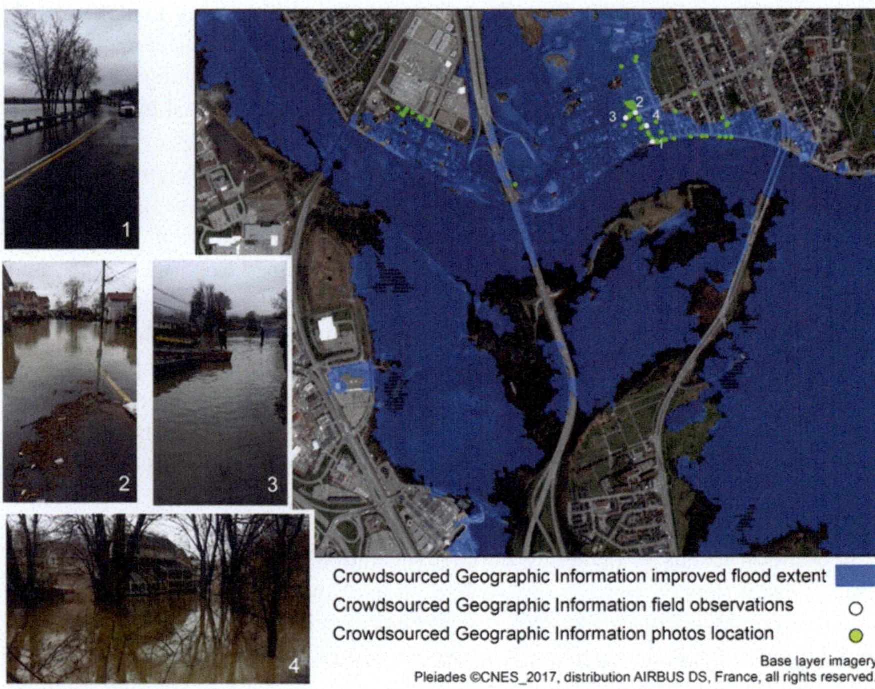

Fig. 12.5 Use of Crowdsourced Geographic Information (CGI) data acquired on 6 April 2017, and a new LIDAR DEM to improve urban flood mapping in Gatineau, Quebec, combined with a RadarSat-2 flood product from 7 April 2017

Limitations and Lessons Learned

Currently available global historical dynamic surface water maps used to train machine learning have known limitations due to input optical Landsat imagery that is the only medium resolution data available with its spatial and temporal coverage. First, optical data does not penetrate cloud and therefore, maximum water extents during cloudy, rainy events including a significant percentage of floods such as the 2017 event are not fully captured. In addition, optical remote sensing has a limited capability to detect water beneath vegetation, while flooded vegetation comprises a high percentage of the overall flood area on many floodplains in Canada. Lastly, because the product is global in scale, it is not optimized to Canada's specific geography or seasonality. All of these factors generally lead to water omission errors. Nevertheless, the new flood mapping methodology is robust enough to overcome these limitations and Pekel's occurrence map proved very useful during the 2017 activation.

The availability of historical radar imagery from Canada's RADARSAT mission going back to 2008 combined with the advantages cloud penetration and the ability to map flooded vegetation suggest that separate dynamic surface water maps should be generated from these data over Canada. While current global products contain error, machine learning used to classify open water is known to be very robust to errors in training data (Ghosh et al. 2016; Olthof and Tolszczuk-Leclerc 2018). Therefore, even if global maps omit the water, errors in training data extracted using these maps should be relatively small compared to the overall training sample size. The activation season was also considered successful using a global historical map product as input, further supporting the use of Pekel's map. However, the more extreme the flood event, the more errors will be introduced when relying on a product that omits water in favour of land. Currently, EGS has no sense of the level of improvement that can be achieved with better historical information. For this reason, work will continue to generate Canadian products tuned to our geography and seasonality from historical optical and radar imagery over select floodplains in Canada. With support, the historical RADARSAT archive can be exploited for this purpose as is ongoing along the Saint John River, NB. Finally, the complete historical and operational surface water mapping approach is iterative as current flood products are used to update and improve historical inundation products, which will then in turn be used to produce flood products next season. As scenes are added to the time-series of water maps, both historical and near-real time water maps should continue to improve.

In addition to falsely detecting flooded vegetation by region growing in urban areas adjacent to open water, a second limitation of the methodology relates to the local topography of riverbanks. Where riverbanks are steep and roughly parallel to the SAR orbit, flooded vegetation will also be falsely detected due to a high backscatter caused by steep local incidence angles. Using ancillary datasets such as slope and aspect based on look direction can help reduce these errors by removing flooded vegetation on slopes above a certain grade. Work is still needed on the best approach to integrate digital elevation data to improve flood maps, particularly in areas of flooded vegetation. Factors such as the quality and resolution of the DEM in relation to the resolution of satellite imagery and decisions on whether to incorporate the DEM directly into region growing, to use it to create a processing mask of flat areas where surface water can pool, or to use it in post-processing still need to be considered.

Editing of flood maps is still required to remove either false detections or those that are not of interest for public safety. However, false detections appear to be relatively rare, and those detections that are not of interest to public safety could be relevant to other potential users. Enhancing flood products with ancillary information to target these users may be an avenue worth pursuing to make flood maps of value to a wider audience. For example, flooding is consistently detected in agricultural fields, while visual observations confirm that this is a common occurrence during the flood season. These detections are ephemeral and are currently being removed from flood maps, but they may be relevant to agricultural production as tilling and seeding are delayed by standing water in fields. Agriculture Canada and crop insurance companies may be interested from a crop inventory perspective to know if crops will be planted late, how much and where. The intersection between flood maps and land cover may be sufficient to characterize different types of flooding in terms of longevity and impacts on different stakeholders.

EGS is generally most active in the response phase of the emergency management cycle that includes mitigation, preparation, response, recovery. However, a request was made to generate maximum 2017 flood extent polygons over the activation region to help evaluate flood risk. Due to the nature of emergency mapping, accuracy assessments are performed on a best effort basis within the 4-h period that the EGS has to generate and distribute its flood extent polygons. Nonetheless, accurate maximum flood extent maps are considered a valuable product to the emergency management and land planning communities during the recovery phase. EGS undertook work to generate a maximum flood extent product that spans all of the affected areas that were mapped at least once by our services. To achieve this, the highest resolution products generated closest to the timing of peak flood were selected, reviewed using all available datasets and merged, including urban flood products in Gatineau and along Pierrefonds Boulevard on the Island of Montreal generated from CGI and the LIDAR DEM. Each product was initially assessed separately against very high-resolution satellite imagery (Worldview, Pleiades) medium resolution optical satellite imagery (Sentinel-2, Landsat), crowdsourced field observation (photography and surveys) and oblique aerial photography acquired by NASP before publishing. Each flood extent polygon was edited and its accuracy improved where errors were noted compared to reference data.

A more formal accuracy assessment of the final maximum flood extent product was subsequently conducted using visually interpreted NASP oblique airphotos as a reference. Interpretation was performed by putting placemarks in Google Earth at locations interpreted as either flood or land at the time of the NASP photo acquisition. Because assigning points in open water under normal non-flood conditions would artificially inflate classification accuracy, reference points were located near the flood margin with flooded points interpreted as locations that were inundated in NASP photos but were interpreted as land in Google Earth during baseline, non-

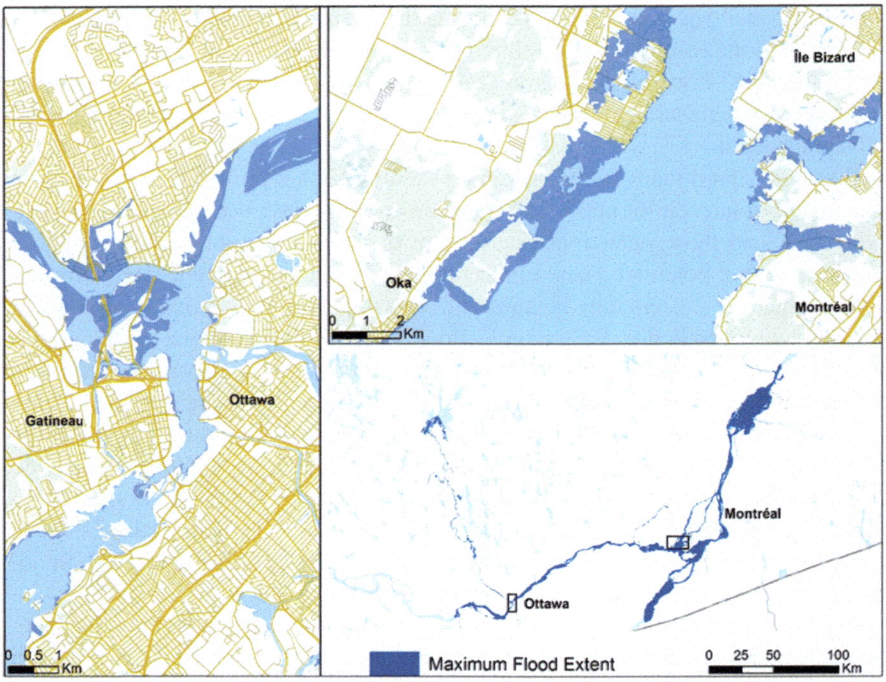

Fig. 12.6 Maximum flood extent of the activation area generated by combining flood maps of different sub-regions

flood conditions. Land points were interpreted just beyond the flood margin as permanent land before and during the flood event. A total of 793 points were collected in this manner from a subset of 1475 photos acquired from west of Ottawa to east of Lac St-Pierre from 7 May 2017 to 16 May 2017. The maximum flood extent is a mosaic of different flood products acquired on different dates that may not correspond precisely to the acquisition dates of co-located NASP. Consecutive dates checked in many locations showed consistent flood margins, suggesting differences in the timing of a few days between NASP and satellite acquisitions likely had a minimal effect on overall accuracy but may have led to error in some cases where flood margins changed significantly over a short period of time. Despite these potential issues, an overall accuracy of 86.2% was obtained for this product using the assessment methodology outlined above. The result is a validated complete and uniform maximum flood extent product shown in Fig. 12.6.

12.3 Conclusions

The 2017 EGS flood activation was considered a success by clients, stakeholders and management at CCMEO. EGS operations adopted new methods to extract open water and flooded vegetation from multiple sensors that were developed the previous year on historical satellite data, which proved to be reliable and robust on a range of sensor data received through the International Disasters Charter. Feedback received from Public Safety Quebec was extremely positive, stating that new products were significantly improved over ones previously generated by EGS during the 2011 Richelieu River activation, particularly in areas of flooded vegetation that were mapped operationally for the first time. Rather than relying on global inundation products from optical data, work on historical inundation frequency integrating radar data should continue for floodplain characterization to better tailor products to the Canadian context and assess the level of optimization that is still possible. Finally, EGS envisions enhancing our flood products through ancillary data integration to further improve them and to make them more relevant to a wider set of users and stakeholders.

References

Bolanos, S., D. Stiff, B. Brisco, and A. Pietroniro. 2016. Operational surface water detection and monitoring using RADARSAT-2. *Remote Sensing* 8: 285. https://doi.org/10.3390/rs8040285.

Ghosh, A., N. Manwani, and P.S. Sastry. 2016. On the robustness of decision tree learning under label noise. *arXiv*: arXiv:1605.06296.

Henry, J.-B., P. Chastanet, K. Fellah, and Y.-L. Desnos. 2006. Envisat multi-polarized ASAR data for flood mapping. *International Journal of Remote Sensing* 27: 1921–1929. https://doi.org/10.1080/01431160500486724.

Hess, L.L., J.M. Melack, and D.S. Simonett. 1990. Radar detection of flooding beneath the forest canopy: A review. *International Journal of Remote Sensing* 11: 1313–1325. https://doi.org/10.1080/01431169008955095.

Latifovic, R., D. Pouliot, and I. Olthof. 2017. Circa 2010 land cover of Canada: Local optimization methodology and product development. *Remote Sensing* 9: 1098. https://doi.org/10.3390/rs9111098.

Li, J., and S. Wang. 2015. An automatic method for mapping inland surface waterbodies with Radarsat-2 imagery. *International Journal of Remote Sensing* 36: 1367–1384. https://doi.org/10.1080/01431161.2015.1009653.

Olthof, I. 2017. Mapping seasonal inundation frequency (1985–2016) along the St-John River, New Brunswick, Canada using the Landsat archive. *Remote Sensing* 9: 143. https://doi.org/10.3390/rs9020143.

Olthof, I., and S. Tolszczuk-Leclerc. 2018. Comparing Landsat and RADARSAT for current and historical dynamic flood mapping. *Remote Sensing* 10: 780. https://doi.org/10.3390/rs10050780.

Pekel, J.-F., A. Cottam, N. Gorelick, and A.S. Belward. 2016. High-resolution mapping of global surface water and its long-term changes. *Nature* 540: 418–422. https://doi.org/10.1038/nature20584.

Townsend, P.A., and S.J. Walsh. 1998. Modeling floodplain inundation using an integrated GIS with radar and optical remote sensing. *Geomorphology* 21: 295–312.

White, L., B. Brisco, M. Dabboor, A. Schmitt, and A. Pratt. 2015. A collection of SAR methodologies for monitoring wetlands. *Remote Sensing* 7: 7615–7645. https://doi.org/10.3390/rs70607615.

White, L., B. Brisco, M. Pregitzer, B. Tedford, and L. Boychuk. 2014. RADARSAT-2 beam mode selection for surface water and flooded vegetation mapping. *Canadian Journal of Remote Sensing* 40: 135–151. https://doi.org/10.1080/07038992.2014.943393.

Chapter 13
The Role of Landsat-8 Multispectral Data in Spill Response: Three Case Studies

Subir Chowdhury, Catherine Evans, and Todd C. Shipman

13.1 Introduction

Over the past three decades, publicly available Landsat multispectral datasets have evolved as a valuable resource for decision makers in various diverse fields such as emergency response, ecological monitoring, agriculture, forestry, land use, water resources, and natural resource exploration (NASA 2013). This chapter exemplifies three case studies in Alberta where Landsat-8 datasets were primarily used to detect early sign of spill and to assess the changing areas of impact over time: a large-scale wastewater spill from a coal mine containment pond due to a structural failure, a release of crude oil emulsion (composed of crude oil and highly saline produced water) from a pipeline leak, and a release of highly saline produced water from a pipeline rupture.

13.2 Case Study 1

On October 31, 2013, a structural failure of the wastewater containment pond in a coal mine, 30 km east of Hinton, Alberta caused 670,000 cubic metres (m^3) of mine wastewater and solids to flow down a local creek and into the Athabasca River downstream of Hinton. Landsat-8 multispectral datasets were used to detect seepage on a suspected location prior to the dam failure. Figure 13.1 shows Landsat-8 imagery acquired before the dam failure on October 25, 2013 and after the dam failure on November 1, 2013. An automated endmember detection and abundance

S. Chowdhury (✉) · C. Evans · T. C. Shipman
Alberta Geological Survey, Alberta Energy Regulator,
Edmonton, Alberta, Canada
e-mail: Subir.Chowdhury@aer.ca

V. Singhroy (ed.), *Advances in Remote Sensing for Infrastructure Monitoring*,
Springer Remote Sensing/Photogrammetry,
https://doi.org/10.1007/978-3-030-59109-0_13

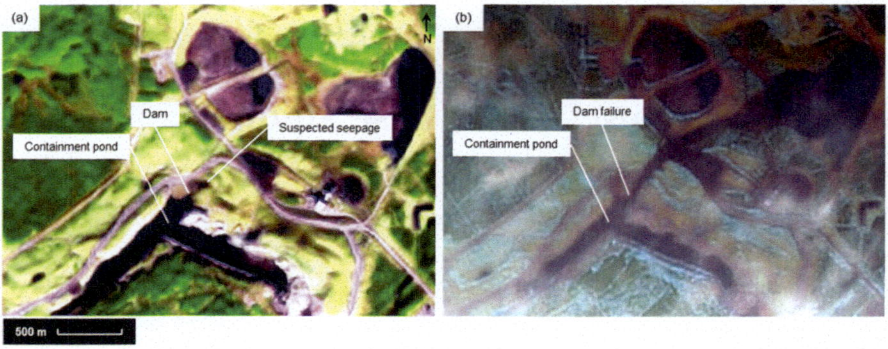

Fig. 13.1 Pan-sharpened Landsat-8 false-colour composite (R: Short Wave Infrared, G: Near Infrared, B: Red) showing the coal mine containment pond and suspected seepage location (**a**) before the dam failure on 2013-10-25 and (**b**) after the dam failure on 2013-11-01 (cloudy [white] and snow cover [cyan])

estimation method, Sequential Maximum Angle Convex Cone (SMACC; Gruninger et al. 2004), was utilized to detect early sign of spill from the coal mine containment pond. SMACC is a convex cone model with constrained oblique projection that automatically extracts image endmembers and offers advantages over highly correlated spectral bands (Harris et al. 2006). SMACC is also a faster and more automated method for finding spectral endmembers compared to the conventional methods such as Pixel Purity Index (PPI) and N-FINDR (Boardman et al. 1995; Winter 1999).

Datasets and Methods

Landsat-8 surface reflectance datasets were obtained from the United States Geological Survey (USGS). Landsat-8 datasets with 30-m spatial resolution and Blue (0.49 μm), Green (0.56 μm), Red (0.66 μm), Near Infrared (NIR) (0.83 μm), Short Wave Infrared (SWIR) 1 (1.65 μm), and SWIR 2 (2.2 μm) bands were used in this study.

Landsat-8 datasets from 2013-08-13, 2013-09-14, 2013-10-16, 2013-10-25 were used as inputs of SMACC to detect endmember materials and to estimate fractional abundances. The number of endmembers was set to 40, where the similar endmembers were coalesced based on the Spectral Angle Mapper (SAM; Kruse et al. 1993) threshold of 0.1 radians. A weakly constrained condition was imposed with the linear mixture model to minimize overestimation or underestimation of abundances (Shang et al. 2008; Chowdhury et al. 2012). Fractional abundances of endmembers present in both containment pond and suspected seepage location were analysed to identify the date of occurrence as shown in Figs. 13.2, 13.3, and 13.4.

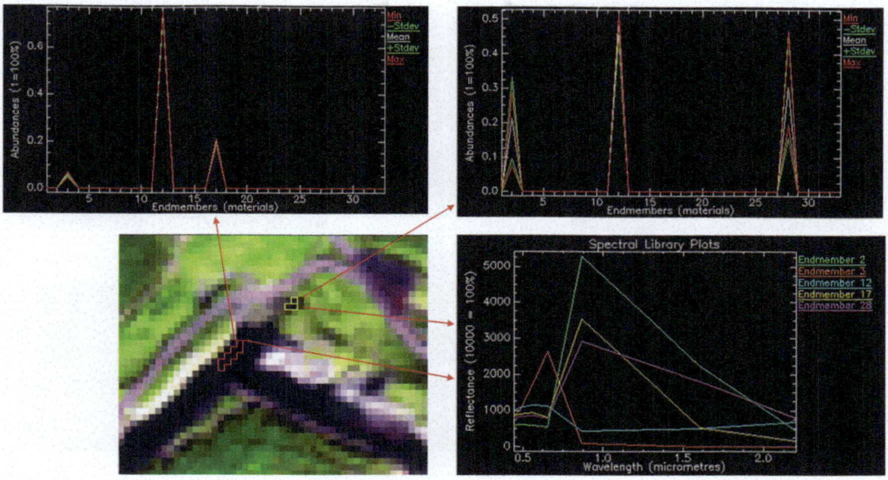

Fig. 13.2 Landsat-8 imagery of 2013-10-25 with selected area of containment pond (red boundary) and suspected seepage area (yellow boundary) showing SMACC estimated fractional abundances of detected endmembers, where Endmember 12 was detected in both areas with up to 0.52 fractional abundance in the suspected seepage location and 0.74 average fractional abundance in the selected area of the containment pond

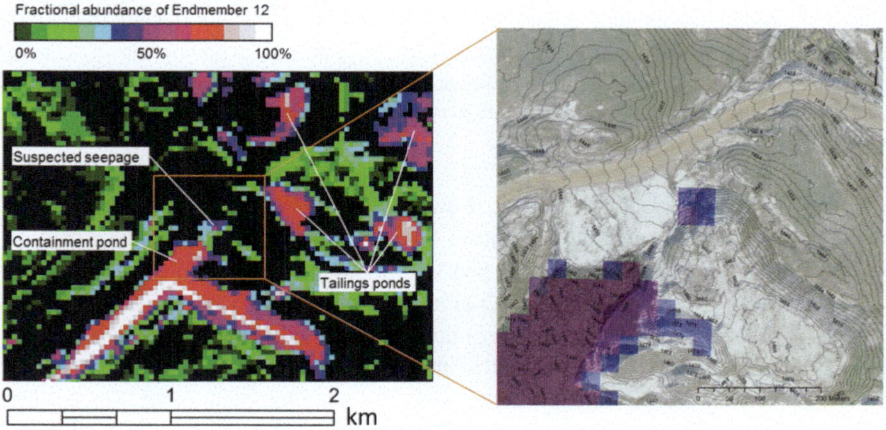

Fig. 13.3 Fractional abundance map showing the proportion of Endmember 12 in the containment pond, suspected seepage location, and other tailings ponds. The inset map shows abundances overlaid on the 2-m elevation (above sea level) contours and a high-resolution aerial photo

Results

SMACC analysis with Landsat-8 data of the coal mining site indicates early sign of seepage on 2013-10-25, with the presence of significant fractional abundances of Endmember 12 on both containment pond and suspected area (Fig. 13.2). The

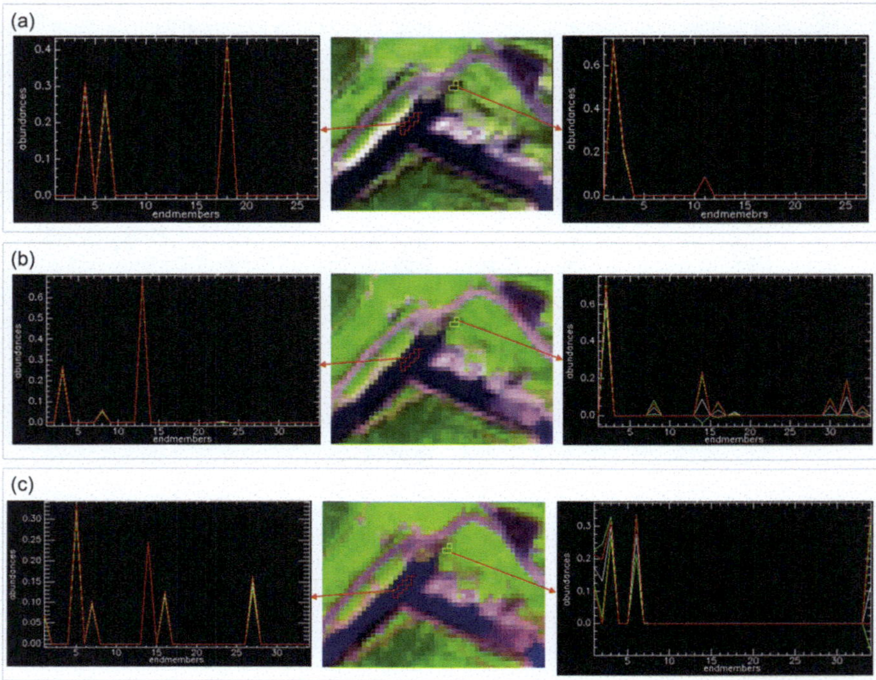

Fig. 13.4 Landsat-8 imagery from (**a**) 2013-10-16, (**b**) 2013-09-14, and (**c**) 2013-08-13, with the SMACC fractional abundances of endmember materials in the containment pond (left) and suspected seepage location (right) showing little to no fractional abundance of common endmember in both locations

spectral signature of Endmember 12 shows an Fe-absorption feature near 0.9 μm that is similar to the generated Landsat-8 spectra of mine tailings described in Mielke et al. 2014. The range of SMACC estimated fractional abundance of Endmember 12 in the 2700 m^2 suspect area was between 0.46 and 0.51 and the one in the 72,000 m^2 selected area of the containment pond was between 0.73 and 0.75, indicating that the 46–51% material in the suspected area is very similar to the 73–75% material present in the selected area of the containment pond. Since other tailings ponds in Fig. 13.3 also show >50% fractional abundances, 2-m elevation contours were used to confirm overflow of the containment pond in the suspect area—these showed a 42 m steep drop in elevation from 1476 m asl (metres above sea level; before the time of the release) to 1434 m asl (after the release). SMACC analysis of 2013-10-16, 2013-09-14, and 2013-08-13 Landsat-8 data found absence of same endmember fractions (>0.05) between containment pond and suspected area as shown in Fig. 13.4. These findings were used to assist the coal mine dam failure incident investigation and prosecution (Alberta Energy Regulator 2019a). Figure 13.5 shows post-remediation state of the coal mine spill site on 2019-05-20 with Sentinel-2 false-colour composite (10-m spatial resolution) and Normalized Vegetation Index (NDVI; Rouse et al. 1974).

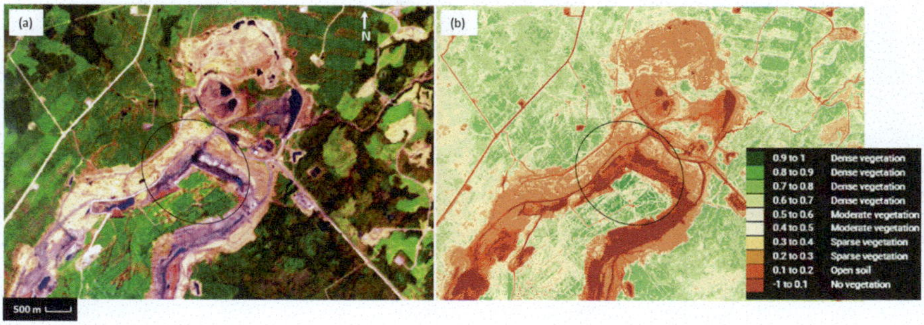

Fig. 13.5 Sentinel-2 imagery of the coal mine spill site (circled), acquired on 2019-05-20 post-remediation, shown as (**a**) false-colour composite (R: B11 [SWIR], G: B8A [NIR], B: B04 [Red]) with (**b**) extracted NDVI (10 m spatial resolution; image courtesy of Land Viewer EOS Platform web interface)

13.3 Case Study 2

On October 6, 2016 release of crude oil emulsion consisting of approx. 50% crude oil and 50% highly saline produced water from a pipeline breach was reported to the Alberta Environmental and Dangerous Goods Emergencies response centre. The release impact area, 15 km northeast of Fox Creek, Alberta was primarily forested peat wetlands, with a flowing creek/beaver fen complex immediately downgradient of the main impact area. When reported, the time elapsed since the start of the release was unknown, but the extent of the impact and the condition of the impacted vegetation suggested that considerable time had elapsed since it began. The release volume was later estimated as 250 m^3. Landsat-8 and Sentinel-2 multispectral datasets were used to find early signs of the release and to establish how the impacted area changed over time. Figure 13.6 shows SPOT-6 imageries (1.5 m spatial resolution) with pre-disturbance land cover on 2015-10-13, stressed vegetation (purple) during the spill on 2016-05-14, and the post-excavation remediation footprint on 2018-10-21. Biomass loss inside the spill boundary was assessed based on the Atmospherically Resistant Vegetation Index (ARVI; Kaufman and Tanre, 1992). ARVI is an improved Normalized Vegetation Index (NDVI; Rouse et al. 1974) that reduces the influence of atmospheric aerosol in polluted areas. The spill material signature and ground-reference data were not available at the time of pipeline rupture, so change detection using the following criteria was used as a proxy for spill detection.

- Surface disturbances are detected multiple times in a chronological order on the same location near the known pipeline break point.
- The extent of surface disturbance expands or remains unchanged over time.
- The intensity of surface disturbance changes over time.

Fig. 13.6 Spill boundary with SPOT-6 true colour composites showing the land cover state before, during, and after the spill on (**a**) 2015-10-13, (**b**) 2016-05-14, and (**c**) 2018-10-21, respectively

Datasets and Methods

Visual analysis of the spill was conducted using the Gram-Schmidt (Laben and Brower, 2000) pan-sharpening of Landsat-8 scenes, where 30-m resolution Red, Green, and Blue bands were pan-sharpened using the 15-m resolution Panchromatic band (0.59 μm).

Change detection maps were produced based on a simple difference in ARVI from initial state to final state, where Landsat-8 scenes with 30-m resolution Blue, Red, and NIR bands were utilized. The level of change detection ranges from −5 to +5, where the range from −5 to −1 represents biomass loss (Chowdhury et al. 2017).

Results

Visual analysis of Fig. 13.7 shows no sign of spill on 2014-06-22, 2014-07-08, and 2014-07-31, early sign of spill on 2015-06-09, 2015-07-18, and 2015-08-28, and appearance of dead vegetation due to spill on 2016-05-17, 2016-07-29, and 2016-08-30.

Landsat-8 imagery from 2014-07-08 was used as a representative state prior to the spill. The gradual increase in the release-impacted area over time was observed in change detection maps. Figure 13.8 shows biomass loss as an indicator of spill from July 2014 to August 2016. The detected release impact area increased gradually from 300 m^2 on 2015-06-09 to 4000 m^2 on 2016-08-30. These findings were used to assist the pipeline spill incident response and investigation (Alberta Energy Regulator 2019b). In addition, Sentinel-2 true colour composite and ARVI (10 m spatial resolution, Figs. 13.9 and 13.10, respectively) show the presence of dead vegetation inside the release area boundary on 2016-05-15, 2016-06-06, 2016-07-27, 2016-09-12, and 2016-06-04; and the footprint from post-spill remediation process on 2017-07-04 and 2019-07-17.

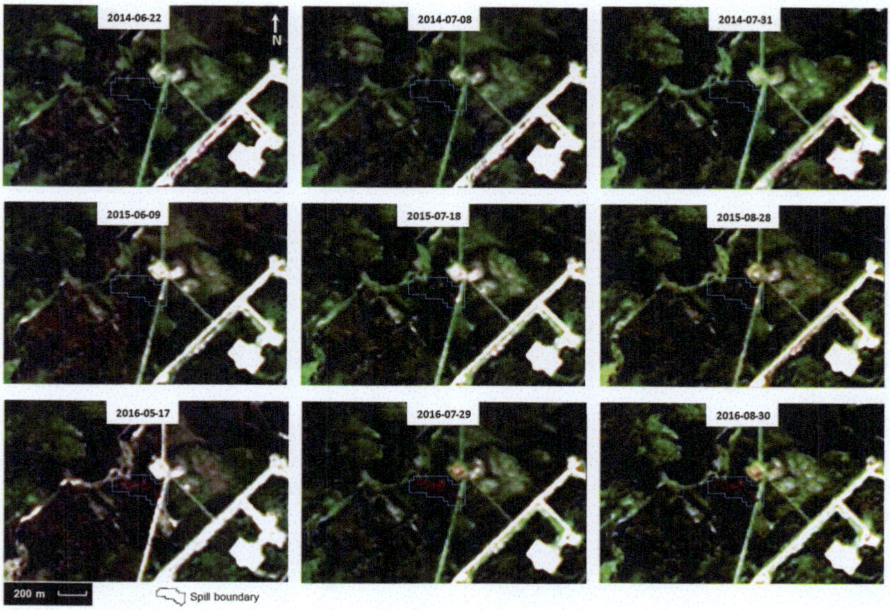

Fig. 13.7 No sign of spill was observed in the Pan-sharpened Landsat-8 scenes from (**a**), (**b**), and (**c**), where early sign of spill was visible in (**d**), (**e**), and (**f**), and dead vegetation resulted from the spill was visible in (**g**), (**h**), and (**i**)

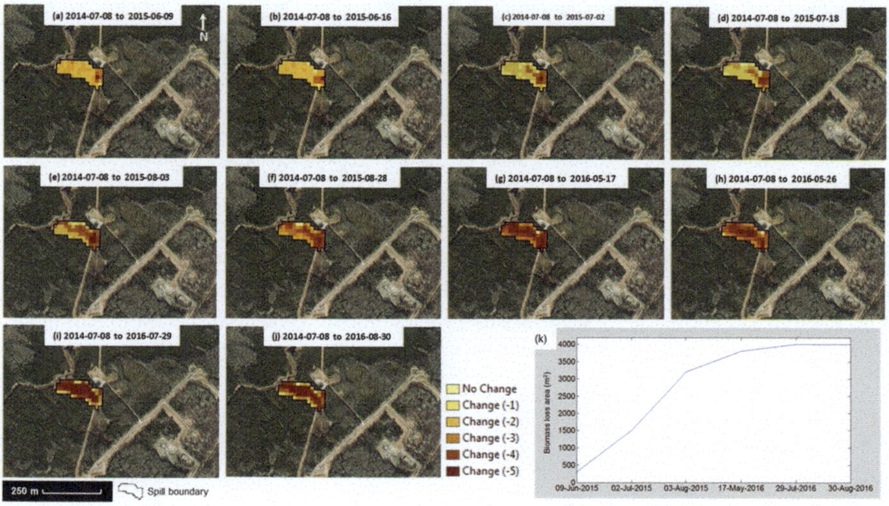

Fig. 13.8 Biomass loss in the spill boundary with change −3 to −5 is observed in (**a–j**), where the release-impacted area over time is shown in (**k**). Change detection results using Landsat-8 data were overlaid on SPOT-6 imagery from 2018-10-21

Fig. 13.9 Sentinel-2 true colour composite shows presence of dead vegetation (purple) in the release area boundary from (**a**) to (**d**) and bare ground/low biomass within the remedial excavation areas in (**e**) and (**f**)

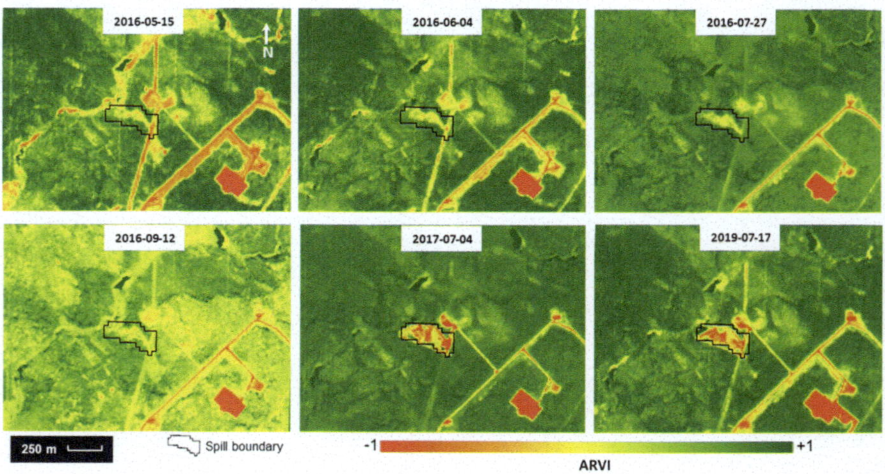

Fig. 13.10 ARVI derived from Sentinel-2 data shows little to no biomass (yellow to red) in the spill boundary from (**a**) to (**d**) and the remedial excavation footprint in (**e**) and (**f**)

13.4 Case Study 3

On June 1, 2013, a release of produced water from a ruptured pipeline was reported in to the Alberta Environmental and Dangerous Goods Emergencies response centre. The volume of the release was later determined to be 15, 363 m^3 of highly saline brines containing traces of dissolved hydrocarbons, which impacted a 42-hectare forest and wetland area 20 km northeast of Zama City, Alberta. The term 'produced water' is used to describe the water that is produced along with the hydrocarbons extracted from an oil- or gas-bearing reservoir. The salinity of produced water varies considerably but can be several times that of the seawater, as it was in this case. The Total Dissolved Solids (TDS) of seawater is usually >35,000 mg/L with a chloride ion concentration of 19,350 mg/L and a sodium ion concentration of 10,760 mg/L (Drever 1997). These produced water brines can be extremely destructive to vegetation, soil, and aquatic environments if released, and releases can often be detected by looking for indications of vegetation death (Government of Alberta 2000). Landsat-8 multispectral datasets were used to estimate the impacts of the spill over time and to ascertain if there were detectable indications of the release before it was reported on June 1, 2013. Figure 13.11 shows an example of the Landsat-8 imagery acquired before the spill on 2013-05-23 and after the spill on 2013-06-24. Changes to the spill footprint over time were estimated from the loss of vegetation water content (dead vegetation/biomass), detected using the Normalized Difference Moisture Index (NDMI; Wilson and Sader 2002). NDMI uses SWIR and NIR bands: NIR is sensitive to internal leaf structure and SWIR is sensitive to leaf moisture content. NDMI is also used for cloudy/hazy scenes because SWIR can penetrate thin cloud/haze. SWIR also differentiates between the moisture contents of soil and vegetation, using the absorbance of leaf moisture and chlorophyll content to detect vegetation. In addition, a potential early indicator of the release was examined using multi-temporal biomass and wetness detection with NDVI and SAM.

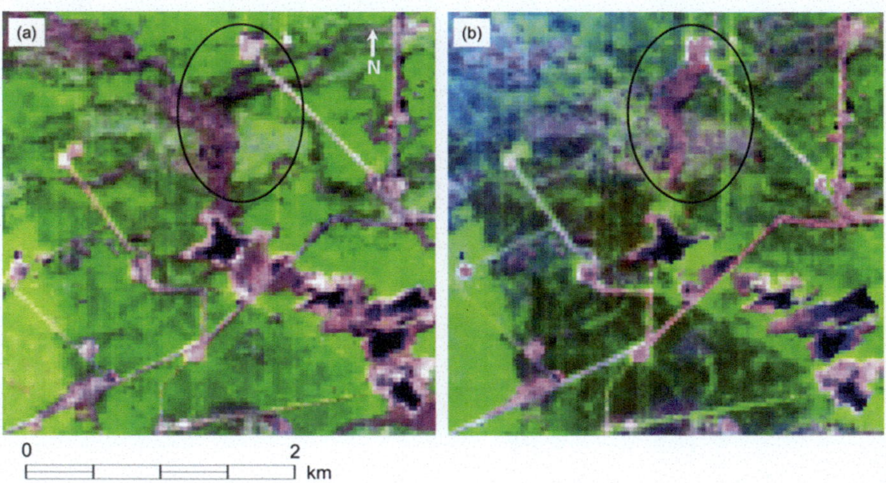

Fig. 13.11 Landsat-8 false-colour composite (R: SWIR, G: NIR, B: Red) showing the pipeline spill site (circled) with (**a**) usual spring water levels before the spill on 2013-05-23 and (**b**) land disturbance due to salt and remediation impacts after the spill on 2013-06-24

Datasets and Methods

NDMI maps were produced using the Landsat-8 datasets from 2013-05-30, 2013-06-08, 2013-06-24, 2013-07-01, and 2013-07-10 to estimate the extent of the pipeline release and to monitor the remediation/clean-up process over time. During the initial containment and clean-up, functional divisions were established by the response teams in order to manage the affected areas. Division 1 included the affected forested area from the ruptured line to the edge of the upper wetland areas. Divisions 2, 3, and 4 were established within the large wetland complex down gradient of the spill to divide the affected wetland areas into zones of decreasing impact, with boundaries set where controls on water flow or key monitoring points

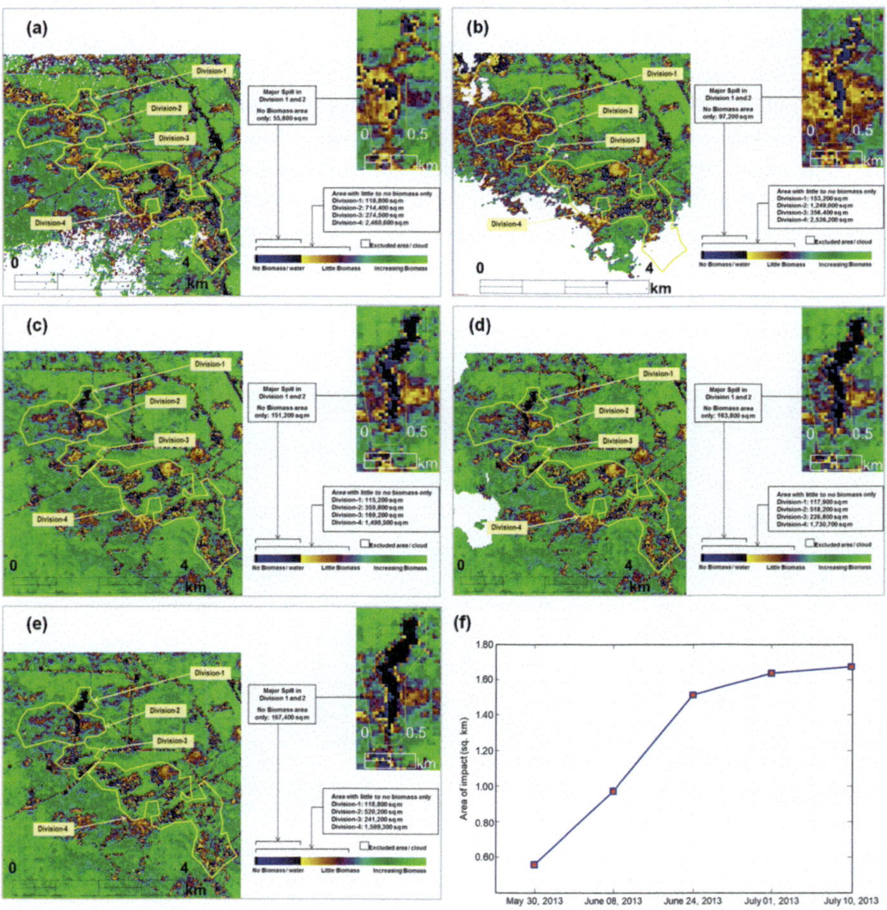

Fig. 13.12 Time-series NDMI from (**a**) 2013-05-30, (**b**) 2013-06-08, (**c**) 2013-06-24, (**d**) 2013-07-01, and (**e**) 2013-07-10 showing estimates of little to no biomass for individual divisions including (**f**) area of impact (no biomass) over time due to release impacts in Divisions 1 and 2

could be established. NDMI ranges from −1 to +1, where NDMI <0 (set through trial and error sensitivity analysis) represents areas with little to no biomass that is identified in the more impacted divisions. In addition, NDMI < −0.5 (set through sensitivity analysis) represents areas with increasing biomass loss over time that was estimated for the major impact areas in division 1 and 2 (Fig. 13.12). A possible early indication of the release was examined with Landsat-8 data from 2010-05-15, 2011-05-09, and 2013-05-23, with wet areas detected using SAM and little-to-no-biomass areas detected using NDVI (Fig. 13.13). A pure water endmember signature was manually selected from the centre of the waterbody in Division 3 and was used in SAM to detect wet areas with spectral similarity threshold of 0.1 radians. Pixels with dark gradient represent little to no biomass with NDVI <0.3.

The size of the release, the terrain (forested upland sloping down to a large wetland complex that included many pocket wetlands connected by stream channels), and the high seasonal movements of water during initial containment and remediation periods meant that managing the release impacts required a complex water management program. Four divisions were established in the impacted areas down gradient from the pipeline break point during the post-spill clean-up process in order to minimize biomass/vegetation disturbance, control further migration of produced water salts, and make effective use of remediation resources. Figure 13.14 shows the containment trench between Divisions 1 and 2 and blocked culverts between Divisions 3 and 4 to prevent or reduce further downstream flow of salt-impacted water. Similar divisions were manually drawn on the time-series NDMI maps as shown in Fig. 13.12 (a) 2013-05-30, (b) 2013-06-08, (c) 2013-06-24, (d) 2013-07-01, and (e) 2013-07-10. Areas with little to no biomass/water were estimated for each division to ensure compliance assurance regarding the blocked culverts and containment trench. The SAM and NDVI results in Fig. 13.12 show spatial similarities in extent of water and biomass contents on 2013-05-23, 2011-05-09, and 2010-05-15; this indicates spring run-off and shows no sign of release impacts prior to 2013-06-01. The area estimates for 2013-05-30, considered as a baseline/pre-disturbance state, show similar spring water levels. A significant increase (534,600 m^2) in little-to-no-biomass/water areas was observed in Division 2 from 2013-05-30 to 2013-06-08. This increase is likely due to

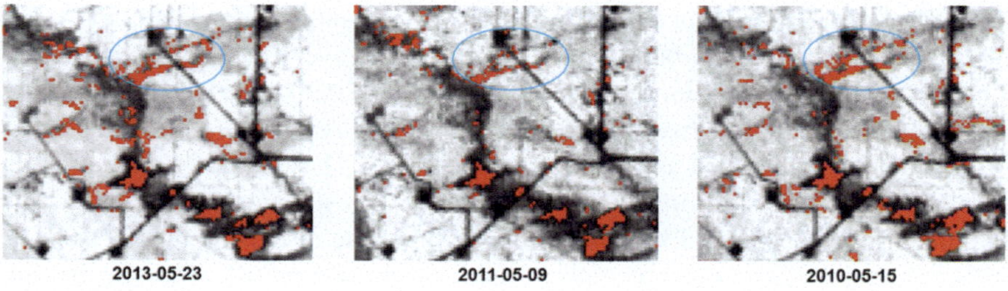

Fig. 13.13 SAM extracted wet areas (red) from Landsat-8 imageries of 2013-05-23, 2011-05-09, and 2010-05-15 are overlaid on corresponding NDVI; white to black gradient represents very high (white) to no (black) biomass; spill site is shown in the blue circled area

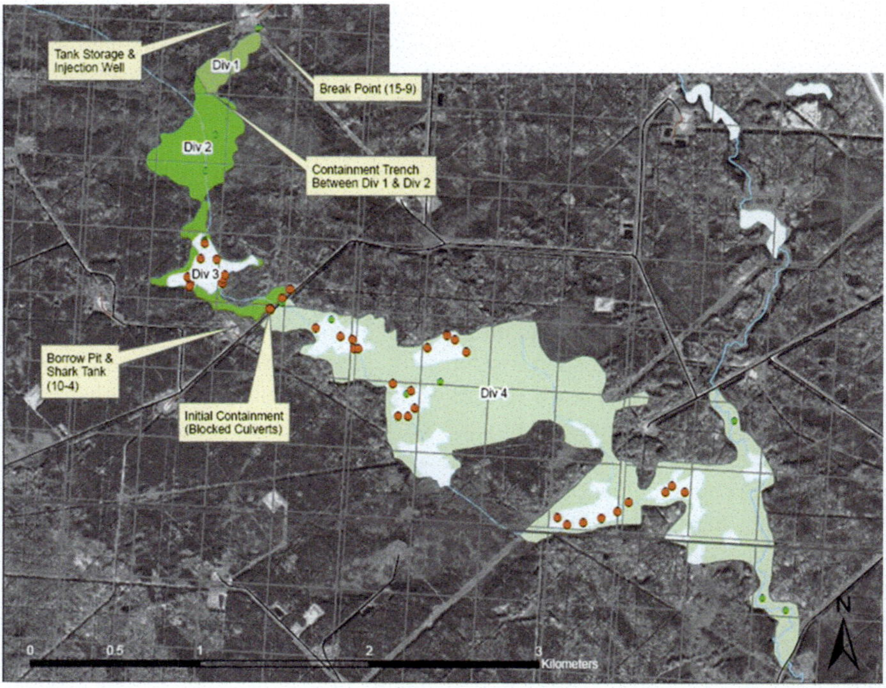

Fig. 13.14 Produced water release incident response map showing the pipeline break point and four divisions (shown as different shades of green) in the wetland system; red and green dots show the water sampling locations in the downgradient divisions; blue lines indicate stream channels and light blue polygons indicate open water areas

vegetation death from salt impacts (Fig. 13.12). Significant decrease (1,037,700 m^2) in little-to-no-biomass/water areas in Division 4 was observed from 2013-06-08, which may reflect an increase in water levels after the culverts were blocked; the south portion of the 2013-06-08 scene with cloud cover was removed from analysis. Fluctuations of the low-biomass area estimates in Divisions 3 and 4 over time do not indicate vegetation loss/biomass disturbance from salt impacts—these are associated with variation in surface water area, flow, and water accumulation. An increasing extent of the no-biomass area was observed over time in Divisions 1 and 2 (Fig. 13.12f), where the total area of impact was 1,120,000 m^2. This live biomass loss is due to salt impacts from the produced water release and the physical impacts of the release containment and remediation activities. These findings were used to assist the pipeline spill incident response and investigation (Alberta Energy Regulator 2019c). Figure 13.15 shows the current state of the site on 2019-05-29 with Sentinel-2 false-colour composite and its derived NDVI. The site is still undergoing long-term reclamation and remediation process with monitoring of natural attenuation (dilution).

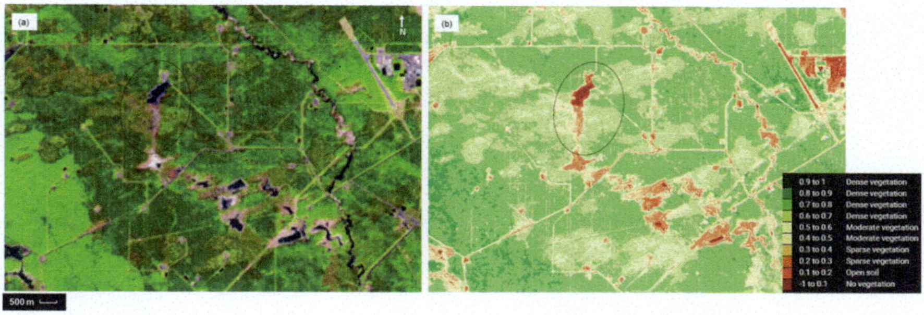

Fig. 13.15 Sentinel-2 imagery of the pipeline spill site (circled), acquired on 2019-05-29, shown as (**a**) false-colour composite (R: B11 [SWIR], G: B8A [NIR], B: B04 [Red]) with (**b**) extracted NDVI (10 m spatial resolution; image courtesy: Land Viewer EOS Platform web interface)

13.5 Conclusions

These case studies demonstrate the operational usage of publicly available Earth Observation data for transparent and efficient compliance assurance regarding spill response and mitigation. The coal mine spill incident investigation (case study 1) using SMACC analysis of Landsat-8 data indicates possible seepage on October 25, 2013 prior to the containment pond dam failure on October 31, 2013. The investigation of a crude oil emulsion release (case study 2) was able to detect early signs of the release in imagery from June 09, 2015 using change detection with ARVI. The release was not discovered and reported until October 06, 2016. For the produced water release investigation (case study 3), NDMI time-series analysis of Landsat-8 data showed how the migration of released salts and physical impacts of remediation affected biomass/vegetation over time, while the SAM and NDVI analysis of Landsat-8 data found no evidence of the release prior to June 01, 2013. These findings were used by the AER incident investigators to establish a starting point for the release events. Data from Earth Observations assisted investigators in reconstructing the incident scenarios, when taken into consideration with other sources of information (e.g. containment pond capacity and overflow factors, water level and precipitation data for the coal mine pond release; pipeline pressure records, corrosion information and pipeline maintenance history for the pipeline releases; statements and other data obtained from witnesses for all cases). Although Landsat-8 has limitation such as the 16-day revisit time and the potential for cloud cover to interfere with the usefulness of some scenes, it still has great potential as an investigation tool. Currently, other publicly available imagery, such as Sentinel-2 multispectral data with improved spatial resolution (10 m) and more frequent imaging (a 6-day revisit time), serves as an additional resource to assist spill incident investigation and to help develop efficient plans for spill prevention, preparedness, and response.

Acknowledgments This study was carried out at the Alberta Geological Survey (AGS), Alberta Energy Regulator (AER) under the Earth Observation (EO) support for incident response project. The authors are indebted to the United States Geological Survey (USGS) and the European Space

Agency (ESA) for providing Landsat-8 and Sentinel-2 datasets, respectively, and are grateful to Cory McLaughlin and Brian Temple from the AER for providing opportunities to assist these spill incident investigations with these datasets. In addition, the authors would like to thank Lisa Atkinson, Dan Palombi, and Andrew Beaton from the AGS/AER for their support.

References

Alberta Energy Regulator. 2019a. Compliance Dashboard, Noncompliance & Enforcement, Reference No. 201706-04. http://www1.aer.ca/compliancedashboard/enforcement.html?searchcol=1&searchstr=201706-04. Accessed 03 June 2019.

———. 2019b. Compliance Dashboard, Investigations, Reference No. 2016-022. http://www1.aer.ca/compliancedashboard/investigations/2016-022_ISR_Paramount Resources Ltd_20180917.pdf. Accessed 03 Sept 2019.

———. 2019c. Compliance Dashboard, Investigations, Reference No. 2013-004. http://www1.aer.ca/compliancedashboard/investigations.html?searchcol=1&searchstr=2013-004. Accessed 03 June 2019.

Boardman, J.W., F.A. Kruse, and R.O. Green. 1995. Mapping target signatures via partial Unmixing of AVIRIS data. In *Summaries of JPL airborne earth science workshop, Pasadena, CA*, 23–26.

Chowdhury, S., J. Zhang, K. Staenz, and D. Peddle. 2012. Spectral mixture analysis of hyperspectral data using Genetic Algorithm and Spectral Angle Constraint (GA-SAC). In *2012 4th Workshop on Hyperspectral Image and Signal Processing: Evolution in Remote Sensing (WHISPERS), Shanghai*, 1–4.

Chowdhury, S., D.K. Chao, T.C. Shipman, and M.A. Wulder. 2017. Utilization of Landsat data to quantify land-use and land-cover changes related to oil and gas activities in west-Central Alberta from 2005 to 2013. *GIScience Remote Sensing* 54 (5): 700–720.

Drever, J.I. 1997. *The geochemistry of natural waters*. Vol. 345. 3rd ed. Englewood Cliffs: Prentice-Hall.

Government of Alberta. 2000 "Interim Salt Contamination Assessment & Remediation Guidelines." Edmonton: Environmental Sciences Division, Environmental Service, 2000. https://open.alberta.ca/dataset/d53c62c1-7dec-4396-aa8a-2a01703d2060/resource/b7bee18b-c7cf-4f85-957d-bcd2dc68a13a/download/2001-saltcontaminationremediationguidelines.pdf. Accessed 15 Oct 2019.

Gruninger, J., A. Ratkowski, and M. Hoke. 2004. The sequential maximum angle convex cone (SMACC) Endmember model. In *Proc. SPIE—Algorithms and Technologies for Multispectral, Hyperspectral, and Ultraspectral imagery X*, vol. 5425, 1–14.

Harris, J.R., P. Ponomarev, S. Shang, P. Budkewitsch, and D. Rogge. 2006. A comparison of automatic and supervised methods for extracting lithological Endmdmbers from Hyperspectral data: Application to lithological mapping in southern Baffin Island, Nunavut. In *Current research, part C, geological survey of Canada, paper 2006–C4*, 1–19.

Kruse, F.A., A.B. Lefkoff, J.W. Boardman, K.B. Heidebrecht, A.T. Shapiro, P.J. Barloon, and A.F.H. Goetz. 1993. The spectral image processing system (SIPS)—Interactive visualization and analysis of imaging spectrometer data. *Remote Sensing of Environment* 44 (2–3): 145–163.

Mielke, C., N. Bosche, C. Rogass, H. Kaufmann, C. Gauert, and M. de Wit. 2014. Spaceborne mine waste mineralogy monitoring in South Africa, applications for modern push-broom missions: Hyperion/OLI and EnMAP/Sentinel-2. *Remote Sensing* 6 (8): 6790–6816.

NASA. 2013. *Landsat: continuing to improve everyday life*. Greenbelt: NASA, Goddard Space Flight Center. https://landsat.gsfc.nasa.gov/wp-content/uploads/2014/09/Landsat_Improve_Life.pdf. Accessed 12 Oct 2019

Rouse, J.W., R.H. Haas, J.A. Scheel, and D.W. Deering. 1974. Monitoring vegetation Systems in the Great Plains with ERTS. In *Proceedings, 3rd Earth Resource Technology Satellite (ERTS) Symposium*, vol. 1, 48–62.

Shang, J., R. Neville, K. Staenz, L. Sun, B. Morris, and P. Howarth. 2008. Comparison of fully constrained and weakly constrained Unmixing through mine-tailing composition mapping. *Canadian Journal of Remote Sensing* 34 (S1): S82–S91.

Wilson, E.H., and S.A. Sader. 2002. Detection of forest harvest type using multiple dates of Landsat TM imagery. *Remote Sensing of Environment* 80: 385–396.

Winter, M.E. 1999. N-FINDR: An algorithm for fast autonomous spectral end-member determination in Hyperspectral data. In *Proc. SPIE Conf. Imaging Spectrom., Pasadena, CA*, 266–275.

Chapter 14
Using EO to Understand the Significance of Cement Production Infrastructure in the Syrian Conflict

Mark H. Bulmer

14.1 Introduction

Minor public protests by the Syrian opposition began almost immediately after the initial protests in Cairo in January 2011 (World Bank 2017a). The first large demonstrations began 2 months later in March, and in the following months escalated as demonstrations spread and increased in size within the country. By the summer of 2011, the armed conflict had started fought between the Ba'athist Syrian Arab Republic led by President Assad supported by domestic and foreign allies against domestic and foreign forces opposing the Assad regime and each other in varying combinations. In 2014, Islamic State of Iraq and Syria (IS) gained control of large parts of Syria and advanced into Iraq. As of early 2020, the conflict remains active but the Assad Government has regained control of much of Syria but not the northern, Kurdish-controlled region[1] part of which has been occupied by Turkish Armed Forces (TAF) and the Turkish-backed Free Syrian Army (TFSA) in Operation Peace Spring (McKernan 2019). Reconstruction is critical to Syria's post-conflict development. The combatants have morphed with IS no longer holding territory, but many longer-term outcomes and political, social, security-related, and institutional effects remain unobservable beyond remote sensing using earth-observing (EO) satellites and anecdotal evidence. In addition, because of the ongoing conflict, the country has remained difficult to access. Given these constraints, the analysis focuses on the effects of the conflict on cement production in northern Syria in four areas: (1)

[1] Also known as Rojava.

M. H. Bulmer (✉)
Roedown Research R2, Davidsonville, MD, USA

Department of Earth Sciences, University College London, London, UK

V. Singhroy (ed.), *Advances in Remote Sensing for Infrastructure Monitoring*, Springer Remote Sensing/Photogrammetry,
https://doi.org/10.1007/978-3-030-59109-0_14

natural resources used in cement production, (2) cement production, (3) cement use in military works, and (4) cement use in reconstruction. Information from actors on the ground and international agencies supplemented remote sensing assessments.

Prior to the conflict, Syria was rapidly urbanizing, with 56% of the population living in urban areas, most of which are in rain-fed agricultural regions, including the basin of the Euphrates River, or along interior trade routes (World Bank 2017a). The two largest cities, Damascus and Aleppo, accounted for nearly 37% of the urban population and 20% of the total population. According to the 2010 Syrian census data, there were 4,128,941 conventional dwelling units across Syria's 14 governorates. Cities like Homs, Aleppo, Damascus, Raqqa, Idlib, and many smaller towns have been battlegrounds for government and rebel offensives. Both civilians, rebels, and IS along with their affiliates, have sought protection in dense urban environments, in steep rugged terrain and underground to overcome barrel bombs used by Syrian Government Forces (SGF) and high technology weapons of the anti-IS coalition (Operation Inherent Resolve militaries) as well as Russia, Iran, and Turkey. Where combatants have sought to hide among civilians in urban areas (World Bank 2017a) this has resulted in both groups seeking protection during fighting on the surface or when aerial and artillery bombardments occur. Bombardments have largely been conventional but have included chemical munitions (CIEDCOR 2016). This has meant that underground spaces such as natural caves and engineered tunnels and facilities have been used for moving and protecting fighters, fighting, hospitals, smuggling, trade, and protecting civilians. In the case of cities like Aleppo, Homs, Damascus, Afrin, Raqqa, Kobane,[2] and Manbij (Fig. 14.1) all of these uses could occur within the same underground space over the course of a single day (Bulmer 2019a, b). As the conflict has continued, it has caused the partial or total breakdown of urban systems in many cities with the destruction of houses and public service-related infrastructure like roads, schools, and hospitals. This has led to economic collapse in many areas. As bridges, water resources, grain silos, quarries, cement production, and other economically significant assets became strategic targets, the physical damage ratios increased (PAX 2015). The longer the conflict continues, the slower the post-conflict recovery will be. The duration of the conflict also affects the rate of recovery. The rate of deterioration has moderated over the course of the conflict as the battlespace has contracted but the effects have become more persistent.

14.2 Cement Industry in Syria

Cement is a high quality, cost-effective building material that is a key component of construction projects throughout the world. Construction of a new dry process cement line represents a significant amount of capital expenditure, depending on the

[2] Ayn Al-Arab is another name for Kabane or Kobane.

location. Beginning in Europe in the 1970s, continuing in the USA during the 1980s, and later in Asia (outside China), the cement industry consolidated worldwide. This has resulted in large international cement companies that include Lafarge (France), Buzzi (Italy), Cemex (Mexico), Cimentos de Portugal SGPS, S.A. (Cimpor, Portugal), HeidelbergCement (Germany), Holcim (Switzerland), Italcementi (Italy), Taiheiyo (Japan), and Votorantim (Brazil). These compete against one another and also against local producers in the various markets around the world. In Syria, a decision to allow private investment in the cement sector was made in 2004 by the Baath Party as part of the countries transition to a 'social market economy', with the first licenses awarded in 2005. This broke the decades-old state monopoly over the cement sector (Lund 2018). A major free trade deal was signed with Turkey in 2007 and northern Syria prepared for a construction boom. Within Syria, Italcementi, Altoun, Lafarge, and Pharoun Groups each have one site with an additional four run by local companies (Fig. 14.1). There are six integrated plants and two grinding

Fig. 14.1 Locations and distributions of cement plants in Syria. Modified from www.cemnet.com

plants.[3] Across Syria, consumption of cement in 2010 was ten million (M) metric tonnes (Mt).

At the start of 2011, Syria was a fast-growing, lower-middle-income country. In aggregate terms, the Syrian economy was improving from an unfavourable base, during the 2000s. Gross domestic product (GDP) grew at an average of 4.3% per year from 2000 to 2010 in real terms, principally driven by growth in the non-oil sectors, and inflation averaged 4.9% (World Bank 2017a, b). The current account was largely in balance and by the end of 2010; Syria was estimated to have international reserves to cover more than 9 months of imports. Foreign investment represented 1.3% of GDP on average during this period and focused mainly on pharmaceuticals, food processing, and textiles (World Bank 2017a, b). In January 2008, Lafarge acquired the cement activity of the Orascom group (Orascom Building Materials Holding S.A.E., Egypt) for $12.8 Bn forming Lafarge Cement Syria (LCS) (Reuters 2010), making it the market leader in the Middle East and Africa (Lafarge 2014). This was backed by President Sarkozy of France who was looking to build a new relationship with Syria. Before being bought by Lafarge, Orascom Cement had signed a deal with the Syrian company MAS[4] to build one of the country's first private cement factories in Aleppo Province. The Syria Cement (al-Jalabiyya) Factory is located in Kharab Eshek Village, south of Kobane in the province of Aleppo, 160 km from Aleppo city and 135 km from Raqqa (Fig. 14.1). When Lafarge took over Orascom Cement, they inherited its joint venture with MAS, which would become known as Lafarge Cement Syria Factory (LCSF). Its production capacity of three million (M) Mt./year was greater than any of Syria's six government-owned cement plants. LCSF began cement production in May 2010. The cement market was largely in the construction sector. On 19 September 2014, the LCSF was evacuated as the surrounding area fell to IS as it advanced to seize Mosul and Tikrit (Fig. 14.2). By late 2014, IS controlled five important cement plants: the LSCF al-Jalabiya plant, the al-Raqqah Guris Cement Plant (1.5 M Mt./year), and the Fallujah, Kubaisa, and al-Qa'im plants in Iraq, with total production of almost 3 M Mt./year. For 2014, the estimated annual income from cement for IS was just under $300 M (Bindner and Poirot 2016). This was reduced three-fold in 2015 to an estimated $100 M with the loss of the LSCF plant and reduced output at the other plants as IS were attacked by anti-IS forces. This decrease would be further accelerated by the loss in March 2016 of the Kubaisa plant (resulting in an annual loss of some $15 M). Between 2010 and 2015, consumption of cement across Syria decreased by 7.8 M Mt. to 2.2 M Mt. This largely occurred from 2013 onward as security decreased and risks related to the transport of cement weakened demand (Taib 2015).

[3] https://www.cemnet.com/global-cement-report/country/syria.

[4] MAS Economic Group through its subsidiaries distributes agricultural products, manufactures metal cans, canning and preserving fruits and vegetables, and real estate development services. The company was founded in 1980 and is based in Damascus, Syria.

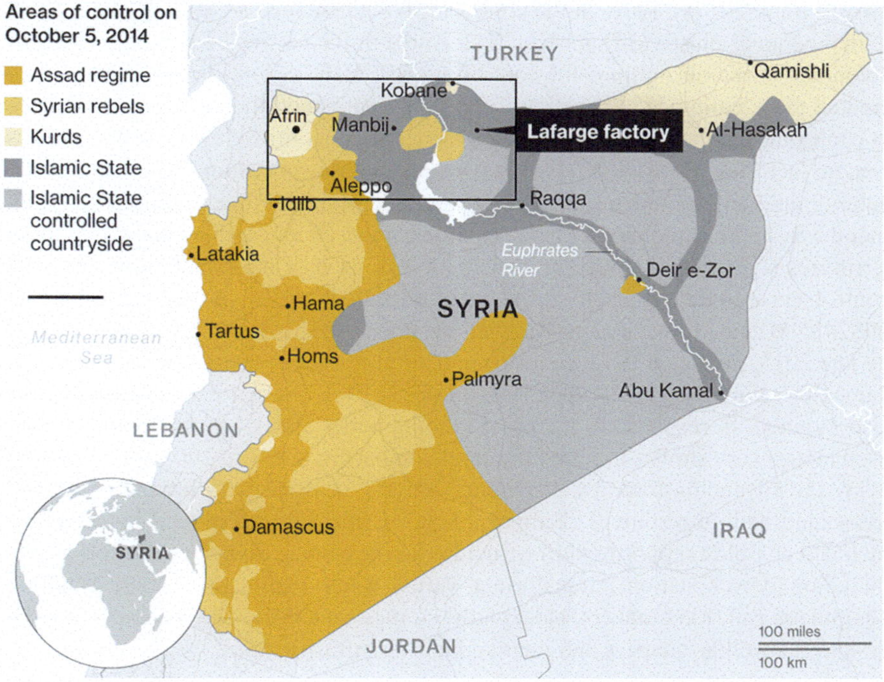

Fig. 14.2 Areas of Syria controlled by different parties to the Syrian conflict on October 5, 2014. Modified from Bloomberg

14.3 Military Use of Natural Resources

Both Syria and Iraq experienced extreme environmental stress from a drought that lasted from 2007 to 2010 (Worth 2010; Gleick 2014; Kelley 2015). This caused large-scale failure of crops and livestock deaths, displacing an estimated 1.5 M people from rural farming communities to urban areas in the south, in search of alternative jobs and livelihoods. This displacement coincided with the arrival of 1.2 M to 1.5 M Iraqi refugees between 2003 and 2007, who settled primarily in the urban regions in the south (UNEP 2003). Water scarcity played a significant but complicated role in creating the conditions that led to the political unrest and violent insurrection in Syria in the spring of 2011 (Mohtadi 2012) that spilled into Iraq (King 2016). The high air temperatures associated with the drought caused an increased demand for fuel and electricity for air conditioning. In Syria, the drought created food insecurity and migration that could not be effectively addressed by the government. This further deepened ethnic and socio-political fractures that IS was able to exploit (Worth 2010; World Bank 2017a). Rural areas in Syria have consistently been poorer than urban centres, and the urban-rural divide has grown since 1997. The poorest part of the country in 2007 was the rural northeast. While 44% of the total population lived in the northeast, it accounted for 56% of the extreme poor

(World Bank 2017a). Early in the civil war SGF abandoned areas in northern Syria to defend areas closer to Damascus. The Kurds in the north slowly created a semi-autonomous region fighting for control against SGF, IS and its affiliates as they created their 'caliphate', as well as Turkey. IS understood the military significance of capturing and then operating, denying or destroying critical infrastructure such as cement factories, electricity, fuel, food, water, and transportation. This included the natural and environmental resources that underpinned them (Bulmer 2018a). Income from the exploitation of commodities, minerals and arable land within these territories (Fig. 14.2), as well as systematic extortion of the populations under its control, ensured financial self-sufficiency for IS. In 2015, 4% of the total IS income of $2435 M came from cement (Bindner and Poirot 2016). Airstrikes, around 10,000 by late 2015, impacted IS revenues from natural resources disrupting transport and damaging refineries and oil pipelines (Bulmer 2019c). Extraction infrastructure was also targeted. In response to IS forces coming under attack by anti-IS forces they used tactics very similar to retreating Iraqi combat engineers in 2003, attempting to delay the advancing force by damaging, denying or destroying infrastructure and associated natural resources (Bulmer 2018a). Critical to the cement industry this included complex supporting infrastructure for the cement manufacturing processes including power stations, transformer yards, water treatment, cranes, handling equipment, rail, road tankers, stockyards, storage tanks, silos, kilns, and conveyor systems as well as quarries and their supporting infrastructure.

Quarries

Cement factories are normally built close to large deposits of raw materials (Taib 2013) and to gain access to long-term quarrying rights. Raw materials for making cement are usually present in limestone, chalk, marl, shale, and clay, and are available in Syria (Syrian Ministry of Industry 1964). Due to favourable geology in the northwest and northeast, quarries are located around Idlib, Aleppo, Manbij, Kobane, and Raqqa. The processes used to recover the rock at a quarry depend upon the geology of the deposit. Often, overburden, not suitable for use as aggregate has to be removed and stored for later use in the restoration process once the extraction has finished. Hard rock deposits will usually be broken out of the ground using explosives, although some weaker rocks can be excavated directly into dump trucks or broken using a 'ripper' attachment on an excavator or on the back of a dozer. The objective is to fragment the rock to a size suitable for loading and transporting to the plant for further processing, and to do it in a way which is safe and minimizes environmental impacts. Once the material has been removed from the ground, it has to be transported to the processing plant via a conveyor belt or dump trucks. Modular, mobile processing equipment enables some or all of the required processing and loading of dump trucks for delivery to take place within the extraction area itself. This equipment is able to follow the extraction area as it develops within a quarry, but can also be moved from quarry to quarry if required. Inhabitants near quarries are subjected to air pollution from fugitive dust,[5,6] noise pollution from trucks and machinery,[7] and the destruction of the landscape.[8] As well as quarries negatively impacting those who live nearby, they have residual detrimental impacts on the environment (European Commission 2009). The most harmful environmental impacts are runoff of chemical pollutants into bodies of water, loss of natural habitats, farmland, and vegetation, and natural resource exhaustion (Wardrop et al. 2001). In Syria, the conflict created a whole new threat, because once quarries were no longer working IS and its affiliates made extensive use of explosives meant for quarrying and the mining and quarry vehicles themselves, to create suicide vehicle-borne improvised explosive devices (SVBIED's).[9] Their size combined with improvised armouring made them difficult to stop (Fig. 14.3).

Water

Dust suppression does not appear to have been standard practice for all quarries in Syria but water was used at LSCF for damping down roads and debris stockpiles, as well as for dust suppression on the actual crushers and screening units and at the debris recycling sites. For site dust management, water trucks are used to spray water and for dust suppression at the crusher and screening units this is provided by sprinklers/nozzles fitted to the input/output points of the recycling process. Once no longer being used for dust suppression, even water trucks were used by IS to make VBIED's (Fig. 14.3d). Table 14.1 shows rates of water consumption, calculated to first order, for crushing operation including site dust management:

Cement factories use demineralized water heated to form steam in the rotary kilns. Water is used in the production processes to cool machinery, or turned into steam to generate electricity for the plant and in boilers used in the raw material processing (Fig. 14.6). Heated water is cooled and recycled requiring on-site water treatment. Reverse osmosis is used to produce demineralized water from a river or

---

[5] PM_{10} tends to settle out within the quarry areas but $PM_{2.5}$ travels into the surrounding areas.

[6] Repeated inhalation of respirable crystalline silica (quartz) may cause silicosis, a fibrosis (scarring) of the lungs. Silicosis is irreversible and may be fatal. Silicosis increases the risk of contracting pulmonary tuberculosis. Some studies suggest that repeated inhalation of respirable crystalline silica may cause other adverse health effects including lung and kidney cancer.

[7] Extraction, loading and unloading, conveying, crushing, screening, loadout, and storage. Typical equipment used at a quarry includes: excavator and hammer, dump trucks, drill machine, compressors, dozers, wheel loaders, fuel tank trucks, and water trucks.

[8] Rock quarrying operations involve extraction of rock through overburden removal, drilling and blasting; and loading and unloading.

[9] In May 2105 IS seized the Khunayfis phosphate mine as well as the one at Al-Sharqiyah in Homs Province. At least six Belaz haul trucks were captured. They captured more of these when they seized the Badush cement plant 20 km NW of West Mosul. IS also controlled production facilities at sulphur extraction plants run by the Mishraq Sulphur State Company 25 km south of Mosul and Syria's main salt mine in Deir Ez-Zor Province.

Fig. 14.3 Range of quarry and mine vehicles used by IS to make SVBIED's in Syria and Iraq. (**a**) A Belaz dump truck used in mining and quarrying with an up-armoured cab driven in Syria August 2015 (Rogoway 2015; Kaaman 2019). (**b**) Dump truck used for moving quarry and mine spoil with an up-armoured cab driven in Anbar October 7, 2015 (Kaaman 2017). (**c**) Front-end loader armoured with steel plates; used as a SVBIED in NW Mosul on 15 March 2017 (Rogoway 2017). (**d**) Water tanker truck SVBIED used by IS in Mosul June 2014 (Kaaman 2017)

Table 14.1 Water consumption per day for small to large crushers calculated using figures from Occupational Knowledge International (2008). Ltr = Litres

Crusher	Water use ltrs/day
Small crusher	2000
Medium crusher	5000
Large crusher	10,000

ground water source such as at LSCF, while distillation or multi-stage flash is used to produce demineralized water from salt water at plants near the coast (Fig. 14.1). IS managed the water usage at the five cement plants they controlled and operated them as part of its management across the 'Caliphate'. Over the period August 2012 to July 2015 in Iraq and Syria, King (2016) identified 44 incidents of water manipulation for political or military advantage. Of these, 21 are attributable to IS showing its systematic use of water to achieve military, economic, and political effect. Control of water has been significant in recruiting both workers and fighters locally and gaining legitimacy.

Cement, Aggregates and Concrete

Concrete is a mixture of cement, water, and aggregates. Prior to the start of the Syrian Civil War, the Syrian Government was keen to develop the stone industry and to attract foreign direct investment. The Natural Dimensional Stones industry was reviewed and assessed in a report funded by European Aid (Damiani and Giovannangeli 2008). A breakdown of the production cost of cement (before distribution and administrative costs) is approximately: energy 30%, raw materials and consumables 29%, labour, maintenance, and other production costs 28%, and depreciation 13%. The cement industry is one of the world's largest emitters of carbon dioxide (CO_2) (Mittelman 2018). Cement is made by crushing and grinding calcium carbonate (limestone), silica (sand), alumina, and iron ore in appropriate proportions and heating the resulting mixture in a rotary kiln to approximately 1500 °C (Lafarge 2014). Common materials used to manufacture cement include limestone, shells, chalk or marl combined with shale, clay, slate, blast furnace slag, silica, and iron ore. These when heated at high temperatures form a rock-like substance that is ground into fine cement powder. Different rock types will produce aggregates with different properties, which will be suitable for a range of different applications. Often a range (or mixture) of sizes are produced, using crushing and screening (Fig. 14.4). The key to achieving a strong, durable concrete rests in the careful proportioning and mixing of the ingredients.

Fig. 14.4 Range of standard stone grades produced by a crusher

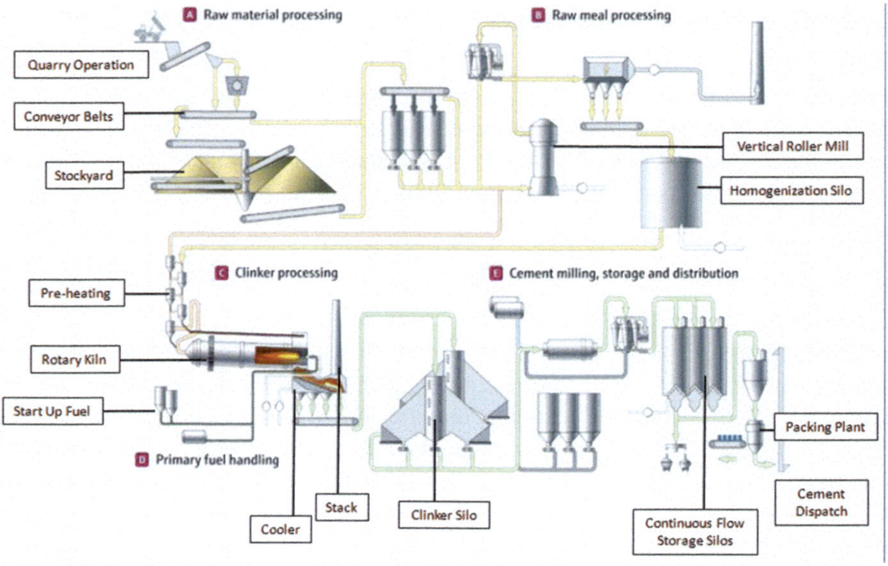

Fig. 14.5 Schematic of cement processing

There is a dry and a wet method to manufacturing cement but each requires raw material processing, raw meal processing, clinker processing with primary fuel handling followed by cement milling, storage, and distribution (Fig. 14.5). In the dry method the quarried rock is crushed to a maximum size of 15 cm, then to 7 mm when it is combined with other ingredients such as iron ore or fly ash. Once mixed this is fed into a pre-heating chamber that consists of a series of near-horizontal cylindrical steel rotary cement kilns mounted on an axis inclined slightly from the horizontal. The ground raw material or slurry enters into the higher end and moves down to a blast furnace burning powdered coal, oil, alternative fuels, or gas. As the material moves through the kiln, gases are driven off (such as carbon dioxide from limestone). When the material reaches the lower part of the kiln it forms clinker. The clinker is red-hot as is emerges from the lower end of the kiln and is brought down to handling temperature using coolers. The heated air from the coolers is cycled back into the kilns. After the clinker is cooled, it is ground in a rotating ball mill where a mix of gypsum and limestone is added to control the setting of the cement. The cement product is then ready for transport. The wet process is alike except the raw materials are ground with water before being fed into the kiln. The slurry's water content is usually between 32 and 36%.

Cement, aggregates, and ready mixed concrete are the finished products for sale. To lower capital costs and where technically available and economically viable, ground blast furnace slag, pozzolan or fly ash may be substituted for certain raw materials when making cement, or mixing slag, pozzolan or fly ash with cement at the end of the process. Ground blast furnace slag is a by-product of steel manufacturing, and fly ash is a product of burning coal in electric thermal utility plants.

Whether and how they are used depends on the physical and chemical characteristics of the slag or ash and on the physical and chemical properties required of the cement being produced.[10] By 2013 as the armed conflict spread, access to slag, pozzolan or fly ash was becoming problematic for all cement factories in Syria.

Lafarge Cement Syria Factory

The LCSF (Jalabiyya 36°32′25.70"N, 38°35′16.12″E) covers 14.6 km² (Fig. 14.6) with quarrying activities, cement factory, and power plant (ELARD 2009). The site was selected due to its close proximity to a limestone deposit (675 ha) situated on the boundary of the cement plant and a basalt deposit (675 ha) 15 km to the north of the plant site. The quarry sites cover 13.3 km² and were state-owned leased from the

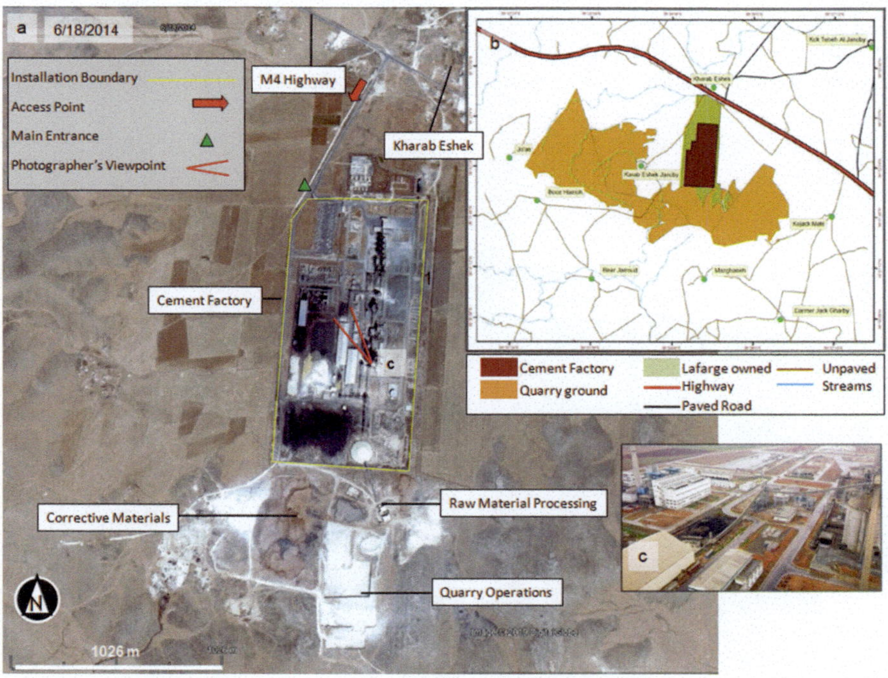

Fig. 14.6 (**a**) WorldView EO image of the Lafarge Syria Cement Plant acquired on 18 June, 2014 four months before IS captured the plant. (**b**) Map showing land owned by Lafarge Cement Syria for the factory, quarry and grounds along with roads and streams. Modified from ELARD 2009. (**c**) View NW over the LSCF from the 80 m tall homogenization silos. The red arrow shows the position of the camera and direction of view. (https://www.glassdoor.co.in/Photos/Lafarge-Cement-Syria-Office-Photos-IMG116658.htm)

[10] Recycling post-industrial material helps to limit CO_2 emissions.

Syrian Government. In 2009, an environmental impact assessment[11] was undertaken with public consultation, looking at the different facilities within the LCSF from raw material quarrying and preparation, through raw processing, cement milling, storage, and distribution (Fig. 14.7). Sources of air emissions were from earthworks (excavation and drilling), machinery and vehicles, cement kiln stack, power plant and combustion facilities as well as open piles. The activities associated with the plant included stripping of the top soil and pollutants entering into soil and groundwater. A feasibility study of water supply was undertaken for the factory based upon a projected initial water requirement of 2000 m^3/day; projected to increase to 4000 m^3/day to meet the future expansion of the facility (ELARD 2009). On the ground, sampling occurred at fuel tanks, waste water treatment plant outlet sites and 1 km downstream of the stacks and quarry sites (ELARD 2009). The soil parameters analysed included pH, clay content (%), organic matter content (%), heavy metals (arsenic, barium, cadmium, chromium (VI), copper, mercury, nickel, lead, and zinc) as well as mineral oil.

Limestone and basalt were quarried using fully mechanized operations comprising of drilling, blasting, loading, and transportation by road to the main crusher to supply about 9500 tons/day of limestone and 2300 tons/day of basalt.

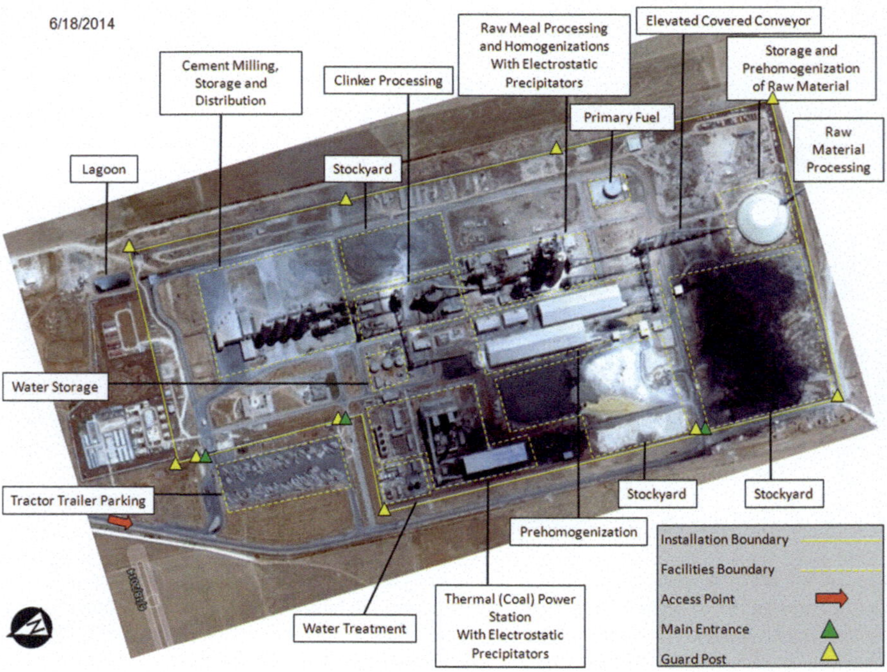

Fig. 14.7 WorldView EO image of the Lafarge Syria Cement Plant acquired on 18 June, 2014 4 months before IS captured the plant. Facilities within the plant are identified

[11] Syrian Environmental Law No. 50 dated 2002.

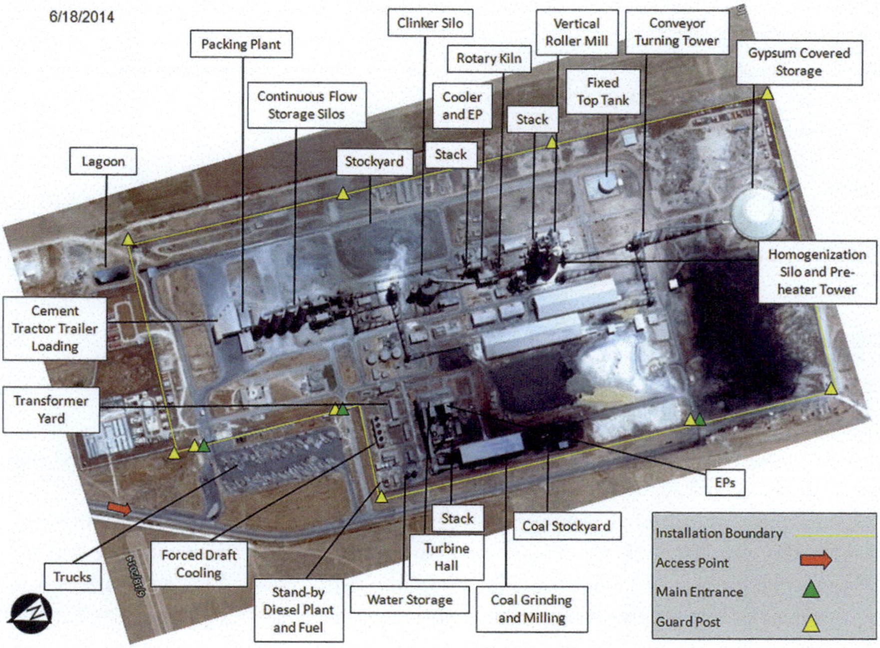

Fig. 14.8 Individual elements within facilities in the plant are identified. WorldView EO image of the Lafarge Syria Cement Plant acquired on 18 June, 2014

Limestone was transported from the quarry to the crusher by off-road dump trucks. Crushed limestone was then transported via conveyor belts to the cement factory. Covered conveyor belts were used to reduce fugitive dust emissions during transportation (Fig. 14.8). Blasting occurred once a week (twice/week at most) with 550 tons of explosive used per month. Ammonium Nitrate mixed with diesel fuel (95%) created ANFO that was combined with Nitroglycerine (5%). ANFO was the main column charge for economic and safety reasons. Additional raw materials required in the cement production were sourced from outside. The annual quantities required to produce 3 M mt/year of cement are shown in Table 14.2.

The main fuel source for the cement plant's kiln and calciner operation (Fig. 14.8) was a mixture of low sulphur (low volatile) bitumen coal and petcoke. The cement plant design was to burn coal or heavy fuel oil (HFO). There is no coal mine in Syria, so the coal was imported either through a Turkish port (Toros or Iskenderun) or through Tartous port in Syria. Due to the size of vessels which can be handled (up to 60 Kt) the preferred option was from Turkey. The origin of the coal was dependent on market conditions from South Africa or Ukraine. Coal from Turkey arrived by truck as there is no railway station close to the plant. The estimated daily requirement of low sulphur coal for the cement plant was 1200 tons, (400,000 mt/year). The only domestic source for petcoke was the Homs refinery (170 kt/year). Imported

Table 14.2 Annual supplies of raw and corrective materials and cement additives needed at LSCF as well as vehicle haulage

Raw Materials	Annual Quantity (mt)	Source of Supply	Distance from Plant km	# of Truck Loads (50 mt Trucks)
Limestone	3,750,000	Koujak and Kharab Eshik Mountains	2	75,000
Basalt	768,240	Kortek Village-Gelabeya area	15	15,364
Pozzolana	700,000	Manakhir quarry	125	14,000
Sand	250,000	Karyateen—Homs governorate	475	5000
Gypsum	150,000	Gureen quarry—Raqqa governorate	170	3000

petcoke would come from US Gulf, Venezuela or India and would be handled like the coal, from Turkey.

The energy required to operate the cement plant at the time of opening was 45 MW/day (ELARD 2009). The same coal for the cement plant was used in the conventional pulverized coal-fired steam plant (Fig. 14.8). The estimated daily requirement of low sulphur coal for the captive power plant (CPP) was 279,725 mt/year, amounting to a total of about 679,725 mt/year. The CPP design used two pulverized coal-fired boilers that generated steam to two steam turbines with one 120 m high stack. Two-diesel oil tanks with a total capacity of 1000 m^3 enabled boiler ignition. Pulverized coal was air-blown into the boiler from fuel nozzles. This burned rapidly, forming a large fireball, heating water circulating through the boiler tubes. Circulating water absorbed the heat forming steam sent to the steam turbine generator for operation. The steam cooled and condensed back into water and returned to the boiler to start the process over again. LCC committed to using the best available technologies for dust and gas emission controls. This included the use of electrostatic precipitators (ESPs)[12] and gas filters in addition to route watering.

LCFS made pre-packaged cement available for pickup at the Jalabiyya warehouses. The plant used four continuous flow storage (CFI) cement silos (Fig. 14.8) enabling simultaneous filling and extracting with a 15,000 mt capacity each. The packing plant used four packing lines with individual capacities of 120 mt/hour. Customers would collect the product using their own drivers and vehicles.[13] In the WorldView EO image acquired 18 June, 2014 the parking area at the northwest entrance contained 127 tractor trailers[14] (Fig. 14.7). The top buyers were construc-

[12] An electrostatic precipitator (ESP) is a filtration device that removes fine particles, like dust and smoke, from a flowing gas using the force of an induced electrostatic charge minimally impeding the flow of gases through the unit.

[13] This meant they often carried large amounts of cash that was later to become a reason for armed groups to stop vehicles to rob them.

[14] A large truck with a long trailer attached to the back of it.

tion companies with distribution networks that could span several cities, though the factory also sold to small, independent customers. The plant employed 1306 personnel while in operation, 85% of them Syrian. When operating at capacity, the LCSF produced cement to fill 160 large trucks each day, ~ 8000 Mt. with a sale value of $500,000. Fifty-kilo (0.0354 m^3) Lafarge cement bags were sold on the market in Manbij, Aleppo, Hassakeh, Qamishli, Kobane, Raqqa, and further afield (Fig. 14.1).

14.4 EO Analysis of Cement Infrastructure

Daily visible images from the moderate-resolution imaging spectroradiometer (MODIS), Visible Infrared Imaging Radiometer Suite (VIIRS), and Operational Land Imager (OLI) passes over Northern Syria were examined for natural and anthropogenic sources of dust (Fig. 14.9) and thermal hotspots associated with fires. The MODIS instrument on the NASA Terra and Aqua satellites captures data in 36 spectral bands ranging in wavelength from 0.4 μm to 14.4 μm. The VIIRS instrument on the Suomi National Polar-orbiting Partnership (Suomi NPP) can collect

Fig. 14.9 MODIS image acquired 7 September, 2015 of a dust source interpreted to be anthropogenic (Ginoux 2012). The white circle shows the location of the Lafarge Cement Syria Factory. A large agricultural region northeast of Raqqa is known as a particularly abundant dust source. In addition, there are several salt flats along the Syria-Iraq border that contribute dust

data in 22 different spectral bands of the electromagnetic spectrum, in the wavelengths between 0.412 μm and 12.01 μm. The Operational Land Imager (OLI) on Landsat eight measures in the visible, near-infrared, and short wave infrared. Resolutions sufficient to observe cement infrastructure on the ground in Northern Syria were examined using EO NASA and ESA satellites such as OLI on Landsat 8,[15] Sentinel-2A,[16] DigitalGlobe World View,[17] and drone footage. Geospatial data were compiled into GIS layers to enable analysis.

Quarries and Factories

Around Aleppo, 54 quarry sites were identified in EO data and compiled into a geodatabase. Identification of quarry machinery as well as quarry dimensional analysis was used to determine activity at the date of image acquisition. Many quarries were also being used to dump fly ash from industrial sites and for trash. Three cement factories were identified in Aleppo city. Thirty four visible-wavelength images were examined over the cement factory on the south side of Aleppo city (Fig. 14.10). Evidence of emission plumes from chimney stacks from 2003 to 2015 was examined to determine their aerial extent and direction (Table 14.2). There are two cement processing lines at the plant and in seven image dates plumes were associated with three stacks related to three near-horizontal rotary kilns at the north cement processing line of the plant (Fig. 14.11). No aerial plumes were identified from 3 July, 2010 to 16 July, 2018 suggesting that the north line of the plant was no longer working. Three plumes were identified on different dates of EO images, each associated with one stack related to two near-horizontal rotary kilns at the south cement processing line of the plant. No aerial plumes were identified from 8 August, 2011 to 16 July, 2018 suggesting that the south line of the plant was no longer working. All plumes were determined to be too small to be visible in MODIS images (Table 14.3).

A Worldview EO image acquired on 8 August, 2011 shows large tractor trailers and rolling stock in the plant siding indicating normal plant operations not observed in subsequent images. The last evidence of quarrying on the east side of the factory is in an EO WorldView image from 8 August 2011. Damage to roofing on the pre-homogenization building and covered conveyors at the plant was first identified in a Worldview image acquired 22 May 2014 (Fig. 14.12). A Worldview image acquired 13 August 2014 shows the main road outside the plant to be blocked. Further damage to fuel tanks and covered conveyors was identified in Worldview image acquired

[15] OLI measures in the visible, near-infrared, and short wave infrared portions of the spectrum.

[16] Sentinel 2-A has a longer repeat period than Landsat8 and uses 13 spectral bands, from the visible and near-infrared to shortwave infrared at resolutions from 10 to 60 m on the ground.

[17] DigitalGlobe's WorldView-2 satellite provides 0.46 m panchromatic (B&W) mono and stereo satellite image data. WorldView-3 satellite provides 31 cm panchromatic resolution, 1.24 m multispectral resolution, 3.7 m short wave infrared resolution (SWIR), and 30 m CAVIS resolution.

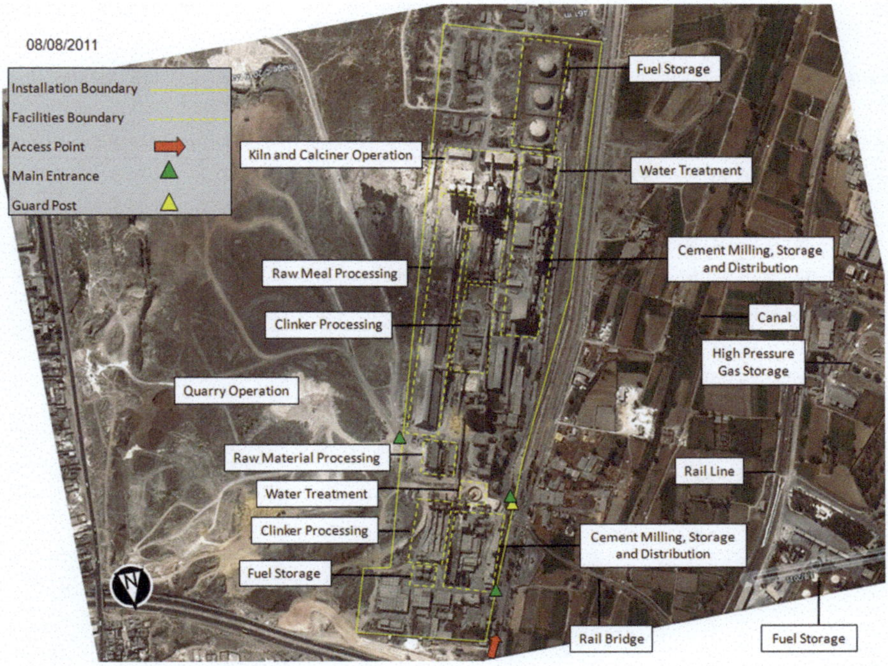

Fig. 14.10 WorldView EO image of the Southern Aleppo Cement Plant acquired on 8 August, 2011. Facilities within the plant are identified

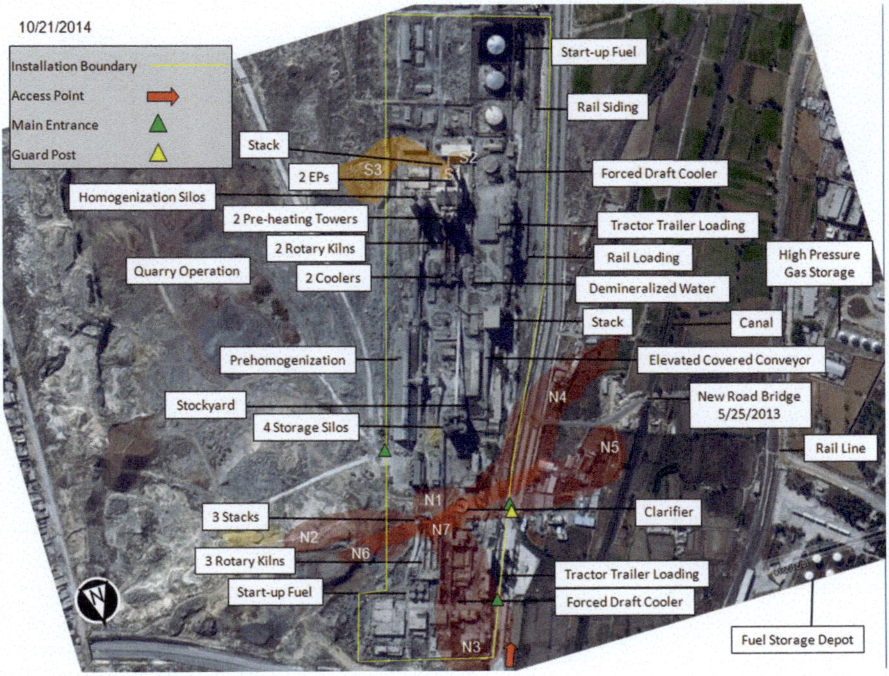

Fig. 14.11 Examination of plumes associated with a cement factory in south Aleppo city from 2003 to 2018 superimposed on a WorldView EO image acquired 21 October, 2014. Individual elements within facilities in the plant are identified. Plumes are associated with three chimney stacks at the north end of the factory and one at the south connected to rotary kilns

Table 14.3 EO analysis of aerial emission plumes at a cement factory in south Aleppo city. P = plume, N is north, and S is south

Site	Latitude	Longitude	Imagery	P Area m^2	P Length m	P Direction
P N1	36° 9′16.94″N	37° 8′41.42″E	9/6/2003	4451	80.2	S
P N2	36° 9′16.94″N	37° 8′41.42″E	8/8/2004	16,016	293.7	E
P N3	36° 9′16.94″N	37° 8′41.42″E	10/12/2006	32,513	308.4	N
P N4	36° 9′16.94″N	37° 8′41.42″E	10/17/2006	31,553	490.2	SW
P N5	36° 9′16.94″N	37° 8′41.42″E	5/29/2007	26,925	398.3	W
P N6	36° 9′16.94″N	37° 8′41.42″E	5/23/2009	5586	194.8	E
P N7	36° 9′16.94″N	37° 8′41.42″E	7/3/2010	376	19.5	Centred
P S1	36° 9′0.59″N	37° 8′36.32″E	7/3/2010	209	18.1	N
P S2	36° 9′0.59″N	37° 8′36.32″E	5/10/2011	508	32.8	E
P S3	36° 9′0.59″N	37° 8′36.32″E	7/31/2011	8631	175.3	E

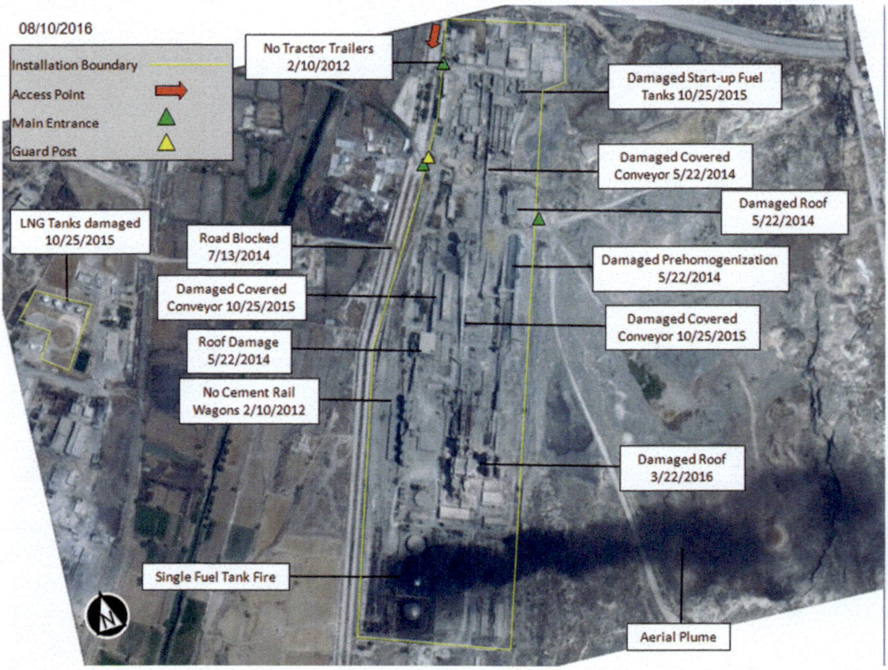

Fig. 14.12 WorldView EO image over the cement plant in south Aleppo city acquired on 10 August, 2016 showing one fuel tank on fire with a plume to the east. Individual elements within facilities in the plant that sustained damage are identified with dates they were identified in images

25 October 2015. The preheater building at the south cement processing line has roof damage in a Worldview image acquired 22 March 2016. A Worldview image acquired on 10 August, 2016 (Fig. 14.12) shows one fuel tank on fire with an aerial plume to the east. A thermal anomaly associated with the fire is visible in Suomi NPP/VIRS. Seen in another Worldview EO image acquired 10 days later (Fig. 14.13a) the same fire spread to the other two fuel tanks. Multiple thermal anomalies associated with the fuel tank fires at the cement plant, as well as at the LNG facility, and oil storage depot to the west, are visible in Suomi NPP/VIRS and MODIS along with an extensive aerial plume that travelled east towards Lake Assad (Fig. 14.13b). In a Worldview EO image acquired on 12 September, 2016 the fires at the tanks had extinguished leaving soot over the surface (Fig. 14.13c). An image dated 23 August, 2018 shows these tanks had not been repaired nor any of the damaged buildings or covered conveyors.

These observations have been compared to official reports. In 2015, the General Organization for Cement and Building Materials (GOCBM), which managed nine state-owned cement companies throughout the country, reported production in only three of its subsidiaries, which were located in relatively safe areas in Damascus, Hamah, and Tartus. The decrease in production was attributed to the deterioration of

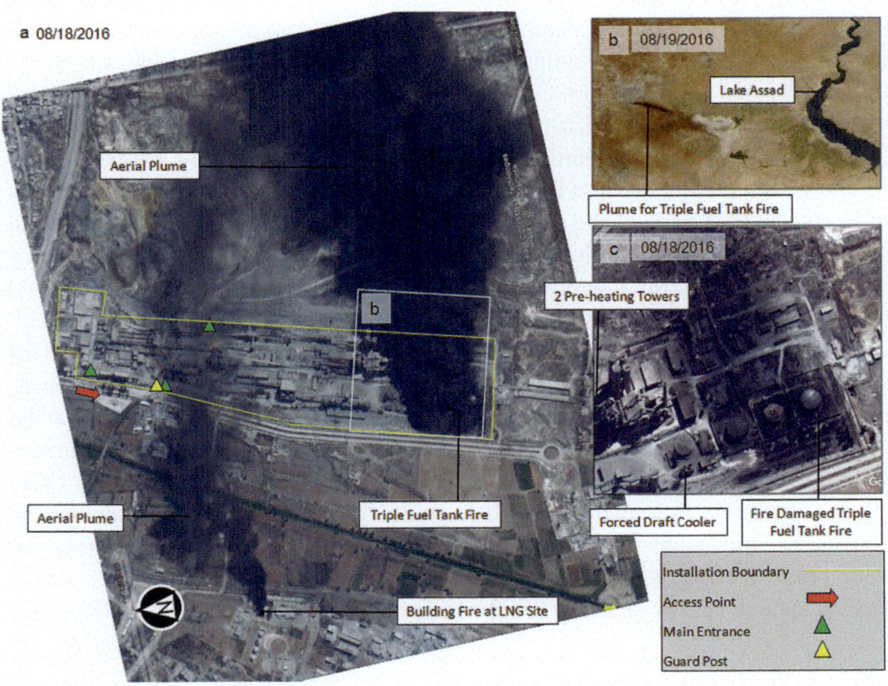

Fig. 14.13 (**a**) WorldView EO image acquired on 18 August, 2016 shows the fire has spread to the other two tanks (Fig. 14.8a). (**b**) A MODIS image acquired 19 August, 2016 showing the plume from the triple fuel tank fire extending east. (**c**) A WorldView EO image acquired on 12 September, 2016 shows the tanks after the fires

security, which forced the closure of six of the company's nine plants (Taib 2015). The Al Badia cement plant,[18] at Abu ash Shamat in southeastern Syria with the capacity to produce 1.6 M Mt./year of cement produced 509,000 Mt. in 2015 compared with 484,000 Mt. in 2014 and 550,000 Mt. in 2013. Production stopped in November 2015 due to fighting in and around the plant area. The Syrian Company for Cement and Building Materials, which owned three plants in Hama Province (Fig. 14.1) with a combined capacity of 2.86 M Mt./year, produced 700,000 Mt. in 2015. Cement output at Tartus Cement and Building Materials Co., which had the capacity to produce 1.8 M Mt./year of cement, was only 570,000 Mt. in 2015. Adra Cement and Building Materials Co., which had the capacity to produce 1.5 M Mt./year of cement, produced 349,000 Mt. in 2015. The GOCBM's combined cement output decreased to about 1.8 M Mt. in 2015 from 2.0 M Mt. in 2014, which was much less than the 5 M Mt./year of cement produced before the armed conflict.

Lafarge Cement Syria Factory

Sixth months after LCSF had started production the protests against the Assad regime started in southern Syria. In August 2011, the USA and European allies called for Assad to step down and the EU imposed sanctions (Portela 2012). These did not include cement so, unlike other international companies that left Syria, Lafarge did not. Fighting started around Aleppo in July, 2012 and the Syrian army withdrew from mostly Kurdish-populated areas including around LCSF but did not leave the plant unguarded until August 2012 when The Kurdistan Workers Party (PKK) gained control because the plant was within the Autonomous Administration of North and East Syria.[19] The Kurdish People's Protection Units (YPG) malitia and Women's Protection Units (YPJ) took over guarding the LCSF which had military significance as a facility that produced cement but also as an installation that could be used to control the M4 Highway running from Aleppo to Raqqa and to Mosul. EO WorldView images acquired between 24 September, 2013 and 22 May, 2014 show enlargement of two quarry sites off the M4, just west of LSFC, and the opening of two new sites covering an area of 0.36 km^2. Investigations into activities at the LSCF at the time have revealed that Lafarge made payments close to \$5 M to armed groups to try to stop the plant being attacked and workers held for ransom (Vely 2017; Jarry 2018; Alderman et al. 2018). In 2012, LCS had signed a deal with the General Establishment of Geology and Mineral Resources, a state body in Damascus, to purchase volcanic ash[20] from a quarry east of Raqqa. In March 2013 anti-Assad rebels captured Raqqa, so a new deal had to be agreed with them (Lund 2018). This

[18] Owned and operated by Al Badia Cement J.S.C.

[19] Also known at Rojava.

[20] Also known as pozzolana. An order for 650,000 tons of volcanic ash worth 216.5 million Syrian pounds (roughly \$3 m at the time) is listed from the al-Manakher quarry east of Raqqa in the 2012 annual report. www.geology-sy.org.

occurred with other suppliers to the LCSF as they were taken over by armed groups (see Table 14.2). During this period LCS worked with a Syrian businessman[21] who had been pro-Assad but then defected and became the chief interlocutor between the government, IS, and the rebels. IS fighters seized LCSF in late September 2014 (Fig. 14.14a, b). During the IS attack the large primary fuel tank for the pre-heater tower and rotary kiln was set on fire (Fig. 14.7) and was filmed for propaganda but retreating YPG/YPJ forces claimed they blew it up to deny IS from taking the oil (Lund 2018). The loss of this primary fuel tank meant that no heating could occur in the cement processing line and that the plant was no longer functional. Once the factory was captured IS constructed additional defensive positions around the

Fig. 14.14 (**a**) View from the approach road to the Lafarge Syria Cement Plant of smoke from the fire at the primary fuel tank in September 2014. Photo from Aron Lund https://twitter.com/aron-lund/status/966600507316989953. (**b**) IS fighter at the LSCP in September 2014. Photo from IS propaganda video (Lund 2018)

[21] Firas Tlass. See Lund 2018 for details.

installation. Silos were used as firing posts, bunkers were dug in the quarries, barbed wire surrounded the perimeter, earthworks were undertaken to protect the gates, and mines were laid around the perimeter and along the roads. The plant workers did not return[22] and the factory was not restarted. Although the LSCF was in IS possession France asked the USA that it not be targeted. YPG/YPJ forces advanced on the factory in February 2015 and captured it on 12 April, 2015. As IS withdrew they blew up parts of the LSCF including the electricity network, and removed some plant equipment to Raqqa for use in making weapons.[23] Eighty percent of the plant was reported damaged (Ara News 2015; International Cement Review 2015). An EO WorldView image acquired 31 August, 2016 (Fig. 14.15a) shows the LSCF with anti-IS coalition helicopters supporting Kurdish YPG/YPJ forces parked next to the

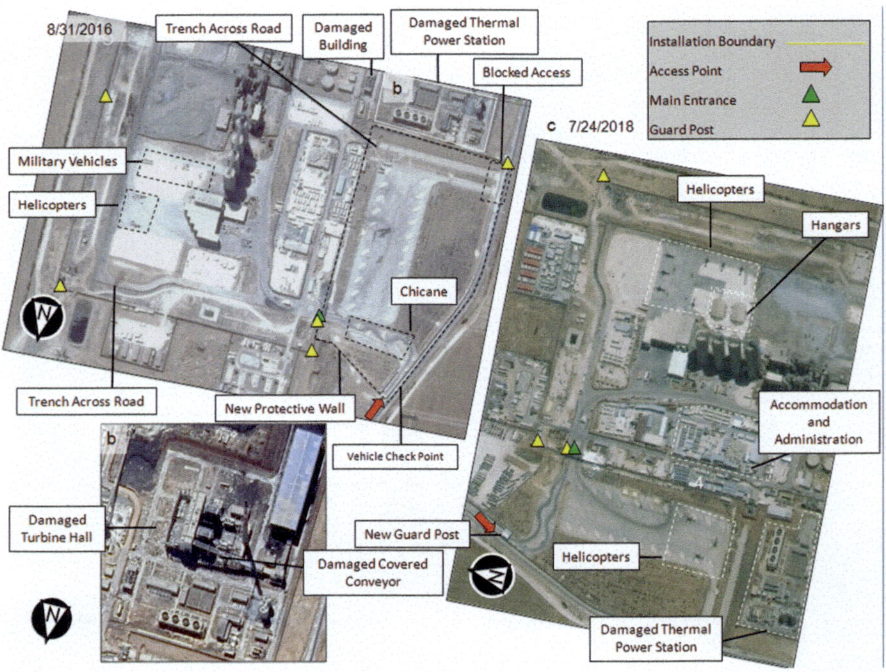

Fig. 14.15 (**a**) WorldView EO image of the Lafarge Syria Cement Plant acquired on 31 August, 2016 after it was captured from IS by YPG and anti-IS forces. (**b**) Enlargement of the conventional pulverized coal-fired steam plant showing extensive damage to the two EPs, boilers and turbine hall. (**c**) WorldView EO image acquired on 24 July, 2018 showing continued use of the LSCP by anti-IS forces. Individual elements within facilities in the plant are identified in (**a**, **b**, and **c**) that were damaged, altered, added or being used for a different purpose

[22] This in no small part was due to IS releasing video of beheading Syrian soldiers captured in Raqqa when they defeated Syrian government forces before advancing on the LCSF.

[23] While this was happening in Syria, Lafarge merged with Swiss company Holcim in 2015 to become LafargeHolcim.

tractor trailer loading facility (Fig. 14.8) along with new force protection measures. The fuel tank (Fig. 14.7) for the pre-heater had not been repaired and extensive damage to the thermal power station is visible (Fig. 14.15a, b). Under the heading of an 'act of war', these damages were not covered by the Lafarge Group insurance policies (Lafarge 2014). An EO WorldView image acquired 27 July, 2018 shows helicopters parked next to the cement truck loading facility, the continuous flow storage silos, and the truck parking area (Fig. 14.15c). Hangers, accommodations and administration areas, as well as improved force protection measures, are also visible showing the LCSF had became a base for operations against IS. In early 2020 the plant has not returned to cement production. Syrian Democratic Forces[24] (SDF and YPG/YPJ) and anti-IS forces had to rapidly abandon the LCSF in October 2019 as TAF and TFSA advanced into Kobane using airstrikes to destroy materiel that could not be moved in time causing further damage to the plant (Sharman 2019). Lafarge has written off the factory and removed it from the company's books (LafargeHolcim 2017).

Border Wall

In 2012, Syrian refugees were attempting to cross the Syrian border into Turkey to flee the advance of SGF against the Free Syrian Army (FSA) in an attempt to block the flow of refugees and supplies for insurgents coming across from Turkey (Oweis and Sezer 2012). After SGF pulled out of Afrin in 2012 (Fig. 14.1), YPG militia took control of the territory. It became Afrin Region, the western-most region of the Autonomous Administration of North and East Syria also known at Rojava. The administrative centre of the region was the town of Afrin. SGF (and possibly Russia), TAF, TFSA,[25] IS and their affiliates all threatened the Afrin Region (Fig. 14.2). In 2016, Turkey began construction of a wall along the border in an attempt to stop IS bombings and then the PKK's /PYD-YPG/YPJ, who had control of the Afrin area, from attacking Turkey (Ozturk 2018). The contract to build 828 km of wall was awarded to TOKI, a Turkish state-owned construction enterprise. Turkey utilized the full resources of the state controlled by President Erdogan in its programme of military works. These military works are visible in EO satellite data at visible (Fig. 14.16a, b), thermal and radar wavelengths and the scale of these works represents significant defensive construction projects (Kaufmann et al. 2011). The wall is composed of 7 tone concrete blocks each 3 m high and 2 m wide (Fig. 14.16c, d, e). By December 2017, 781 km of wall had been constructed requiring 390,500 concrete blocks. In spring 2018, the wall was finished comprising

[24]The SDF was founded in October 2015 and is composed primarily of Kurdish, Arab, and Assyrian/Syriac militias, as well as some smaller Armenian, Turkmen, and Chechen forces led by the YPG. The updated December 2016 constitution of the Autonomous Administration of North and East Syria (NES) named the SDF as the official defence force.

[25]The formation of the National Army was officially announced on 30 December 2017 in Azaz. The official aims of the group are to assist the Republic of Turkey in creating a "safe zone" in Syria.

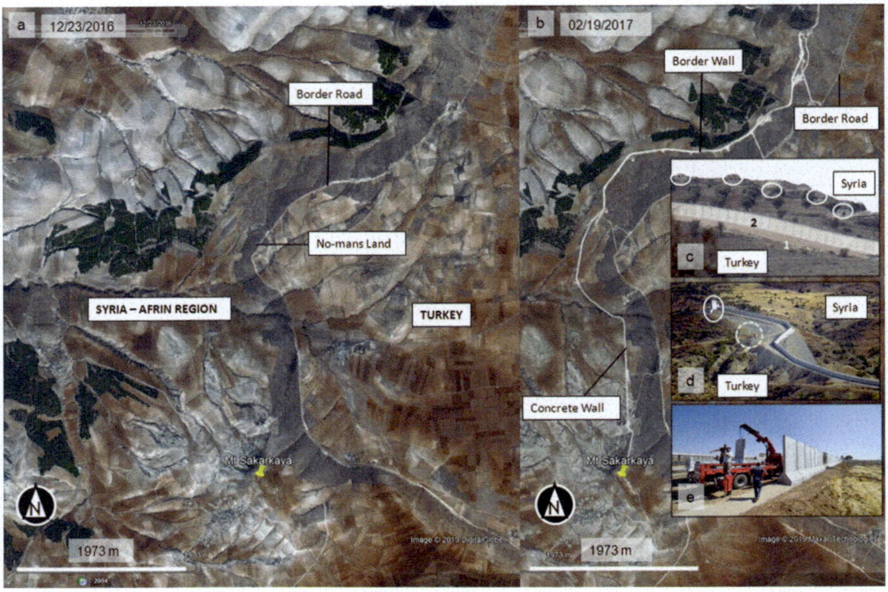

Fig. 14.16 (**a**) WorldView EO image acquired 23 December, 2016 over the border between Syria and Turkey around Mount Sakarkaya in NW Syria. On the Turkish side a road can be seen following the border boundary. (**b**) WorldView EO image acquired 19 February, 2017 over the same area as 'a' showing the new wall along the border. (**c**) A view from the Turkish side across the Syrian border (Turkish News 2016). One shows a security road running along a concrete wall 2. On the Syrian side defensive positions are shown in white circles. (**d**) A view of a section of concrete wall built on a gravel embankment (Ozturk 2018). The white circle shows an automated watch tower overlooks the valley opposite. The dashed white line shows a flat-bed tractor trailer with a crane used for lifting concrete wall sections. (**e**) Flat-bed tractor trailer offloading a concrete wall section (Daily Sabah 2016)

Table 14.4 Total volume of concrete in 414,000 blocks calculated for a range of thicknesses from 0.1 m to 0.5 m. The blocks flare at the base to provide stability (Fig. 14.10e). V is volume m^3 and mt is metric tonnes. t is thickness m

Number of blocks used	V m^3 using $t = 0.1$ m	V m^3 using $t = 0.2$ m	V m^3 using $t = 0.3$ m	V m^3 using $t = 0.4$ m	V m^3 using $t = 0.5$ m
414,000	248,400	496,800	745,200	993,600	1,242,000
	Mt $t = 0.1$ m	Mt $t = 0.2$ m	Mt $t = 0.3$ m	Mt $t = 0.4$ m	Mt $t = 0.5$ m
	597,782	1,195,564	1,793,346	2,398,348	2,988,910
Cost of concrete @$120 /0.76$m^3$	39,221,053	78,442,105	117,663,157	156,884,211	196,105,263
Cost of concrete @$200 /0.76$m^3$	65,368,421	130,736,842	196,105,263	261,473,684	326,842,105

414,000 blocks. To first order, the total volume of concrete used to make these blocks was calculated for a range of thicknesses (Table 14.4).

Comparisons of the metric tonnage required show that with a wall thickness of 0.3 m the demand exceeds the total annual pre-war output of the al-Raqqah Guris Cement Plant (1.5 M Mt./year). With a wall thickness of 0.4 m, the demand would have exceeded the total pre-war output of the LCSF (2 M Mt./year) and at 0.5 m thickness, they would have used the combined pre-war total production of the Fallujah, Kubaisa, and al-Qa'im plants in Iraq (3 M Mt./year). To first order, costs for 414,000 blocks using concrete pricing of \$120 / 0.76 m^3 ranged from \$39.2 M for 0.1 m thickness to \$196.1 M for 0.5 m thickness. This rises to \$65.3 M for 0.1 m thickness to \$326.8 M for 0.5 m thickness calculated using concrete pricing of \$200/0.76 m^3. Included in the border wall construction were 120 border towers (Fig. 14.16d), each 15 m high with a diameter of 7 metres, each computerized with an alarm and automated firing systems (Turkey News 2016).

Trenches, Tunnels, and Underground Facilities

Initially, defensive positions on the Syrian side constructed by the Syrian Kurds on the crest of the mountain slopes (Fig. 14.16c) were much less sophisticated than on the Turkish side but underwent rapid design and material improvements to incorporate lesson learned from weapons systems used by TAF, TFSA, SGF, IS and their affiliates (Bulmer 2020). The YPG/YPJ was a light infantry force with no combat engineers and no airpower to counter TAF, SGF or Russia. To defend against TAF in the Afrin region, along the border with Turkey and in the interior the Syrian Kurds constructed concrete reinforced trenches (Turkey News 2018a), observations posts, tunnels, and underground facilities (UGFs). The aim was to improve survivability against high technology weapons used by TAF (Bulmer 2019a, b). Syrian Kurdish military works were constructed in mountains of limestone, sandstones, and marls and in valleys in alluvial terraces and conglomerates (Syrian Ministry of Industry 1964). Construction of military works in the Afrin region may have been undertaken by PKK, KCK, PYD-YPG/YPJ, and IS all of whom have experience fighting from them (Bulmer 2018b, 2019a, b). Trenches were dug using hand tools and mechanical excavators working along the line of the trench (Bulmer 2020). Trenches excavated in the horizontally layered limestone with marls show them to be coherent enough to support near-vertical walls (Fig. 14.17a). Within machine excavated trenches 5–7 m deep and 3.5 wide, reinforced concrete was used to create cut and cover hardened tunnels (CCHTs) 1.5 m wide, 2 m high with all sides 0.9 m thick that ran for tens, hundreds, and thousands of metres (Fig. 14.17b, and c). These were then covered back over with spoil (cut and cover) to provide additional overhead protection (Fig. 14.17d). In several locations 'double-deck' CCHTs were

Fig. 14.17 (**a**) View in a trench dug by / Syrian Kurds at Mt. Sakarkaya 1 March, 2018. 1 shows the near-vertical trench wall in white limestone with marls reported to be 5 m deep, 2 the trench floor, 3 spoil, and 4 seepage. The dig appears to be fresh with only minimal water seepage and no ponding. (**b**) View of excavated Syrian Kurdish trench at Mt. Sakarkaya (Sarkaya) 1 March 2018 (Turkish News 2018). This was dug using a mechanical excavator in white limestone with marls evidenced by striation marks (6). 1 shows the near-vertical wall with evidence of seepage (4) and ponding (5). 2 concrete floor showing where the walls were to be poured. 7 shows a concrete wall with firing point from a previous trench that has been deepened. An observation post (8) can be seen on the horizon. c View of excavated Syrian Kurdish trench at Sakarkaya (Sarkaya) Hill 1 March 2018 (Murat Kula—Anadolu Agency). 1. Trench walls and floor (2) in white chalky limestone with spoil (3) on the edges. The white oval at 4 shows a tracked excavator in digging position with a two-part hinged arm. The bucket appears to be small (305 mm). 5 Rebar on the walls and floor partially wired into a grid. 6 Form boards ready for concrete pour. d CCHT on Mt. Bursaya (Sozcu News Agency 2018). 1 Concrete wall to the trench, 2 concrete floor, 3 made slope with rocks embedded in concrete, 4 rebar grid on the roof over form boards (5) and shoring (6) ready for the concrete pour

constructed. Observation towers and pill-boxes[26]constructed of reinforced concrete were connected to CCHTs and had ventilation, electrical wiring, and living areas (Bulmer 2020). CCHTs have been identified linked to UGFs with command centres, hospitals, dormitories, weapons, ammunition, and vehicle storage areas reminiscent of World War I (Jones 2010; Doyle 2017). On January 20, 2018 at the start of the Operation Olive Branch, TAF and TFSA advanced across the Syrian border into Afrin. The aim was to defeat the YPG/YPJ and SDF who had been fighting IS with the backing of the anti-IS coalitions but whom Turkey now saw again as 'terrorists'.

[26] Small concrete fort.

Observation towers, pill boxes, CCHTs, and UGFs were the primary targets for Turkish artillery and airstrikes (Bulmer 2020).

Steel Rebar

Using images such as Fig. 14.17 and satellite images such as in Fig. 14.16 it has been possible to determine how Syrian Kurdish trenches were constructued. Once a trench had been excavated rebar was installed and tied on the floor, walls, and lastly ceilings with concrete poured at each stage. To first order, the amount of steel single bar rebar used in the walls (2 m high inside +0.9 m thick for floor +0.9 m thick for ceiling = 3.8 m) and ceilings (1.5 m wide +0.9 m thick for floor +0.9 m thick for ceiling = 3.3 m) along different trench lengths, were calculated (Table 14.5). The spacing of rebar every 0.3 m was based upon examination of images in news reporting and social media with examples in Fig. 14.17a-d. A 300 m long trench is highlighted because one was reported up on Mount Sakarkaya (Turkey News 2018b).

This shows total lengths of single bar rebar for walls, ceiling, and floor range from 2367 m for a 50 long CCHT to 14,200 m for a 300 m reinforced length. One kilometre of CCHT required 47,333 m of single bar rebar. This would have had to be transported using flatbed tractor trailers with either precut lengths or cut to requirement on-site. Table 14.5 was used to calculate first-order costs of 25.4 mm diameter single bar rebar and also a rebar grid-mesh that would have been stronger improving survivability (Table 14.6). The rebar grid-mesh would likely be tighter on the ceiling to provide additional strength against artillery and airstrikes.

Using lengths from Tables 14.5 and 14.6 shows costs of 25.4 mm diameter single bar rebar ($0.75 / 0.3 m) range from $5917 for a 50 m long single bar reinforced CCHT and $35,500 for a 300 m reinforced length. One kilometre of reinforced

Table 14.5 Calculated lengths of steel single bar rebar used in CCHTs observed in Fig. 14.17. *l* = length m

Trench length (*l*) m	Total *l* of single bar rebar for 2 walls along tunnel lengths with rebar every 0.3 m	Total *l* of single bar rebar for floor and ceiling along tunnel lengths with rebar every 0.3 m	Total *l* of single bar rebar for walls, ceiling and floor m
50	1267	1100	2367
100	2533	2200	4733
200	5067	4400	9467
300	*7600*	*6600*	*14,200*
400	10,133	8800	18,933
500	12,667	11,000	23,667
600	15,200	13,200	28,400
700	17,733	15,400	33,133
800	20,267	17,600	37,867
900	22,800	19,800	42,600
1000	25,333	22,000	47,333

Table 14.6 First-order costs of 25.4 mm diameter steel single bar and rebar mesh grid (at $0.75 / 0.3 m) based on the CCHT length ranges from Table 14.5 between 50 m and 1000 m

Cost of single bar rebar for wall, ceiling and floor at $0.75/0.3 m	Total *l* of rebar mesh grid (0.3 m spacing) for 2 walls, ceiling and floor m	Cost for rebar mesh grid (0.3 m spacing) at $0.75/0.3 m for 25.4 mm diameter
5917	4733	11,833
11,833	9467	23,667
23,667	18,933	47,333
35,500	*28,400*	*71,000*
47,333	37,867	94,667
59,167	47,333	118,333
71,000	56,800	142,000
82,833	66,267	165,667
94,667	75,733	189,333
106,500	85,200	213,000
118,333	94,667	236,667

CCHT cost $118,333 for the single bar rebar. The total lengths of rebar required to make a mesh grid with a 0.3 m spacing between vertical and horizontal for a 50 m long CCHT are 4733 m, 28,400 m for 300 m of CCHT and 94,667 for 1000 m of CCHT. Costs of rebar ($0.75/0.3 m) used to make a mesh grid (0.3 m spacing) range from $11,833 for a 50 m long CCHT and $71,000 for a 300 m grid-mesh reinforced length. One kilometre of grid-mesh reinforced CCHT cost $236,667. The calculated demands for rebar require a robust supply chain both for suppliers and flat-bed tractor trailers for delivery.[27] Domestic sources for rebar have all been impacted by the war. In 2010, Syria was a net importer of crude, semi-finished, and finished steel mostly imported from Ukraine and Russia (S&P Global 2013). In 2013, billet[28] producer Hadeed Metal Manufacturing Co planned to commission a new 600,000 Mt./year capacity 8–40 mm diameter rebar mill following the start-up of its 800,000 Mt./year electric arc furnace-based billet plant in 2012. The plan was for rebar to be sold both domestically and exported to regional markets but the war interrupted these plans. The works, in Adra Industrial City near Damascus, sourced scrap locally and also experienced interruptions. EO WorldView images over the factory 29 May, 2012 show construction of large factory buildings completed in the image from 18 August, 2012 but show no aerial plumes associated with production from 2012 to 2018. Syrian rolling mills continued to source billet from Hadeed despite

[27] When power hoisting equipment is not available, bundles of reinforcing bars may be skidded from the truck over timbers or rails extending from the truck bed to the ground. Timbers (dunnage) are normally located on the ground to support and elevate the bundles, and keep them free from mud on the jobsite.

[28] Steel Billets is the second stage product produced in the process of making steel bars. The raw steel cannot be used in its pure form and has to be cast into a shape before being used. The freshly made steel, which is still in the form of a metal bar or rectangle, is called steel billet.

the civil war. In 2013, Hmisho Steel opened a 800,000 Mt./year meltshop in the Hassia industrial area, near Homs that was impacted by fighting. Hmisho's 500,000 Mt./year rebar mill in the port city of Latakia suspended production in January 2013 owing to a lack of billet feedstock. The firm sought suppliers of coke, iron ore, and refractories, and in 2013 had 3000 metric tons of pig iron available for sale. Related to international sanctions, Syria's central bank stopped financing billet imports in early January 2012 by refusing to exchange Syrian pounds into dollars or euros (World Bank 2017a). Nevertheless, re-rollers continued importing by procuring dollars on the black market. Rolling mills in Aleppo, such as Orient Co., ceased operating due to fighting in May 2013, at which time 1 M Mt./year of the total 2 M Mt./year domestic rebar capacity was active. Frequent power shortages restricted Syrian steel output to around 40% of capacity.

Concrete

News reporting, social media, and drone footage from early 2018 in Northern Syria shows concrete lined trenches, CCHTs partially and fully completed. To first order, the volumes of concrete used in the CCHT walls (2 m high and 0.9 m thick) and ceiling/floor (1.5 m wide and 0.9 m thick) were calculated along different trench lengths for two trench wall slabs and also for floor and ceiling slabs over lengths ranging from 50 m to 1000 m (Table 14.7).

Using Table 14.7, the combined volumes of concrete used in the CCHT walls (2 m high and 0.9 m thick) and ceiling/floor (1.5 m wide and 0.9 m thick) along different trench lengths, costs at two price points, along with the number of cement mixers required for transportation, were calculated (Table 14.8).

The total volume of concrete needed to complete two walls, ceiling, and floor for a 50 m CCHT is 315 m^3, for 300 m is 1890 m^3, and for 1000 m is 6300 m^3. The costs

Table 14.7 Calculations of concrete volumes required in the CCHT walls (2 m high and 0.9 m thick) and ceiling/floor (1.5 m wide and 0.9 m thick) for length ranges from 50 m and 1000 m. *V* = volume m^3

Concrete wall slab length m	*V* m^3	2 slabs *V* m^3	Floor and Ceiling Slab length m	*V* m^3	2 slabs *V* m^3
50	90	180	50	67.5	135
100	180	360	100	135	270
200	360	720	200	270	540
300	*540*	*1080*	*300*	*405*	*810*
400	720	1440	400	540	1080
500	900	1800	500	675	1350
600	1080	2160	600	810	1620
700	1260	2520	700	945	1890
800	1440	2880	800	1080	2160
900	1620	3240	900	1215	2430
1000	1800	3600	1000	1350	2700

Table 14.8 Combined volumes (m^3 and mt) of concrete needed to complete CCHTs with two walls, floor, and ceiling over lengths ranging from 50 m to 1000 m along with required costs and the number of cement mixers needed for transportation

2 wall and 2 ceiling/floor concrete slab lengths m	V m^3	mt	Cost of concrete @$120/0.76 m^3	Cost of concrete @$200/0.76 m^3	Number of cement mixers each 6.1 m^3
50	315	758	49,737	82,895	52
100	630	1516	99,474	165,789	103
200	1260	3032	198,947	331,579	207
300	*1890*	*4548*	*298,421*	*497,368*	*310*
400	2520	6064	397,895	663,158	413
500	3150	7580	497,368	828,947	516
600	3780	9096	596,842	994,737	620
700	4410	10,612	696,316	1,160,526	723
800	5040	12,128	795,789	1,326,316	826
900	5670	13,645	895,263	1,492,105	930
1000	6300	15,161	994,737	1,657,895	1033

of concrete range from $49,737 for 50 m length of CCHT, $298,421 for 300 m length, and $994,737 for 1000 m length at a price of $120/0.76m^3. At a concrete cost of $200/0.76m^3, 50 m of CCHT cost $82,895, 300 m costs $497,368 and 1000 m $1.65 M. Reports in the Turkish media were that the concrete used by IS, the Syrian Kurds to make trenches, bunkers, and tunnels was being supplied from the LCSF (Ahval News 2018). Discussion above (Sect. 4.2) of events at the LCSF combined with EO image analysis (Figs. 14.14 and 14.15) has shown that by 2018 the installation had not been operational since October 2014 refuting the claim of Turkish media.

The three closest large cement factories to Afrin are in Aleppo city (Fig. 14.2). The cement factory on the south side of Aleppo has been discussed previously (Sect. 4.1 and Figs. 14.10, 14.11, 14.12, and 14.13). Two, 4 km of each other on the north side of the city, are 6 km from the Aleppo Industrial City—Sheikh Najjar. An EO WorldView image acquired 30 January, 2014 shows the cement plant west of the Aleppo Industrial City—Sheikh Najjar (36°18′15.92"N, 37°10′34.13″E) which was connected by road and rail (Fig. 14.18). This has one cement production line with an oil-fired power station. The cement factory was targeted in May 2014 and one fuel tank was set alight resulting in an aerial plume (Fig. 14.19). Damage to roofs on three buildings where munitions went through is visible in EO WorldView images acquired June 2014. Damage to the roof of the power station is visible in EO WorldView image acquired 17 September, 2014. An EO WorldView image acquired 21 July, 2014 shows the second fuel tank alight. The loss of these primary fuel tanks meant that no heating could occur in the cement processing line and that the plant was no longer functional. Damage to one of the water tanks is visible in an EO WorldView image acquired 14 December, 2014. The same image shows new fighting trenches 3 km south of the factory. An EO Worldview image acquired 21

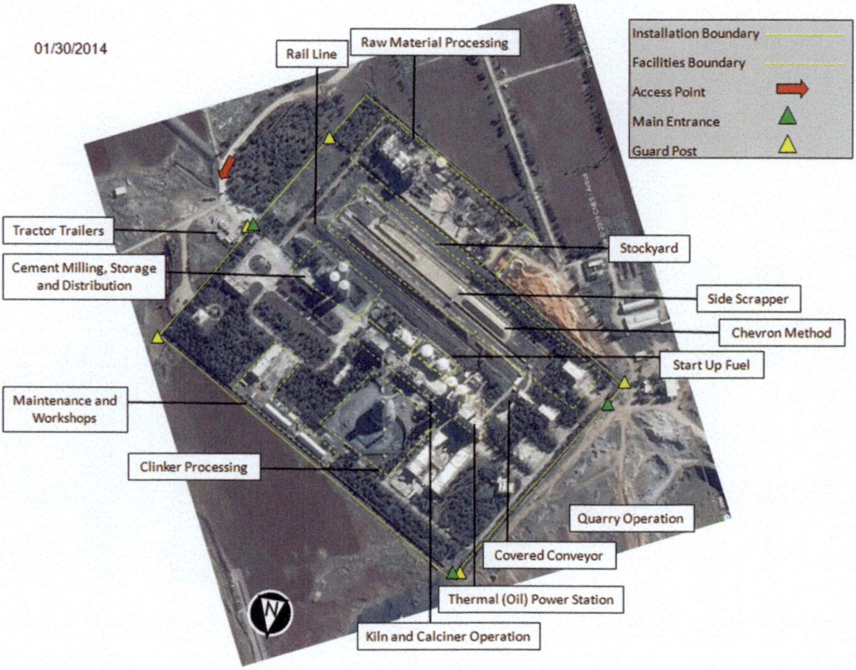

Fig. 14.18 WorldView image acquired on 30 January 2014 over a cement plant on the north side of Aleppo city and west of Aleppo Industrial City—Sheikh Najjar. Facilities within the plant are identified

November, 2015 shows these trenches had extended and were freshly dug around the southern side of the plant indicating that fighting continued. The first repair work on one roof was observed in an image acquired 19 February 2017, with further repairs to three others visible in an image acquired 13 August 2017. Repair to the damaged water tank is also visible in this image. Comparison of two EO WorldView images between 9 February, 2018 and 26 March, 2018 shows that the two damaged fuel tanks were replaced. There is no indication that the plant was restarted but repairs were being carried out. In EO WorldView images, the quarry on the west side of the plant was last active on 5 October, 2011 but continued to be used for periodic small-scale dumping.[29]

The larger nearby cement plant north (Fig. 14.20 of the Aleppo Industrial City—Sheikh Najjar (36°20′18.85"N, 37°12′33.07″E) was connected by road and rail (Fig. 14.20). This had one cement production line with two near-horizontal kilns and another line with one kiln. One production line was for the dry process and the other for the wet process. All were powered by fuel oil and electricity plus steam

[29] A large number of new piles appear in the western quarry in an EO WorldView image acquired 28 January, 2018.

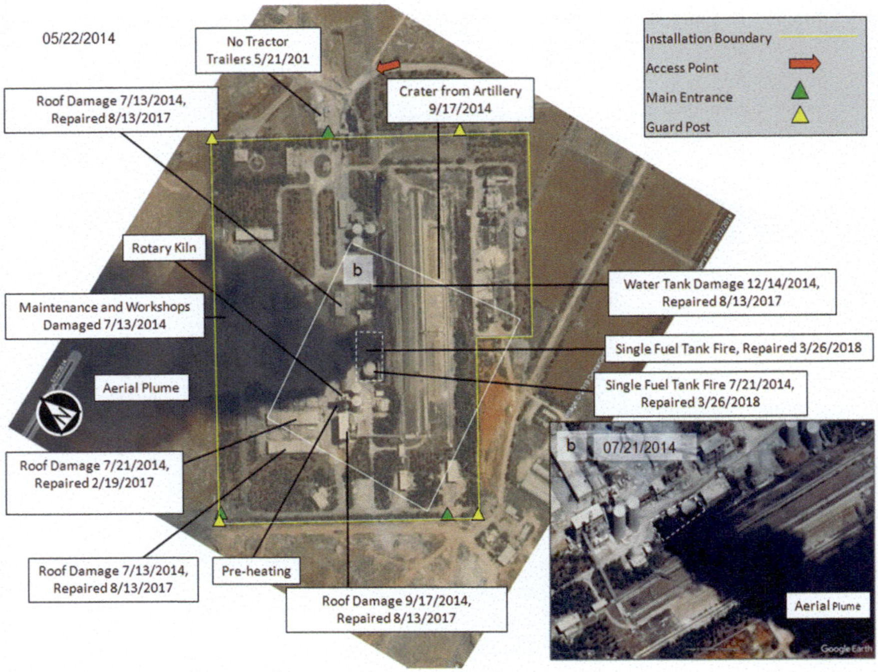

Fig. 14.19 (**a**) WorldView image acquired on 22 May 2014 over the same plant in Fig. 14.18. Individual elements within facilities in the plant are identified along with damage sustained in the conflict with dates they were identified in images. (**b**) WorldView image acquired on 21 July 2014 show two fuel tanks on fire

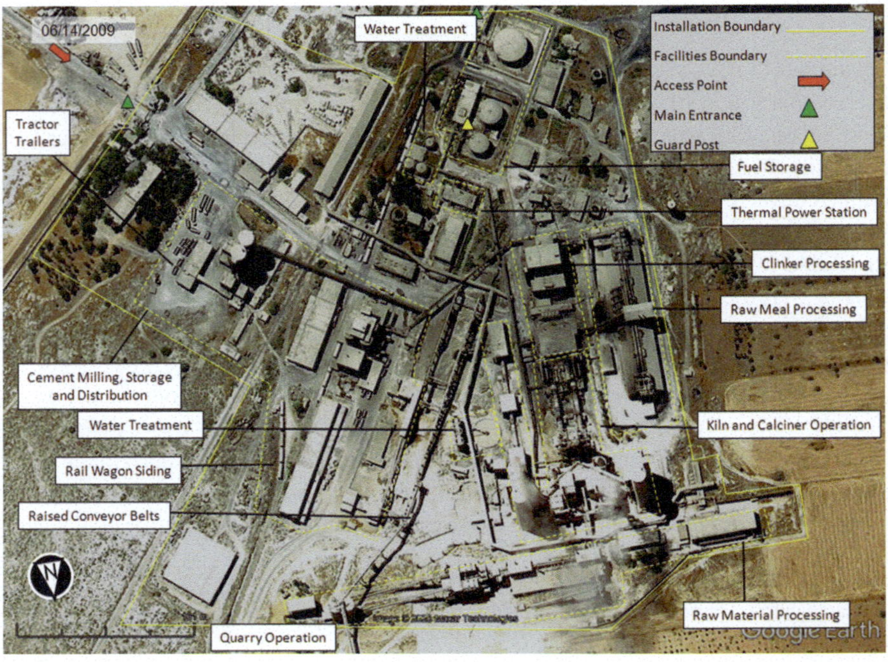

Fig. 14.20 WorldView image acquired on 14 June 2009 over a cement plant on the north side of Aleppo city and north of Aleppo Industrial City—Sheikh Najjar. Facilities within the plant are identified

was produced at an on-site oil-fired power station. The plant appears to have been damaged over successive periods of fighting (Fig. 14.21a). An EO image from 4 June 2012 shows no tractor trailers or rail wagons loading in the plant compared to February that same year (Fig. 14.21a, b) indicating an interruption to normal plant operations. Damage to a workshop building can be identified in EO images acquired 8 June, 2013, to the prehomogenization building plus raw material processing buildings acquired 22 November, 2013, and to maintenance buildings acquired 22 May, 2014. Damage to the prehomogenization building meant that no raw meal processing could occur in the cement processing line and that the plant was no longer functional. An EO image from 29 June 2015 shows that three fuel tanks had been targeted and destroyed (Fig. 14.22a). The two cement production lines had been severely damaged with one near-horizontal kiln no longer in place, the other two in pieces, and the two pre-heater towers partially collapsed (Fig. 14.22b). An EO image from 28 March 2016 shows one of the pairs of homogenization silos collapsed and one of the other pair damaged. Examination of EO WorldView images over the quarry on the north side of the plant show no activity after 25 February 2013. An EO WorldView image acquired 23 July, 2018 shows that unlike at the

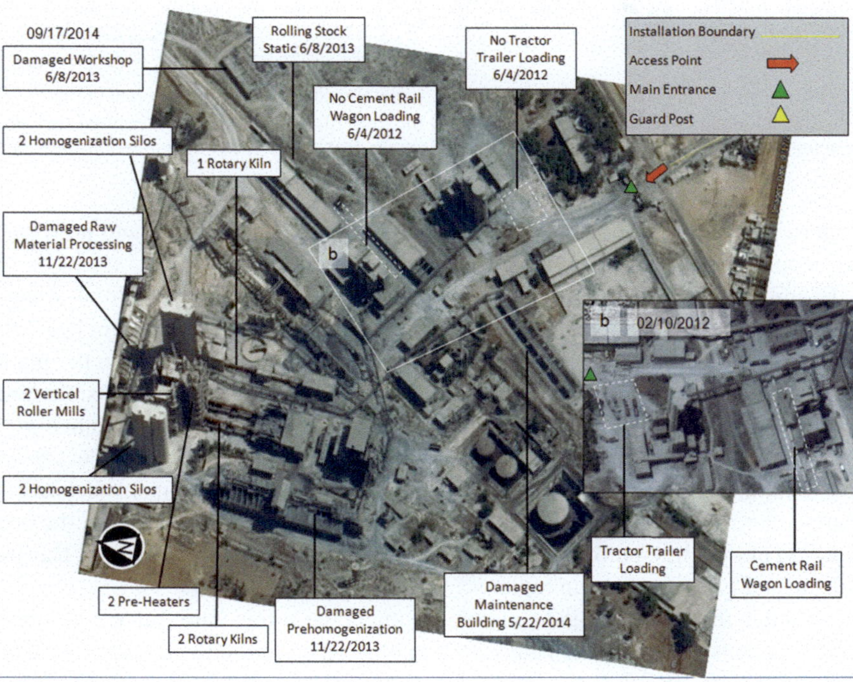

Fig. 14.21 (**a**) WorldView image acquired on 17 September 2014 over the same plant in Fig. 14.20. Individual elements within facilities in the plant are identified along with damage sustained in the conflict with dates they were identified in images. (**b**) WorldView image acquired on 10 February 2012 over the cement milling, storage, and distribution showing the presence of tractor trailers and cement rail wagons

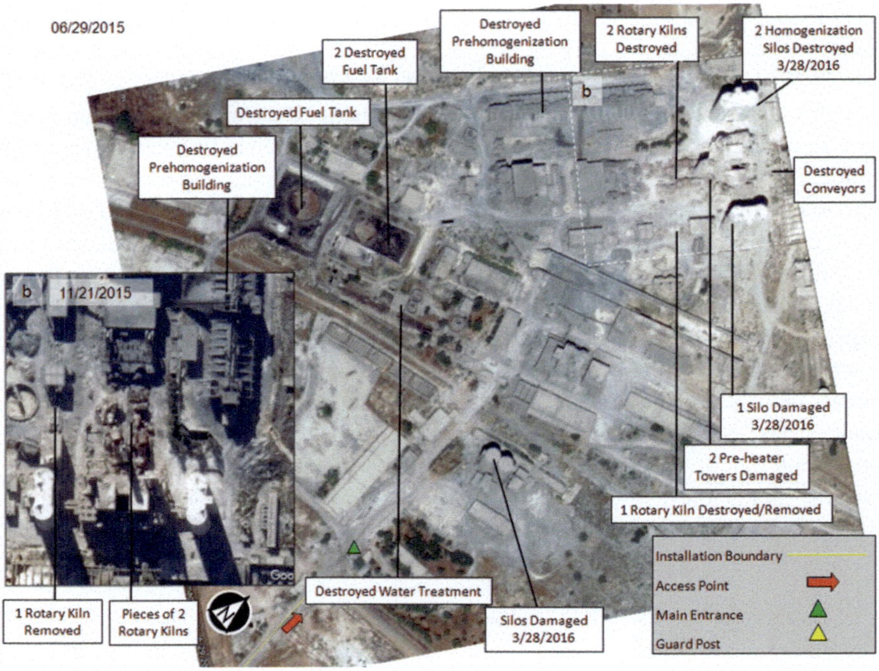

Fig. 14.22 (**a**) WorldView image acquired on 17 September 2014 over the same plant in Fig. 14.21. Individual elements within facilities in the plant showing further damage are identified including three fuel tanks, (**b**) water treatment plant, three near-vertical rotary kilns and homogenization silos

western plant, the three destroyed fuel tanks had not been replaced and no repairs had occurred to the cement production lines.

Examination of the two cement factories to the west and north of Aleppo Industrial City—Sheikh Najjar shows that they were part of the conflict in late 2013 to 2016, similar to the cement factory in the south of Aleppo city. Once the plants were no longer working, they like the LCSF, had tactical value as defensible positions and places where vehicles, machines, pipes, and steel could be obtained to be used by rebels, IS and affiliates to make weapons and up-armour vehicles. Based upon examination of the EO images, it appears that the three cement factories in Aleppo were not working in 2017 to produce cement or concrete used in the construction of trenches, bunkers, observation posts, CCHTs, and UGFs by IS, Kurdish malitia, and FSA in the Afrin region.

Although the sources of the concrete are unknown at this time, Table 14.8 shows that 52 cement mixers would be needed to deliver the required volume for a CCHT 50 m long, 310 for a 300 m long CCHT and 1033 for 1000 m long CCHT. Once on-site, slabs would likely have been poured in section to handle concrete pumping, filling, and water needs. Drone footage released by Turkish media (TRTWorld 2018) shows the YPG/YPJ/SDF pouring concrete to make CCHTs in Kobane in December, 2018 (Fig. 14.23). Similar to the characteristics identified around Afrin

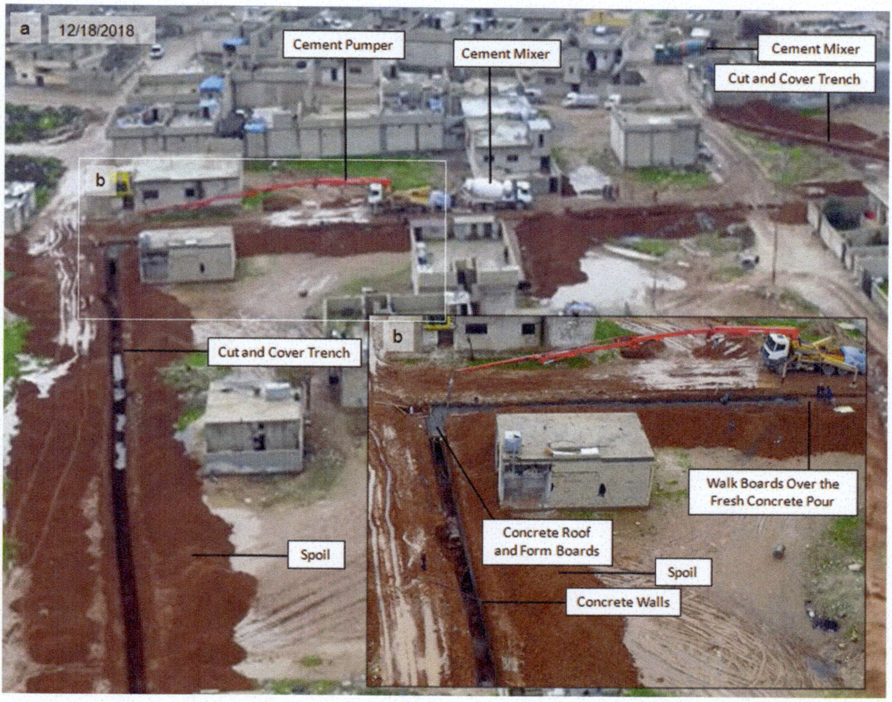

Fig. 14.23 (**a**) Drone footage released by Turkish media of the YPG/YPJ/SDF held city of Kobane showing construction of CCHTs in December, 2018 (TRTWorld 2018). (**b**) Higher resolution portion of still frame from 'a' shows the concrete walls of the trench and the new section of concrete roof. The new section of roof extends back to the pumper truck and a team is working the concrete pour before laying down boards in preparation for the spoil to be moved to cover the tunnel. Daily Sabah https://twitter.com/DailySabah/status/1075122277589626881

(Fig. 14.17), the trenches were mechanically excavated in the terrace alluvial pebble beds and conglomerates. Two cement mixers are visible with the one transferring concrete into a pumper truck (Fig. 14.23a). The arm of the pumper is shown at full extension delivering concrete to make the concrete roof to a CCHT. The concrete walls and form boards can be seen in Fig. 14.23b. The calculated volumes of concrete required for CCHTs (Table 14.8) require a robust supply chain both for suppliers, cement mixers for deliver, and pumper trucks as well as work crews.

Concrete Mix

The concrete mix design used in trenches, bunkers, CCHTs, and UGFs is unknown but first-order calculations can be made for a concrete ratio for cement, fine aggregate, and coarse aggregates of 1:2:4. Using a volume of one cement bag = 0.0354 m^3 so the volume of fine aggregates is

$$0.0354\,\text{m}^3 \times 2 = 0.07\,\text{m}^3 \tag{14.1}$$

The volume of coarse aggregate is

$$0.0354\,\text{m}^3 \times 4 = 0.141\,\text{m}^3 \tag{14.2}$$

The combined volume of aggregates is

$$0.07 + 0.141 = 0.211\,\text{m}^3 \tag{14.3}$$

The water requirement for one bag of cement is 11.667 ltrs. The water for aggregates is 5% giving

$$(5\% / 100) \times 0.211\,\text{m}^3 = 0.01055\,\text{m}^3 * 1000 = 10.55\,\text{ltrs} \tag{14.4}$$

The total quantity of water is

$$11.667 + 10.55 = 22.217\,\text{ltrs} \tag{14.5}$$

These figures can be used with calculations of total volumes of concrete (Table 14.8) to determine total volumes of cement, fine and coarse aggregates for CCHTs ranging from 50 m to 1000 m (Table 14.9). The volume of cement ranges from 45 m^3 for a 50 m long CCHT, 270 m^3 for 300 m and 900 m^3 for 1000 m. Converting these numbers into bags of cement (0.0354 m^3/bag), a 50 m long CCHT requires 1271 bags, 300 m 7627 and 1000 m 25,424 bags. This, along with the volumes of fine and

Table 14.9 Volumes of cement, fine and coarse aggregates for CCHT for a concrete ratio 1:2:4. V = volume m^3

2 wall and 2 ceiling/floor concrete slab lengths m	V m^3	Cement m^3	Bags of cement @ 0.0354 m^3	Fine aggregates m^3	Coarse aggregates m^3	Total V of aggregates m^3
50	315	45	1271	90	180	270
100	630	90	2542	180	360	540
200	1260	180	5085	360	720	1080
300	*1890*	*270*	*7627*	*540*	*1080*	*1620*
400	2520	360	10,169	720	1440	2160
500	3150	450	12,712	900	1800	2700
600	3780	540	15,254	1080	2160	3240
700	4410	630	17,797	1260	2520	3780
800	5040	720	20,339	1440	2880	4320
900	5670	810	22,881	1620	3240	4860
1000	6300	900	25,424	1800	3600	5400

coarse aggregates, quantifies the level of outputs needed from quarrying and from cement factories.

As discussed in Sect. 3.3, the water content in concrete is critical to its quality and strength. The combined quantity of water required for a concrete ratio 1:2:4 (Table 14.10) is 28,331 ltrs for a 50 m length of CCHT, 169,986 ltrs and 566,619 ltrs for a 300 m and 1000 m length, respectively. This shows the amount of water cement plants required to be able to make a concrete mix ratio 1:2:4. The volumes required to build CCHTs, along with the immediate need to improve survivability from TAF airstrikes and artillery, demanded that concrete be mixed in volume and transported on-site in cement mixers rather than be mixed on-site.

Table 14.10 Calculations of water required for a concrete ratio 1:2:4 for CCHTs over a range of lengths

2 wall and 2 ceiling/floor concrete slab lengths m	Water for aggregates m^3	Water for aggregates ltr	Water requirement for bags of cement ltr	Total quantity of water ltr
50	13.5	13,500	14,831	28,331
100	27	27,000	29,662	56,662
200	54	54,000	59,324	113,324
300	*81*	*81,000*	*88,986*	*169,986*
400	108	108,000	118,647	226,647
500	135	135,000	148,309	283,309
600	162	162,000	177,971	339,971
700	189	189,000	207,633	396,633
800	216	216,000	237,295	453,295
900	243	243,000	266,957	509,957
1000	270	270,000	296,619	566,619

Table 14.11 Total costs of rebar mesh grid and concrete for CCHTs calculated using two price points

2 wall and 2 ceiling/floor concrete slab lengths m	Total cost of rebar mesh grid and concrete @$120/0.76 m^3	Total cost of rebar mesh grid and concrete @$200/0.76 m^3
50	61,570	144,465
100	123,140	288,930
200	246,281	577,860
300	*369,421*	*866,789*
400	492,561	1,155,719
500	615,702	1,444,649
600	738,842	1,733,579
700	861,982	2,022,509
800	985,123	2,311,439
900	1,108,263	2,600,368
1000	1,231,404	2,889,298

Combining the analysis above it is possible to calculate the combined costs of rebar (0.3 m intervals) used in the wall (2 m high and 0.9 m thick) plus ceiling and floor (1.5 m wide and 0.9 m thick) along the different CCHT lengths as well as the costs of concrete at two price points (Table 14.11).

The costs range from $61,570 to $144,465 using $120/0.76 m and $200/0.76 m, respectively, for a 50 m long CCHT, $369,421 to $866,789 for 300 m and $1,231,404 to $2,889,298 for 1000 m. Additional costs include purchase/rental of excavator, operators time, concrete mixers, concrete reservoir and pumps, water, wood for forms, and work crew. The use of concrete in bunkers, observations posts, pill boxes, CCHTs, and UGFs is costly requiring massive quantities of financing, plant, materials, and work crews. The costs, when applied across the construction of military works reveal millions of dollars in expenditure. Significant cost reductions could occur if resources for making concrete were owned locally or existed nearby.

14.5 Damage and Reconstruction

The conflict in Syria is having a continued and sustained negative impact on the built and natural environment with wide-scale damage throughout the country. The environmental degradation will burden the country for many centuries to come (PAX 2015, 2016; World Bank 2017a, b; Bulmer 2018a, 2019c). The generated debris, some of it hazardous, will impact the return of residents and industrial activities, reconstruction works, hinder access and negatively impact public health and the natural environment. The conflict-related damage has led to large-scale debris accumulation in Syrian cities much of which is concrete, which will require massive amounts of clearance, transportation, and processing or disposal before new construction. Remote-sensing-based analysis was used to analyse the extent and concentration of debris accumulation in Aleppo city up to 2017 (Fig. 14.24). Damage data from Courage Services Inc., World Bank, and UNOSAT was cross referenced with structural built environment data, providing an understanding of the proportion of different types of buildings damaged in each neighbourhood. By integrating the expertise of on-the-ground Syrian engineering expertise, estimates of the quantities of debris created per damaged household of different types were arrived at. Extrapolating these data across the whole area allowed for neighbourhood wide debris quantities to be calculated. The results show that about 14.9 M mt of debris.[30] Table 14.12 shows three scenarios have been examined for residential debris management (DWR et al. 2017).

[30] In comparison, in Homs, 5.3 M Mt. have accumulated up to 2017. It would take about 2.5 years and 2.3 M truck-kilometers to clear.

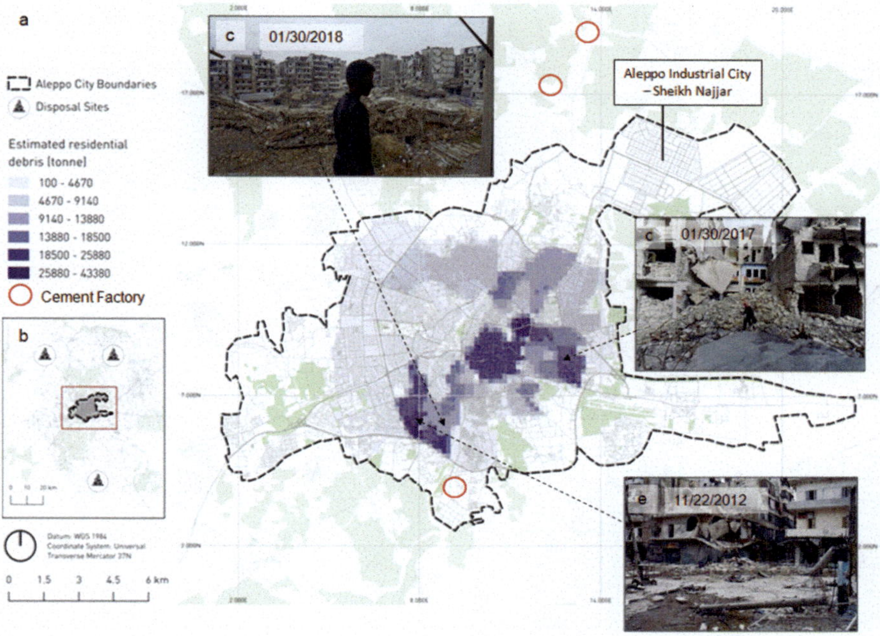

Fig. 14.24 (**a**) Calculated debris accumulations from the start of the Syrian Conflict to February 2017. The red ovals show the locations of the three cement factories discussed in the text. (**b**) The black triangles show three pre-conflict designated disposal sites outside Aleppo city. (**c**) Resident looks from his home at damage in Sukkari district, Aleppo on 21 January, 2018 (CTV News 2018. AP Photo/Mstyslav Chernov). (**d**) Damaged buildings in Aleppo 30 January, 2017 (Associated Press 2018. Reuters/Omar Sanadiki). (**e**) Damaged apartment buildings in the Salahudeen district in Aleppo November, 2012 (ABC News Ed Giles)

Table 14.12 Scenarios for residential debris management in Aleppo city

Scenario	Action
Scenario 1:	All debris is taken to the closest of the three disposal sites as identified from Aleppo's waste management plan for disposal with no recycling. Reconstruction will use new building materials.
Scenario 2:	A network of 43 small and six medium capacity debris crushing sites are established. The small sites are mobile and move through the streets of the most densely impacted neighbourhoods, while the medium crushers are stationary on the city boundaries. These sites all produce a recycled aggregate with 70% of the produced debris and the remaining 30% is sent for disposal to the same sites as identified in scenario 1. Logistics are provided by a fleet of 200 trucks of 16 tonne capacity each. The recycled aggregate is used in reconstruction in Aleppo.
Scenario 3:	A network of five large capacity crushing sites is set up on the outskirts of the city. These process the debris in the same way as in scenario 2 and logistics are provided by the same fleet. The recycled aggregate is used in reconstruction in Aleppo.

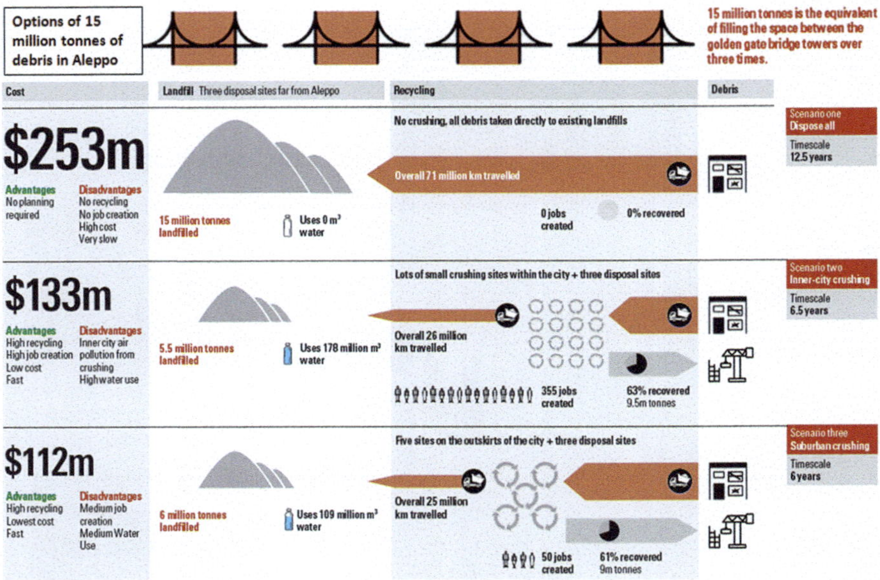

Fig. 14.25 Infographic displaying three options for addressing the 15 million tonnes of debris in Aleppo. Modified from DWR 2017

The scenarios have been modelled using expertise from DWR[31] and URP[32] to include disposal sites as collected from various national or governorate level planning documents, and potential recycling sites located within or around the city based on the most likely and achievable scenarios in the Syrian context, considering access to resources, equipment, and expertise. The hourly capacities, costs, job creation opportunities, and other factors of the processing sites are inputs based on equipment most commonly used in the industry for humanitarian purposes. Figure 14.25 shows the results of the three scenarios in Aleppo city. If there is no recycling of conflict debris, then it will be trucked out the city to landfill sites. Pre-conflict there were three designated landfill sites shown in Fig. 14.24. It would take 12.5 years[33] to move 15 M mt to the three landfills requiring over 71 M truck km,[34]

[31] Disaster Waste Recovery is a UK-based not-for-profit organisation responding to global crises, providing vital debris recovery and waste management support to strengthen disaster-affected communities through to rebuilding their livelihoods. https://www.disasterwaste.org/.

[32] Urban Resilience Platform is a not-for-profit organisation preparing for large-scale debris and waste management works with expertise in post-disaster waste, tunnel spoil management, and offshore decommissioning. https://www.urplatform.eu.

[33] Work weeks are 7 days long and working days are 10 h each.

[34] Two hundred trucks, each with a capacity of 16 metric tonnes, are used throughout the debris processing program from day 1 until the end day. This assumption does not therefore allow for the reality of fluctuating access to resources by the debris program nor access into some of the smaller

using 18.6 M ltrs of fuel, releasing 72.6 M mt of CO_2[35] at a cost of $253 M. The cost of trucking will be $178.7 M. This will require little planning but has the disadvantage that there will be no recycling, no job creation around recycling, high cost and slow progress. In the second scenario, it would take 6.5 years to recover 63% of the debris (9.5 M mt)[36] using small and medium crushers, with 5.5 M mt trucked to landfill requiring 26 M km of truck travel, using 6.8 M ltrs of fuel, releasing 26.6 M mt of CO_2, using 178 M m^3 of water (114,000 m^3/day) for dust suppression at a cost of $133 M. The cost of trucking will be $65.5 M. This creates 355 jobs, lowers cost, and is faster than scenario one but creates inner city air pollution and requires high water usage. In scenario three, it would take 6 years to recover 61% of the debris (9 M mt) using five large capacity crushing sites set up on the outskirts of the city, with 9 M mt trucked to landfill requiring 25 M truck km, using 6.7 M ltrs of fuel, releasing 26.4 M mt of CO_2, using 109 M m^3 of water (50,000 m^3/day) at a cost of $112 M. This has the lowest cost and fastest rate with lower water usage than scenario two but only creates 50 jobs. Debris removal and processing poses significant organizational challenges: it is costly; it takes time to clear; there is unexploded ordinance, and the operations impose further environmental risks, including dust, CO_2 emissions, and water resource pollution. Thus, locating crushing sites often faces resistance from locals. All three scenarios require the availability of cement and concrete to enable reconstruction. Scenario one will create the biggest demand and when matched against the pre-war outputs of cement factories in Syria it is clear that there is a dramatic shortfall. The cement factories are therefore critical but given the damage they have sustained, as well as availability of skilled workers, the shortfall must be met by recycling conflict debris. Should 1000 km of roads around Aleppo require reconstruction, with a 40 cm sub-base layer of recycled debris and a road width of 6 m then the demand for recycled debris would be approximately 30% of that produced in scenarios two and three.

When considering the requirement for natural raw materials for the reconstruction demand, the pre-conflict (2011), total production (extraction) of sand, gravel, and crushed rock from all quarries in Syria were recorded as being 1.84 M mt (Taib 2013). The total quantity of conflict debris in the city of Aleppo up to 2017 is 15 M mt,[37] and assuming reconstruction to the same degree as before the conflict (possibly a wrong assumption), then the amount of gravel, sand, and crushed rock will be

---

roads and streets. Indeed the competition with other sectors in terms of access to transportation resources is not taken into account.

[35] The emissions have been calculated for truck movements within the scenarios, where both trucks carrying debris and empty trucks are included. The algorithm relies on distance travelled by the trucks (kilometres) multiplied by typical construction truck CO_2 emissions as provided by the UK Government GHG Conversion Factors for Company Reporting.

[36] The debris which is created by the conflict is not 100% recyclable, it will require separation, and a portion of the debris will need to go to disposal in all cases, even after processing at the recycling sites.

[37] This number will have increased due to continued fighting in Aleppo over 2017–2019.

similar to that originally built. This will thus require ca. 7.5 times the annual output of all quarries in Syria, just for Aleppo and not including all the other damaged cities in Syria and Iraq (World Bank 2017b). This will place unprecedented demand on the quarrying and cement industry for skills, resources, and access to geology further supporting the demand and need for recycling debris into a reconstruction material.

In the context of the conflict in Syria, new quarries are likely to be opened and used as locations to take conflict-related debris from cities such as Aleppo. There are important lessons here from Lebanon, which witnessed increased use of quarries post 2006 reconstruction (DWR et al. 2017). This demonstrated the environmental impact of this increased use of quarries and caused groundwater problems, leading to tensions with local communities that had to deal with the environmental degradation (Rose 2019). Examination of EO images show domestic waste dumped in disused quarries around Aleppo. This biodegrades, generating methane, CO_2, and leachate. In the permeable sand, gravel, and fractured rock this may enter and pollute ground water (Wardrop et al. 2001; Singleton et al. 2013). At individual sites gas or leachate can be prevented from escaping by sealing the quarry with clay or a tough impermeable plastic but no evidence has been found for such practices in Syria. Even with a lining, leachate and gas are still produced that have to be removed.[38] Using quarries around Aleppo for conflict-generated debris there will be no requirement for lining if the debris is inert.[39] This can occur by sorting at the clearance site to make sure that only concrete and stone is loaded onto trucks that travel to the designated quarry sites. Dust suppression measures (WRAP 2006) will need to be carried out on all trucks moving from the debris loading site to the quarry unloading site[40] and along the route.[41] Once at the designated quarries, the debris can be tipped or moved using heavy equipment.

Remediating quarries provides an important opportunity for reconciliation in a nation divided by the conflict. Land rehabilitation is an essential part of quarrying and done well can make disturbed areas suitable for new sustainable land uses (Neri and Sánchez 2010; Loudes et al. 2012; Clemente et al. 2004). It increases public acceptance of quarrying that will be critical in reconstruction and shows that former quarry sites are not merely degraded areas, but can have alternative land use able to act as a catalyst for development (Damigos and Kaliampakos 2003; McCandles 2012). A variety of sustainable uses incorporating climate change adaptation strategies could create sites of greater social and environmental benefit (Jim 2001). Much work has been undertaken for sustainable redevelopment of disused quarries where they have become public and private spaces for research and education, aquaculture,

---

[38] Landfill gas can be used to generate electricity but leachate has to be pumped out and treated elsewhere.

[39] Not chemically reactive.

[40] Covering loads, speed limits, and prohibit activities during high wind.

[41] Paved roads, purchase PM_{10} efficient sweeper, chemical dust suppressant, apply water, air quality monitoring, limit speeds.

recreation, storage, industry, and housing[42] (Carmona and Freeman 2005). The abandonment of no longer operational quarries, wastes land, pollutes neighbouring areas, and allows for the deterioration of natural habitats. In the case of Aleppo city, there are clear opportunities to connect quarry extraction for reconstruction, disused quarries for conflict-related debris management and for quarries as sites of opportunity for sustainable alternative land use that can contribute to public and environmental health as well as contribute to the aesthetic recovery of the post-conflict landscape.

14.6 Conclusions

The ongoing conflict in Syria has destroyed cities with widespread damage to industrial sites and critical infrastructure including cement production. Analysis of the fighting, which is based on EO imagery, social media monitoring and the reports of International agencies, has identified that the cement industry has sustained severe damage but has been critical in the production of military works built by the Assad government, the rebels and IS. It will be critical in reconstruction for which cost estimates are >$200 Bn (Karasapan 2017). The conflict has disastrously impacted the environment and public health. As well as killing 400,000, internally displacing six million, and creating five million refugees outside Syria (UNOCHA 2016), it has destroyed productive factors; severely diminished economic connectivity, reduced incentives to pursue productive activities, and broken economic and social networks and supply chains. The war economy reflects territorial, religious, political, and ethnic divisions of the armed conflict with middlemen being critical to enabling trade between warring factions. In January, 2017 the Assad government signed an agreement with Iran to reconstruct phosphate mining, farm land, oil and gas terminals, and a port (Karasapan 2017). Turkey, Russia, and China are looking to develop further agreements. The UN has to determine the role and goal of continued sanctions. Syrian and Iranian government efforts to repopulate key areas with mostly Shia populations, replacing displaced Sunnis, is exacerbating divisions rather than attempting reconciliation (Chulov 2017). Turkey, working with the TFSA is building towns within Northern Syria that it cleared so that it can move Syrian refugees from Turkey back into Syria. Those areas that have remained under Kurdish control have attempted to consolidate using rapid reconstruction but this is being constrained by the lack of domestic construction materials, the closure of the border with Turkey, and barriers to trade with Iraq's Kurdistan Regional Government. The U.S. Administration's position on supporting partners to stop further Syrian government gains backed by Russian airpower in the NE has become uncertain. Even with these attempts at guaranteeing security through political and religious

[42] Examples include Brownstone Park, Connecticut; Quarry Falls, San Diego; Bellwood Quarry, Atlanta; Butchart Gardens, Victoria Canada; Groundscraper Hotel, Shanghai, China.

manipulation of the population, the fact remains that airstrikes and artillery bombardment of urban areas have caused the destruction of a major part of Syria's housing stock. This has continued into 2020 (Reuters 2020). An assessment by the World Bank in 2017 put the figure at 7% of the housing stock physically destroyed and 20% partially damaged with cities such as Aleppo and Raqqa having sustained greater destruction and damage. This has increased with an additional 3 years of fighting. Millions of tonnes of conflict debris, some of which contains hazardous materials such as asbestos, heavy metals, and the toxic residues from conventional weapons have been created. The need to clear and reconstruct is a massive undertaking. In Aleppo alone, the demand for cement and aggregate will be over 7.5 times the pre-war annual output of all quarries in Syria. This will place unprecedented demand on the quarrying industry for skills, resources, and access to suitable geology. Given the assessment of the damage to the cement sector and the scale of reconstruction, there will not be enough to meet the demands of rebuilding in all the damaged cities at the same time. This will require recycling and repurposing of conflict debris into a reconstruction material as well as large quantities of new cement. Use of EO image analysis will enable progress to be monitored.

Acknowledgements The author thanks the reviewers for their comments. I would like to thank Martin Bjerregaard of Disaster Waste Recovery (DWR), and Aiden Short of Urban Resilience Platform (URP), for their discussions regarding cement production, quarrying and debris management.

References

Ahval News. 2018. *Foreign cement company providing material for YPG fortifications*. 10 February. https://ahvalnews.com/afrin/foreign-cement-company-providing-material-ypg-fortifications-islamist-newspaper. Accessed 10 July 2019.

Alderman, L., E. Peltier, and Saad, H. (2018) 'ISIS Is Coming!' How a French Company Pushed the Limits in War-Torn Syria. The New York Times March 10. https://www.nytimes.com/2018/03/10/business/isis-is-coming-how-a-french-company-pushed-the-limits-in-war-torn-syria.html. Accessed 07 July 2019.

Ara News. 2015. Islamic state builds its economy on Syrian ashes: Ara news, February 26. http://aranews.net/2015/02/islamic-state-builds-its-economy-on-syria-ashes. Accessed 07 July 2019.

Associated Press. 2018. Damage to Aleppo 'Much Bigger' Than Expected, U.N. Official Says. 1 February. https://www.pbs.org/newshour/world/damage-aleppo-much-bigger-expected-u-n-official-says. Accessed 14 July 2018.

Bindner, L., and G. Poirot. 2016. ISIS Financing 2015. Centre for the Analysis of Terrorism. May 2016. https://cat-int.org/wp-content/uploads/2016/06/ISIS-Financing-2015-Report.pdf. 33.

Bulmer, M.H. 2018a. Military Use of Environmental Degradation by Islamic State, Northern Iraq. *Scientia Militaria. South African Journal of Military Studies* 46 (1). https://doi.org/10.5787/46-1-1228.

———. 2018b. Geological Considerations of Contemporary Military Tunnelling near Mosul, Northern Iraq. In *Military Aspects of Geology: Fortification, Excavation and Terrain Evaluation*, ed. T. Rose. London: GSL Special Publication.

———. 2019a. Contemporary military use of Subterranea. *The British Army Review.* 175: 106–113.

———. 2019b. Contemporary Use of Subterranea to Increase Survivability. In *Paper presented at the 13th International Conference on Military Geosciences, University of Padova, Italy, 24–28 June 2019.*

———. 2019c. Environmental degradation in the Battlespace. In *Culture in Conflict. The Learning from Conflict Series. BAR Special Report*, 48–59.

———. 2020. Geological considerations for military works in the Afrin Battlespace, Syria. In *Military Geosciences in Peace and War*, ed. A. Bondesan and J. Yehlen. Amsterdam: Elsevier. In Review.

Carmona, M., and J. Freeman. 2005. The Groundscraper: Exploring the Contemporary Reinterpretation. *Journal of Urban Design* 10 (3): 309–330.

Chulov, M. 2017. Iran repopulates Syria with Shia Muslims to help tighten Regime's control. The Guardian. 13 January. https://www.theguardian.com/world/2017/jan/13/irans-syria-project-pushing-population-shifts-to-increase-influence. Accessed 15 July 2019.

Clemente, A.S., C.C. Werner, C.C. Máguas, M.S. Cabral, M.A. Martins-Loução, and O.O. Correi. 2004. Restoration of a Limestone Quarry: Effect of Soil Amendments on the Establishment of Native Mediterranean Sclerophyllous Shrubs. *Restoration Ecol* 12 (1): 20–28.

CIEDCOR. 2016. *Chemical IEDs. Current status and trend threat.* November. https://ciedcoe.org/index.php/reports/analysis-reports/54-0007-chemical-ieds-current-status-and-trend-threat-november-2016/file. 18.

CTV News. 2018. *Life Slowly Returns to Aleppo, Shattered by Syrian Civil War.* 26 January. https://www.ctvnews.ca/world/life-slowly-returns-to-aleppo-shattered-by-syrian-civil-war-1.3776401. Accessed 14 July 2019.

Cunningham, E. 2014. *Islamic state jihadists are using water as a weapon in Iraq*. The Washington Post, last modified October 7, 2014, https://www.washingtonpost.com/world/middle_east/islamic-state-jihadists-are-using-water-as-a-weapon-in-iraq/2014/10/06/aead6792-79ec-4c7c-8f2ffd7b95765d09_story.html. Accessed 10 Feb 2019.

Daily Sabah. 2016. *Turkey-Syria Border Wall to be Completed by Mid-2017, Defense Minister says.* 2 November. https://www.dailysabah.com/turkey/2016/11/02/turkey-syria-border-wall-to-be-completed-by-mid-2017-defense-minister-says. Accessed 7 July 2019.

Damiani, E., and P. Giovannangeli. 2008. *Sector Study of the Natural Dimensional Stone Sector in Syria. European Aid. SME Support Programme. Project No (SSP) MEDA/2005/17542.* EuropeanAid/122282/C/SV/SY. http://www.ncosyria.com/assets/files/studies/Natural%20Dimensional%20STONE%20sector.pdf. 35.

Damigos, D., and D. Kaliampakos. 2003. Assessing the benefits of reclaiming urban quarries: A CVM analysis. *Landscape and Urban Planning* 64: 249–258.

Disaster Waste Recovery, Roedown Research and Urban Resilience Platform. 2017. *Aleppo: Debris and environmental Management for Recovery. Phase 1 report. 10 February*. To Courage Services Inc: Arlington. pp. 87.

Doyle, P. 2017. *Disputed Earth: Geology and Trench Warfare on the Western Front 1914–1918*, 208. London: Unicorn Publishing Group.

Environmental and Social Impact Assessment of Construction and Operation of Syria Cement Plant and Associated Activities, Syria (2009) Earth Link and Advanced Resources Development (ELARD) in association with Golders Associates Africa (GAA). https://www.eib.org/attachments/pipeline/20090033_eis_en.pdf. 267.

European Commission: Delegation Damascus (2009) Country Environmental Profile for the Syrian Arab Republic. Max Kasparek & Marwan Dimashki. Agreco consortium. Final Report. Brussels. http://eeas.europa.eu/delegations/syria/documents/content/eu_syria/cep_syria_en.pdf. 93.

Geological map of NW Syria. 1:500,000. Sheet 1. 1964. Syrian ministry of industry. *Dept of Geological and Mineral Research.*

Giles E (2012) Damaged Apartment Buildings in Aleppo. ABC News, 22 November. https://www.abc.net.au/news/2012-11-23/damaged-apartments-aleppo-syria/4388846. Accessed 13 July 2019.

Ginoux, P. 2012. Global-scale attribution of anthropogenic and natural dust sources and their emission rates based on MODIS Deep Blue aerosol products. *Reviews of Geophysics* 50 (3): 3005.
Gleick, P.H. 2014. Water, drought, climate change, and conflict in Syria. *Am Meteorol Soc* 6 (3): 331–340. https://doi.org/10.1175/WCAS-D-13-00059.1.
Guidance for Controlling Silica Dust from Stone Crushing with Water Spray Technology. 2008. *Occupational Knowledge International*. http://www.okinternational.org/docs/Guidance%20for%20Controlling%20Silica%20Dust%20from%20Stone%20Crushing%20with%20Water%20Spray%20Technology%20-%20for%20Employers.pdf. 7.
Jarry, E. 2018. *Exclusive: France Asked U.S. not to Bomb Lafarge Factory in Syria in 2014*. Reuters. April 25. https://www.reuters.com/article/us-lafargeholcim-syria-france-exclusive/exclusive-france-asked-u-s-not-to-bomb-lafarge-factory-in-syria-in-2014-emails-idUSKBN1HW2DE. Accessed 7 July 2019.
Jim, C.Y. 2001. Ecological and landscape rehabilitation of a quarry site in Hong Kong. *Restoration Ecology* 9 (1): 85.
Jones, S. 2010. *Underground Warfare 1914–1918*, 288. Barnsley: Pen & Sword.
Kaaman, H. 2017. *The History and Adaptability of the Islamic State Car Bomb*. Open Source Research on SVBIED's. https://hugokaaman.com/2017/02/14/the-history-and-adaptability-of-the-islamic-state-car-bomb/. Accessed 07 July 2019.
———. 2019. Mining Haul Trucks – The Largest SVBIEDs in the World. https://hugokaaman.com/2019/03/25/mining-haul-trucks-the-largest-svbieds-in-the-world/. Accessed 07 July 2019.
Karasapan O (2017) Rebuilding or redefining Syria? Brookings Institution, February 13. https://www.brookings.edu/blog/future-development/2017/02/13/rebuilding-or-redefining-syria. Accessed 15 July 2019.
Kaufmann, J.E., H.W. Kaufmann, A. Jankovic-Potocnik, and P. Lang. 2011. *The Maginot line. 792 history and guide*. Barnsley: Pen & Sword Books Ltd.
Kelley, C.P. 2015. Climate change in the Fertile Crescent and implications of the recent Syrian drought. *Proceedings of the National Academy of Sciences of the United States of America* 112: 3241–3246. https://doi.org/10.1073/pnas.1421533112.
King, M.D. 2016. The Weaponization of water in Syria and Iraq. *The Washington Quarterly* 38 (4): 153–169.
Lafarge. 2014. Annual Report. Registration Document. https://www.companyreporting.com/sites/default/files/annual-report-index/lafarge-annual-report-2014.pdf. 332.
LafargeHolcim. 2017. Building for Growth. Annual Report. https://www.lafargeholcim.com/annual-interim-reports?field_year_tid_i18n_exposed=1165.
Lund, A. 2018. The Factory: A Glimpse Into Syria's War Economy. Twitter 22 February. https://twitter.com/aronlund/status/966204849414078464. Accessed 11 July 2019.
Luodes, H.H., P.P. Kauppila, N.N. Luodes, S.S. Aatos, J.J. Kallioinen, S.S. Luukkanen, and J.J. Aalto. 2012. Characteristics and the environmental acceptability of the natural stone quarrying waste rocks. *Bull Eng Geol Environ* 71 (2): 257–261.
McCandles, C. 2012. No Longer just a hole in the Ground. The Adaptive Re-Use of Resource Depleted Quarries. http://www.mit.edu/people/spirn/Public/Ulises-11-308/Quarrying.pdf.
McKernan, B. 9 October 2019. Turkey launches military operation in northern Syria. The Guardian. Accessed 30 Oct 2019.
Mittelman, E. 2018. The Cement Industry, One of the World's Largest CO2 Emitters, Pledges to Cut Greenhouse Gases. Yale Environment 360. https://e360.yale.edu/digest/the-cement-industry-one-of-the-worlds-largest-co2-emitters-pledges-to-cut-greenhouse-gases. Accessed 13 July 2019.
Mohtadi, S. 2012. Climate change and the Syrian uprising. Bulletin of the Atomic Scientists. http://thebulletin.org/web-edition/features/climate-change-and-the-syrian-uprising. Accessed 01 July 2019.
Neri, A., and L. Sánchez. 2010. A procedure to evaluate environmental rehabilitation in limestone quarries. *Journal of Environmental Management* 91 (11): 2225–2237. https://doi.org/10.1016/j.jenvman.2010.06.005.

Oweis, K.Y., and C. Sezer. 2012. Syrian Refugees Brave Mines, Machineguns to reach Turkish Sanctuary. Reuters, April 6. https://www.reuters.com/article/syria-turkey-border/syria-refugees-brave-mines-machineguns-to-reach-turkish-sanctuary-idUSL6E8F60LY20120406. Accessed 08 July 2019.

Ozturk, B. 2018. Turkey's Great Wall and its Archenemy on the Syrian Border. SETA 24 January. https://www.setav.org/en/turkeys-great-wall-and-its-archenemy-on-the-syrian-border/. Accessed 05 July 2019.

Portela C (2012) The EU's sanctions against Syria: Conflict management by other means. Security policy brief no. 38, Egmont Royal Institute for international relations, September., aei.pitt.edu/39406/1/SPB38.pdf.

PAX. 2015. Amidst the Debris. A Desktop Study on the Environmental and Public Health Impact of Syria's Conflict. https://www.paxforpeace.nl/publications/all-publications/amidst-the-debris. 84.

———. 2016. Scorched Earth and Charred Lives. https://www.paxforpeace.nl/publications/all-publications/scorched-earth-and-charred-lives. 27.

Reuters. 2010. Lafarge to Buy Orascom Cement for $12.8 Billion. 5 August. https://www.cnbc.com/id/22178287. Accessed 07 July 2019.

———. 2020. Syrian Government Forces Capture Towns in Advance on Rebel-Held Idlib. January 26. https://www.reuters.com/article/us-syria-security/syrian-government-forces-capture-towns-in-advance-on-rebel-held-idlib-idUSKBN1ZP0FC. Accessed 30 Jan 2020.

Rogoway, T. 2015. ISIS used this huge crudely modified truck to attack a key Syrian air base. Foxtrot Alpha, 24 August. https://foxtrotalpha.jalopnik.com/isis-used-this-huge-crudely-modified-truck-to-attack-a-1726246835. Accessed 08 July 2019.

———. 2017. This ISIS armored loader just broke through Iraqi army lines and blew up. The Drive, 15 March 2017. http://www.thedrive.com/the-war-zone/8358/this-isis-armored-bulldozer-just-broke-throughiraqi-army-lines-and-blew-up. Accessed 08 July 2019.

Rose, S. 2019. In Lebanon, pro and Anti-Syria Camps Face off Over a Cement Factory. National World. 30 June. https://www.thenational.ae/world/mena/in-lebanon-pro-and-anti-syria-camps-face-off-over-a-cement-factory-1.881081. Accessed 15 July 2019.

S&P Global. 2013. Special Report: Syria Steel Investments Continue Despite War. https://www.steelbb.com/it/?PageID=157&article_id=120692.

Sharman, J. 2019. US Bombs its own Military Base in Syria to 'Reduce Usefulness'. Independent. 17 October. https://www.independent.co.uk/news/world/middle-east/syria-war-trump-turkey-invasion-us-base-bomb-lafarge-cement-factory-manbij-russia-video-a9159851.html. Accessed 20 Jan 2020.

Singleton, V., B. Jacob, M. Feeney, and J. Little. 2013. Modeling a proposed quarry reservoir for raw water storage in Atlanta. *Georgia. J. Environ. Eng.* 139 (1): 70–78.

Sozcu News Agency. 2018. Positions on Mount Burseya have been viewed! Details that stand out. 28 January. https://www.sozcu.com.tr/2018/gundem/burseya-dagindaki-mevziler-goruntulendi-dikkat-ceken-detaylar-ortaya-cikti-2193127/. Accessed 29 June 2019.

Taib, M. (2013). The Mineral Industry of Syria. In: US Geological Survey 2011 Minerals Yearbook. http://minerals.usgs.gov/minerals/pubs/country/2011/myb3-2011-sy.pdf.

———. 2015 The Mineral Industry of Syria. In: US Geological Survey 2011 Minerals Yearbook. https://prd-wret.s3-us-west-2.amazonaws.com/assets/palladium/production/atoms/files/myb3-2015-sy.pdf. 7.

TRTWorld. 2018. Exclusive: Footage Shows YPG Terrorists Digging Trenches in Northern Syria. 18 December. https://www.trtworld.com/middle-east/exclusive-footage-shows-ypg-terrorists-digging-trenches-in-northern-syria-22589. Accessed 10 July 2019.

Turkey News. 2016. Security Wall Partially Completed on Turkey's Syrian Border. News Service 19 October. https://www.yenisafak.com/en/gundem/security-wall-partially-completed-on-turkeys-syrian-border-2550353. Accessed 09 July 2019.

———. 2018a. PKK Digs Trenches, Tunnels in Futile Attempt to Escape Turkey's Afrin Op. News Service 1 March. https://www.yenisafak.com/en/world/pkk-digs-trenches-tunnels-in-futile-attempt-to-escape-turkeys-afrin-op-3135794. Accessed 02 July 2019.

———. 2018b. Terrorist PYD/PKK Tunnel Used to Attack Turkish Troops. News Service 29 January. https://www.yenisafak.com/en/world/terrorist-pydpkk-tunnel-used-to-attack-turkish-troops-3033549. Accessed 04 July 2019.

UNEP. 2003. *Desk study of the environment in Iraq*, 98. Nairobi: United Nations Environment Programme. https://postconflict.unep.ch/publications/Iraq_DS.pdf.

UNOCHA. 2016. UN Office for the Coordination of Humanitarian Affairs 16_02_2016. https://www.unocha.org/syria

Vely, Y. 2017. Le groupe Lafarge aurait versé plus de 500 000 dollars à l'Etat islamique. Paris Match, November 23. www.parismatch.com/Actu/International/Le-groupe-Lafarge-aurait-verse-plus-de-500-000-dollars-a-l-Etat-islamique-1400795. Accessed 14 July 2019.

Wardrop, D.R., C.C. Leake, and J.J. Abra. 2001. Practical techniques that minimize the impact of quarries on the water environment. *Applied Earth Science: Transactions Of The Institution Of Mining & Metallurgy, Section B* 110 (1): 5.

Western Regional Air Partnership's (WRAP). 2006. Fugitive Dust Handbook. September 7, 2006. 244.

World Bank. 2017a. *The toll of war. The economic and social consequences of the conflict in Syria*, 148. Washington DC: The World Bank Group.

———. 2017b. *Syria Damage Assessment of Selected Cities Aleppo, Hama Idlib*. Phase III, March. http://documents.worldbank.org/curated/en/530541512657033401/pdf/121943-WP-P161647-PUBLIC-Syria-Damage-Assessment.pdf.

Worth R.F. 2010. *Earth is parched where Syrian farms thrived*. New York edition, A1. New York: New York Times. 13 October 2010.

Index

V. Singhroy (ed.), *Advances in Remote Sensing for Infrastructure Monitoring*, Springer Remote Sensing/Photogrammetry,
https://doi.org/10.1007/978-3-030-59109-0

T

The manufacturer's authorised representative in the EU is Springer Nature Customer Service Centre GmbH, Europaplatz 3, 69115 Heidelberg, Germany. If you have any concerns regarding our products, please contact ProductSafety@springernature.com

Printed and bound by CPI Group (UK) Ltd, Croydon, CR0 4YY

30/09/2026

02225247-0009